GEORGE BEST

THE THREE VOYAGES OF MARTIN FROBISHER

IN SEARCH OF A PASSAGE TO

CATHAY AND INDIA BY THE NORTH-WEST

A.D. 1576-8

Elibron Classics
www.elibron.com

Elibron Classics series.

ISBN 1-4021-9548-6 (paperback)
ISBN 1-4021-2636-0 (hardcover)

This Elibron Classics Replica Edition is an unabridged facsimile of the edition published in 1867 by the Hakluyt Society, London.

REPORT FOR 1867.

THE HAKLUYT SOCIETY has now reached the twentieth year of its existence. Since the second year, two volumes have regularly been delivered to Members; and their value is established by the price they fetch, when, on rare occasions, a complete set gets into the market. At a recent sale, the whole series was purchased for £16. 16s., the price charged to Members being then only £15. 15s.

But, besides attaining the main objects of the Society, namely, the supply of carefully edited editions of rare or unpublished voyages, travels, and other geographical records, to subscribers, the Council may congratulate the Members on the high position that many of their volumes have attained in English literature, both as regards intrinsic merit and practical usefulness. The Society's labours have been pretty equally distributed over the great divisions of the world. Out of the thirty-eight volumes forming the series, twelve relate to discoveries and explorations in the New World, twelve to the continent and islands of Asia, six to the Arctic Regions, three to Russia, one to Australia, and four to general voyages or circumnavigations. Most of these volumes are of such a character that no future students or authors who write upon the subjects to which they relate can fail to consult them. Thus, Sir Robert Schomburgk's Guiana is essential to any author who treats of Sir Walter Raleigh's life and acts; the "World

Encompassed" to the biographer of Drake; the "Select Letters" to any future life of Columbus. The historian of Arctic exploration will find the only carefully edited editions of the voyages of Barentz, Hudson, and Frobisher in the Society's series. Mr. Major's "Voyages to Terra Australis" is the main source whence histories of the future wealthy states of the southern Continent will take their rise. And Colonel Yule's laborious and admirably edited work will in future be the standard authority on all questions relating to the early intercourse between Europe and the far East.

With regard to practical usefulness the works of the Hakluyt Society stand equally high. In the Report for 1865 it will be remembered that the great practical value of one of the Society's volumes in navigating a ship through intricate channels, was noticed. Captain Penny, who is well known as the commander of one of the expeditions which went in search of Sir John Franklin, reported that, when Parry's chart failed him, he was enabled to guide his ship through the Savage Group into Fox's Channel by the aid of the volume on Hudson's Bay, edited by Mr. Barrow. Explorers in the Amazon Valley and the region of the Andes, are furnished, by Mr. Markham, with portable editions of the leading discoverers of former times, whose footsteps they are following. Equally important to the traveller in the East are the editions of Galvano, Middleton, Jordanus, Varthema, Conti, and others.

The Society's volumes are quoted by historians, such as Helps and others; by those who write the lives or portions of the lives of great explorers, of Hawkins or Drake, Columbus or De Soto, Raleigh or Frobisher or Hudson; by travellers and explorers; and by those who study the antiquities of India or South America; and the use of these volumes is essential to the adequate treatment of their respective subjects.

These are facts which may be verified by any one, and they afford just cause for congratulation to the Members of the Hakluyt Society. The results of twenty years of labour have, we believe, been alike satisfactory to subscribers and practically useful both to students and to explorers.

In looking forward, the Council are happy to be able to report that several valuable works have been undertaken by editors. Many others have been suggested for publication, and indeed the mine from which these rich ores are extracted is practically inexhaustible.

The two volumes of Colonel Yule's "Cathay and the Way Thither", have been delivered to Members since the last General Meeting.

The following work is nearly ready and will shortly be delivered:—"The Three Voyages of Sir Martin Frobisher," with a selection from his letters now in the State Paper Office. Edited by Rear-Admiral R. Collinson, C.B.

Several other works have been undertaken by editors, but the next that will be ready for delivery will probably be—"Events in the Philippine Islands," by the Doctor Antonio de Morga, Alcalde of the Royal Audience of New Spain, published at Mexico in 1609. To be translated and edited by the Honble. Henry Stanley.

Mr. Stanley has lately discovered that the description of the coast of East Africa and Malabar (the volume issued in 1866) which was attributed, in the title-page, to Duarte Barbosa, on the authority of Ramusio, was in reality written by Magellan. A brief notice on the subject by Mr. Stanley will be issued to Members with the next delivery, for insertion into their copies of the volume in question.

The Council regret to have to announce that, owing to the lamented death of the well known traveller Dr. Barth, the important work by Leo Africanus, the great repository of African geographical knowledge, is again without an editor.

The following Six Members retire from the Council, viz.—

1. R. W. Grey, Esq.
2. John W. Kaye, Esq.
3. Count de Lavradio.
4. Thomas K. Lynch, Esq.
5. Sir C. Nicholson, Bart.
6. Major-Gen. Sir Henry Rawlinson, K.C.B., M.P.

Of this number, the three following are proposed for re-election, viz.—

1. Sir C. Nicholson, Bart.
2. R. W. Grey, Esq.
3. Major-Gen. Sir Henry Rawlinson, K.C.B., M.P.

And the names of the following gentlemen are proposed for election—

1. Sir Walter Elliot, K.S.I.
2. Rear-Admiral Alfred Ryder, R.N.
3. W. E. Frere, Esq.

STATEMENT OF THE ACCOUNTS OF THE SOCIETY FOR THE YEAR 1866-67.

	£	s.	d.		£	s.	d.
Balance at Banker's at last Audit.	£468	12	0	Mr. Richards, for Printing	£331	15	0
Received by Bankers during the year........................	241	19	0	Mr. Weller, for Maps............	61	6	6
				Transcriptions..................	6	0	0
				Woodcut for "Frobisher's Voyages"	8	8	0
				Gratuity to Agent's Foreman	5	0	0
				Expended in Petty Cash	3	14	3
					416	3	9
				Present Balance at Banker's	293	1	6
				Present Balance in Petty Cash ..	1	5	9
	£710	11	0		£710	11	

Examined and approved May 23rd, 1867.

CHARLES BAGOT PHILLIMORE.
WILLIAM NEVILLE STURT.

THE

HAKLUYT SOCIETY.

President.

SIR RODERICK IMPEY MURCHISON, Bart., K.C.B., G.C.St.S. F.R.S., F.R.G.S., D.C.L.
Mem. Imp. Acad. Sc. St. Petersburg, Corr. Mem. Inst. Fr., &c. &c.

Vice-Presidents.

Rear-Admiral C. R. DRINKWATER BETHUNE, C.B.
The Right Hon. SIR DAVID DUNDAS.

Council.

The Right Hon. H. U. ADDINGTON.
Rev. G. P. BADGER, F.R.G.S.
J. BARROW, Esq., F.R.S.
E. H. BUNBURY, Esq.
Rear-Admiral R. COLLINSON, C.B.
Sir WALTER ELLIOT, K.S.I.
Sir HENRY ELLIS, K.H., F.R.S.
General C. FOX.
W. E. FRERE, Esq.
R. W. GREY, Esq.
JOHN WINTER JONES, Esq., F.S.A.
R. H. MAJOR, Esq., F.S.A.
Sir CHARLES NICHOLSON, Bart.
Captain SHERARD OSBORN, R.N., C.B.
Major-General Sir HENRY C. RAWLINSON, K.C.B., M.P.
Rear-Admiral ALFRED RYDER, R.N.
Viscount STRANGFORD.

Honorary Secretary—C. R. MARKHAM, Esq.

Bankers—Messrs. RANSOM, BOUVERIE, and Co., 1, Pall Mall East.

The Hakluyt Society, which is eſtabliſhed for the purpoſe of printing rare or unpubliſhed Voyages and Travels, aims at opening by this means an eaſier access to the ſources of a branch of knowledge, which yields to none in importance, and is ſuperior to moſt in agreeable variety. The narratives of travellers and navigators make us acquainted with the earth, its inhabitants and productions; they exhibit the growth of intercourſe among mankind, with its effects on civilization, and, while inſtructing, they at the ſame time awaken attention, by recounting the toils and adventures of thoſe who firſt explored unknown and diſtant regions.

The advantage of an Aſſociation of this kind, conſiſts not merely in its ſyſtem of literary co-operation, but alſo in its economy. The acquirements, taſte, and diſcrimination of

1866.

a number of individuals, who feel an intereſt in the ſame purſuit, are thus brought to act in voluntary combination, and the ordinary charges of publication are alſo avoided, ſo that the volumes produced are diſtributed among the Members (who can alone obtain them) at little more than the coſt of printing and paper. The Society expends the whole of its funds in the preparation of works for the Members; and ſince the coſt of each copy varies inverſely as the whole number of copies printed, it is obvious that the members are gainers individually by the proſperity of the Society, and the conſequent vigour of its operations.

Gentlemen deſirous of becoming Members of the Hakluyt Society ſhould intimate their intention to the Secretary, MR. CLEMENTS R. MARKHAM, 21, *Eccleſton Square, S.W.*, or to the Society's Agent for the delivery of its volumes, MR. RICHARDS, 37, *Great Queen Street, Lincoln's Inn Fields;* when their names will be recorded, and, on payment of their ſubſcription of £1 : 1 to Mr. Richards, they will receive the volumes iſſued for the year.

New Members have, at preſent (1867), *the privilege of purchaſing the complete ſet of the publications of the Society for previous years for sixteen guineas, but have not the power of ſelecting any particular volume.*

The Members are requeſted to bear in mind that the power of the Council to make advantageous arrangements, will depend, in a great meaſure, on the prompt payment of the subscriptions, which are payable in advance on the 1st of January, and are received by MR. RICHARDS, 37, Great Queen Street, Lincoln's Inn Fields. Poſt Office Orders ſhould be made payable to MR. THOMAS RICHARDS, at the *Weſt Central Office, High Holborn.*

WORKS ALREADY ISSUED.

1—The Observations of Sir Richard Hawkins, Knt.

In his Voyage into the South Sea in 1593. Reprinted from the edition of 1622, and edited by Capt. C. R. DRINKWATER BETHUNE, R.N., C.B.
Iſſued for 1848.

2—Select Letters of Columbus.

With Original Documents relating to the Diſcovery of the New World. Tranſlated and Edited by R. H. MAJOR, Esq., of the Britiſh Muſeum.
Iſſued for 1849.

3—The Discoverie of the Empire of Guiana,

By Sir Walter Ralegh, Knt. Edited, with copious Explanatory Notes, and a Biographical Memoir, by SIR ROBERT H. SCHOMBURGK, Phil. D., etc.
Iſſued for 1850.

4—Sir Francis Drake his Voyage, 1595,

By Thomas Maynarde, together with the Spaniſh Account of Drake's attack on Puerto Rico, Edited from the Original MSS., by W. D. COOLEY, Esq.
Iſſued for 1850.

5—Narratives of Early Voyages

Undertaken for the Diſcovery of a Paſſage to Cathaia and India, by the Northweſt, with Selections from the Records of the worſhipful Fellowſhip of the Merchants of London, trading into the Eaſt Indies; and from MSS. in the Library of the Britiſh Muſeum, now firſt publiſhed; by THOMAS RUNDALL, Esq.
Iſſued for 1851.

6—The Historie of Travaile into Virginia Britannia,

Expreſſing the Coſmographie and Commodities of the Country, together with the manners and Cuſtoms of the people, gathered and obſerved as well by thoſe who went firſt thither as collected by William Strachey, Gent., the firſt Secretary of the Colony; now firſt Edited from the original manuſcript in the Britiſh Muſeum, by R. H. MAJOR, Esq., of the Britiſh Muſeum.
Iſſued for 1851.

7—Divers Voyages touching the Discovery of America

And the Islands adjacent, collected and publiſhed by Richard Hakluyt, Prebendary of Briſtol in the year 1582. Edited, with Notes and an introduction, by JOHN WINTER JONES, Esq., of the Britiſh Muſeum.
Iſſued for 1852.

8—A Collection of Documents on Japan.

With a Commentary by THOMAS RUNDALL, ESQ.
Iſſued for 1852.

9—The Discovery and Conquest of Florida,

By Don Ferdinando de Soto. Tranſlated out of Portugueſe by Richard Hakluyt; and Edited, with notes and an introduction, by W. B. RYE, Esq., of the Britiſh Muſeum. *Iſſued for* 1853.

10—Notes upon Russia,

Being a Tranflation from the Earlieft Account of that Country, entitled Rerum Muscoviticarum Commentarii, by the Baron Sigifmund von Herberftein, Ambaffador from the Court of Germany to the Grand Prince Vafiley Ivanovich, in the years 1517 and 1526. Two Volumes. Tranflated and Edited, with Notes and an Introduction, by R. H. MAJOR, Esq., of the Britifh Mufeum. Vol. 1. *Iffued for* 1853.

11—The Geography of Hudson's Bay.

Being the Remarks of Captain W. Coats, in many Voyages to that locality, between the years 1727 and 1751. With an Appendix, containing Extracts from the Log of Captain Middleton on his Voyage for the Difcovery of the North-west Passage, in H.M.S. "Furnace," in 1741-2. Edited by JOHN BARROW, Esq., F.R.S., F.S.A.

Iffued for 1854.

12—Notes upon Russia. Vol. 2.

Iffued for 1854.

13—Three Voyages by the North-east,

Towards Cathay and China, undertaken by the Dutch in the years 1594, 1595, and 1596, with their Difcovery of Spitzbergen, their refidence of ten months in Novaya Zemlya, and their safe return in two open boats. By Gerrit de Veer. Edited by C. T. BEKE, Esq., Ph.D., F.S.A.

Iffued for 1855.

14-15—The History of the Great and Mighty Kingdom of China and the Situation Thereof.

Compiled by the Padre Juan Gonzalez de Mendoza. And now Reprinted from the Early Tranflation of R. Parke. Edited by SIR GEORGE T. STAUNTON, Bart. With an Introduction by R. H. MAJOR, Esq. 2 vols.

Iffued for 1855.

16—The World Encompassed by Sir Francis Drake.

Being his next Voyage to that to Nombre de Dios. Collated, with an unpublifhed Manufcript of Francis Fletcher, Chaplain to the Expedition. With Appendices illuftrative of the fame Voyage, and Introduction by W. S. W. VAUX, Esq., M.A. *Iffued for* 1856.

17—The History of the Tartar Conquerors who Subdued China.

From the French of the Père D'Orleans, 1688. Tranflated and Edited by the EARL OF ELLESMERE. With an Introduction by R. H. MAJOR, Esq.

Iffued for 1856.

18—A Collection of Early Documents on Spitzbergen and Greenland,

Confifting of: a Tranflation from the German of F. Martin's important work on Spitzbergen, now very rare; a Tranflation from Isaac de la Peyrère's Relation de Greenland; and a rare piece entitled "God's Power and Providence fhowed in the miraculous prefervation and deliverance of eight Englifhmen left by mifchance in Greenland, anno 1630, nine months and twelve days, faithfully reported by Edward Pelham." Edited, with Notes, by ADAM WHITE, Esq., of the Britifh Mufeum.

Iffued for 1857.

19—The Voyage of Sir Henry Middleton to Bantam and the Maluco Islands.

From the rare Edition of 1606. Edited by BOLTON CORNEY, Esq.
Issued for 1857.

20—Russia at the Close of the Sixteenth Century.

Comprising "The Russe Commonwealth" by Dr. Giles Fletcher, and Sir Jerome Horsey's Travels, now first printed entire from his manuscript in the British Museum. Edited by E. A. BOND, Esq., of the British Museum.
Issued for 1858.

21—The Travels of Girolamo Benzoni in America, in 1542-56.

Translated and Edited by ADMIRAL W. H. SMITH, F.R.S., F.S.A.
Issued for 1858.

22—India in the Fifteenth Century.

Being a Collection of Narratives of Voyages to India in the century preceding the Portuguese discovery of the Cape of Good Hope; from Latin, Persian, Russian, and Italian Sources, now first translated into English. Edited, with an Introduction by R. H. Major, Esq., F.S.A.
Issued for 1859.

23—Narrative of a Voyage to the West Indies and Mexico,

In the years 1599-1602, with Maps and Illustrations. By Samuel Champlain. Translated from the original and unpublished Manuscript, with a Biographical Notice and Notes by ALICE WILMERE.
Issued for 1859.

24—Expeditions into the Valley of the Amazons

During the Sixteenth and Seventeenth Centuries: containing the Journey of Gonzalo Pizarro, from the Royal Commentaries of Garcilasso Inca de la Vega; the Voyage of Francisco de Orellana, from the General History of Herrera; and the Voyage of Cristoval de Acuna, from an exceedingly scarce narrative written by himself in 1641. Edited and Translated by CLEMENTS R. MARKHAM, Esq. *Issued for* 1860.

25—Early Indications of Australia.

A Collection of Documents shewing the Early Discoveries of Australia to the time of Captain Cook. Edited by R. H. MAJOR, ESQ., of the British Museum, F.S.A. *Issued for* 1860.

26—The Embassy of Ruy Gonzalez de Clavijo to the Court of Timour, 1403-6.

Translated, for the first time, with Notes, a Preface, and an Introductory Life of Timour Beg. By CLEMENTS R. MARKHAM, Esq.
Issued for 1861.

27—Henry Hudson the Navigator.

The Original Documents in which his career is recorded. Collected, partly Translated, and Annotated, with an Introduction by GEORGE ASHER, LL.D.
Issued for 1861.

28—The Expedition of Ursua and Aguirre,

In search of El Dorado and Omagua, A.D. 1560-61. Translated from the "Sexta Noticia Historical" of Fray Pedro Simon, by W. BOLLAERT, Esq.; with an Introduction by CLEMENTS R. MARKHAM, Esq.
Issued for 1862.

29—The Life and Acts of Don Alonzo Enriquez de Guzman.

Tranflated from a Manufcript in the National Library at Madrid, and edited, with Notes and an Introduction, by CLEMENTS R. MARKHAM, Esq.

Iffued for 1862.

30—Discoveries of the World by Galvano

From their firft original unto the year of our Lord 1555. Reprinted, with the original Portuguefe text, and edited by VICE-ADMIRAL BETHUNE, C.B.

Iffued for 1863.

31—Marvels described by Friar Jordanus,

Of the Order of Preachers, native of Severac, and Bifhop of Columbum; from a parchment manufcript of the Fourteenth Century, in Latin, the text of which has recently been Tranflated and Edited by COLONEL H. YULE, C.B., F.R.G.S., late of H.M. Bengal Engineers.

Iffued for 1863.

32—The Travels of Ludovico di Varthema

In Syria, Arabia, Perfia, India, etc., during the Sixteenth Century. Tranflated by J. WINTER JONES, Esq., F.S.A., and edited, with Notes and an Introduction, by the REV. GEORGE PERCY BADGER.

Iffued for 1864.

33—The Travels of Cieza de Leon in 1532-50

From the Gulf of Darien to the City of La Plata, contained in the firft part of his Chronicle of Peru (Antwerp 1554). Tranflated and edited, with Notes and an Introduction, by CLEMENTS R. MARKHAM, Esq.

Iffued for 1864.

34—The Narrative of Pascual de Andagoya.

Containing the earlieft notice of Peru. Tranflated and edited, with Notes and an Introduction, by CLEMENTS R. MARKHAM, Esq.

Iffued for 1865.

35—The Coasts of East Africa and Malabar

In the beginning of the Sixteenth Century, by Duarte Barbofa. Tranflated from an early Spanifh manufcript by the HON. HENRY STANLEY.

Iffued for 1865.

36—Cathay and the Road Thither.

A Collection of all minor notices of China, previous to the Sixteenth Century. Tranflated and edited by COLONEL H. YULE, C.B. Vol. 1.

Iffued for 1866.

37—Cathay and the Road Thither. Vol. 2.

Iffued for 1866.

38—The Three Voyages of Sir Martin Frobisher.

With a Selection from Letters now in the State Paper Office. Edited by REAR-ADMIRAL COLLINSON, C.B.

Iffued for 1867.

OTHER WORKS UNDERTAKEN BY EDITORS.

Events in the Philippine Iſlands by Doctor Antonio de Morga, Alcalde of the Royal Audience of New Spain, publiſhed at Mexico in 1609. Tranſlated and edited by the HON. HENRY STANLEY.

Journeys of Caterino Zeno and other Italians to Perſia in the Fifteenth and Sixteenth Centuries. Tranſlated and edited by CHARLES GREY, Esq.

The Travels of Joſafa Barbaro and Ambrogio Contarini in Tana and Perſia. Tranſlated from Ramuſio by E. A. ROY, Esq., and edited, with an Introduction, by VISCOUNT STRANGFORD.

The Royal Commentaries of the Ynca Garcilaſſo de la Vega; tranſlated and edited by CLEMENTS R. MARKHAM, Esq.

The Voyages of Davis and Baffin in ſearch of a North Weſt Paſſage, together with the "Seaman's Secrets" of Davis. Edited by Captain SHERARD OSBORN, R.N., C.B.

Hans Stade. Adventures in Brazil in the Sixeeenth Century. Tranſlated and edited by Captain R. F. BURTON, H.M. Conſul at Santos.

Pigafetta's Narrative of the Voyage of Magalhaens, from the Italian text of Amoretti. Tranſlated and edited by M. FREDERICK DE HELLWALD.

"The Seyyeds of the Al Boo Said"; an account of Oman from the immigration of the Azdites from Mareb in A.D. 118; containing information reſpecting Oman, the coaſt of Mekran, the Perſian Gulf, and the eaſt coaſt of Africa during the 16th and 17th centuries; to be tranſlated from the Arabic, and edited by the REV. GEORGE P. BADGER, F.R.G.S.

The Fifth Letter of Hernan Cortes, describing his Voyage to Honduras in 1525-26. Tranſlated and edited by DON PASCUAL GAYANGOS.

Roſmital's Embaſſy to England, Spain, etc., in 1466. Edited by R. C. GRAVES, Esq.

WORKS SUGGESTED TO THE COUNCIL FOR PUBLICATION.

Voyages of Alvaro de Mandana and Pedro Fernandez de Quiros in the South Seas, to be tranflated from Suarez de Figueroa's "Hechos del Marques de Cañete," and Torquemada's "Monarquia Indiana."

Inedited Letters, etc., of Sir Thomas Roe during his Embaffy to India.

John Huigen van Linschoten. Difcourfe of a Voyage unto the Eaft Indies; to be reprinted from the Englifh tranflation of 1598,

The Voyage of John Saris to India and Japan in 1611-13, from a manufcript copy of his Journal, dated 1617.

The Topographia Chriftiana of Cosmas Indicopleuftes.

Bernhard de Breydenbach, 1483-84, A.D. Travels in the Holy Land.

Felix Fabri, 1483. Wanderings in the Holy Land, Egypt, etc.

Voyage of Du Quefne to the Eaft Indies in 1692, from a manufcript Journal by M. C. * * * *

El Edrifi's Geography.

Narrative of Giovanni da Verrazzano, a Florentine, concerning the land called New France, difcovered by him in the name of his Majefty: written at Dieppe, A.D. 1524.

Voyage made by Captain Jaques Cartier in 1535 and 1536 to the ifles of Canada, Hochlega, and Saguenay.

Nicolo and Antonio Zeno. Their Voyages to Frifland, Eftotiland, Vinland, Engroenland, etc.

Ca da Mofto. Voyages along the Weftern Coaft of Africa in 1454: tranflated from the Italian text of 1507.

The Difcovery and Conqueft of the Canary Iflands, by Bethencourt in 1402-25.

Leo Africanus.

J. dos Santos. The Hiftory of Eaftern Ethiopia. 1607.

Joam de Caftro. Account of a Voyage made by the Portuguefe in 1541, from the city of Goa to Suez.

Bethencourt. The Difcovery and Conqueft of the Canary Iflands, A.D. 1402-25.

John and Sebaftian Cabot. Their Voyages to America.

Willoughby and Chancellor. Their Voyages to the North-east.

Icelandic Sagas narrating the Difcovery of America.

The Voyage of Vafco de Gama round the Cape of Good Hope in 1497. Tranflated from a contemporaneous manufcript, accompanied by other documents, forming a monograph on the life of De Gama.

La Argentina. An account of the Difcovery of the Provinces of Rio de la Plata from 1512 to the time of Domingo Martinez de Irala; by Ruiz Diaz de Guzman.

LAWS OF THE HAKLUYT SOCIETY.

I. The object of this Society ſhall be to print, for diſtribution among its members, rare and valuable Voyages, Travels, Naval Expeditions, and other geographical records, from an early period to the beginning of the eighteenth century.

II. The Annual Subſcription ſhall be One Guinea, payable in advance on the 1st January.

III. Each member of the Society, having paid his Subſcription, ſhall be entitled to a copy of every work produced by the Society, and to vote at the general meetings within the period ſubſcribed for; and if he do not ſignify, before the cloſe of the year, his wiſh to reſign, he ſhall be conſidered as a member for the ſucceeding year.

IV. The management of the Society's affairs ſhall be veſted in a Council conſiſting of twenty-one members, viz., a Preſident, two Vice-Preſidents, a Secretary, and ſeventeen ordinary members, to be elected annually; but vacancies occurring between the general meetings ſhall be filled up by the Council.

V. A General Meeting of the Subſcribers ſhall be held annually. The Secretary's Report on the condition and proceedings of the Society ſhall be then read, and the Meeting ſhall proceed to elect the Council for the enſuing year.

VI. At each Annual Election, ſix of the old Council ſhall retire, of whom three ſhall be eligible for re-election.

VII. The Council ſhall meet every month, excepting Auguſt, September October, and November, for the diſpatch of buſineſs, three forming a quorum, including the Secretary, and the Chairman having a caſting vote.

VIII. Gentlemen preparing and editing works for the Society, ſhall receive twenty-five copies of ſuch works reſpectively, and an additional twenty-five copies if the work is alſo tranſlated.

RULES FOR THE DELIVERY OF THE SOCIETY'S VOLUMES.

I. The Society's productions will be delivered without any charge, within three miles of the General Poſt Office.

II. They will be forwarded to any place beyond that limit, the Society paying the coſt of booking, but not of carriage; nor will it be anſwerable in this caſe for any loſs or damage.

III. They will be delivered by the Society's agent, Mr. Thos. Richards, 37, Great Queen Street, Lincoln's Inn Fields, to perſons having written authority of ſubſcribers to receive them.

IV. They will be ſent to the Society's correſpondents or agents in the principal towns throughout the kingdom; and care ſhall be taken that the charge for carriage be as moderate as poſſible.

LIST OF MEMBERS

OF

THE HAKLUYT SOCIETY.

Addington, The Right Hon. H. U., 78, Eaton-place, S.W.
Admiralty (The), 2 *copies*.
All Souls College, Oxford.
Allport, Franklin, Esq., 156, Leadenhall-street.
Anderson, H. L., Esq., 5, Leinster-gardens, Bayswater.
Antiquaries, the Society of.
Army and Navy Club, 13, St. James's-square.
Arrowsmith, John, Esq. 35, Hereford-square, South Kensington.
Astor Library, New York
Athenæum Club, The, Pall Mall.
Athenæum Library, Boston, U.S.

Badger, Rev. George Percy, F.R.G.S., 7, Dawson-place, Bayswater.
Bain, James, Esq., 1, Haymarket.
Bank of England Library and Literary Association.
Barrow, J., Esq., F.R.S., F.S.A., 17, Hanover-terrace, Regent's Park.
Batho, J. A., Esq., 49, Upper Charlotte-street, Fitzroy-square.
Bell, Rev. Thomas, Berbice.
Benzon, E. L. S., Esq., Sheffield.
Berlin, The Royal Library of.
Bethune, Admiral C. R. Drinkwater, C.B., 4, Cromwell-road.
Bibliothèque Impériale, Paris.
Birmingham Library (The)
Birmingham, The Central Library.
Blackie, Dr. Walter G., Villafield, Glasgow.
Bowring, Sir John, LL.D., Athenæum Club.
British Museum (*copies presented*)

Brodhead, J. R., Esq., New York.
Broughton, Lord, 42, Berkeley-square.
Brown, J. A., Esq., Newcastle-place, Clerkenwell.
Brown, J. M., Esq., Portland, U.S.
Brown, R., Esq., King's Langley, Herts.
Brown, W. H., Esq., Chester.
Brownrigg, Sir Robert C., Bart., 12, Eaton-place.
Bruce, John, Esq., F.S.A., 5, Upper Gloucester-street, Dorset-square.
Bunbury, E. H., Esq., 35, St. James's-street.
Burton, Captain Richard F., H.M. Vice-Consul, Santos, Brazil.

Cambridge University Library.
Campkin, Henry, Esq., F.S.A., Reform Club, Pall Mall.
Canada, The Parliament Library.
Canada (Upper), Department of Public Instruction for.
Cannon, Charles, Esq., British Museum.
Carlton Club, Pall Mall.
Cartwright, Henry, Esq., Barbican Lodge, Gloucester.
Cautley, Sir Proby, K.C.B., India Office.
Christie, Jonathan Henry, Esq., 9, Stanhope-street, Hyde-park-gardens
Churchill, Lord Alfred S., F.R.G.S., 16, Rutland Gate.
Collinson, Rear-Admiral, C.B., The Haven, Ealing.
Colonial Office (The), Downing-street.
Congress, Library of, United States.
Cooper, Lieut.-Colonel E. H., 5, Bryanstone-square.
Cotton, R. W., Esq., Barnstaple.

Delft, Royal Academy of.
Dilke, Sir C. Wentworth, Bart, 76, Sloane-street.
Ducie, Earl of, 1, Belgrave-square, S.W.
Dundas, Rt. Hon. Sir David, 13, King's Bench Walk, Temple.
Dundas, George, Esq., 9, Charlotte-square, Edinburgh.
Dundas, John, Esq., 25, St. Andrew's-square, Edinburgh.
Duprat, M. B., Paris.
Duprat, Chevalier Alfredo.

École Normale, Montreal.
Elliot, Sir Walter K.S.I., Wolflee, Edinburgh.
Ellis, Sir Henry, K.H., F.R.S., 24, Bedford-square.
Ely, Miss, Philadelphia.

Forbes, Captain Charles Stuart, R.N., K. 3, Albany.
Foreign Office (The).
Forster, John, Esq., Palace Gate House, Hyde Park Gate, W.
Fox, General, 1, Addison-road, Kensington.
Franklin, Lady, Upper Gore-lodge, Kensington-gore.
Frere, W. E., Esq.
Fuller, Thomas, Esq., 119, Gloucester-terrace, Hyde Park

Gayangos, Don Pascual de, Madrid.
Gladdish, William, Esq., Gravesend.
Glasgow College.
Goodenough, Capt. J. G., R.N., F.R.G.S., H.M.S. Minotaur.
Grey, Charles, Esq., India Office, S.W.
Grey, R. W., Esq., 47, Belgrave-square.
Griffith and Farran, Messrs., 21, Ludgate-street.
Grinnell, Cornelius, Esq., F.R.G.S., 180, Piccadilly.
Grote, A., Esq., Calcutta.

Hall, Rear Admiral Sir William H., K.C.B., 48, Phillimore-gardens, Campden Hill.
Harcourt, Egerton, Esq., Whitwell Park, York.
Hardinge, Captain E., R.N., F.R.G.S., 32, Hyde Park Square.
Harker, Turner James, Esq., 10, Northampton Park, Islington.
Harris, Captain H., R.N., 35, Gloucester-terrace, Bayswater.
Hellwald, Frederick de, 9, Türkenstrasse, Vienna.
Holmes, James, Esq., 4, New Ormond-street.
Home Office (The), Whitehall.
Horner, Rev. J. S. H., Wells Park, Somersetshire.
Hull Subscription Library.

India Office, 20 *copies*.

Johnson, W., Esq., R.N., F.R.G.S., North Grove House, Southsea.
Jones, J. Winter, Esq., F.S.A., British Museum.

Knowles, John, Esq., 42, Moorgate-street.

Lavradio, His Excellency the Count de, 12, Gloucester-pl., Portman-sq.
Liverpool Free Public Library.
London Institution, Finsbury Circus.
London Library, 12, St. James's-square.
Lott, Capt. E. G., 159, Parliament-street, Liverpool.
Loyes, Edw., Esq., 33, Paternoster-row.
Lucas, Samuel, Esq., 6, Cork-street, W.
Lynch, Thomas Kerr, Esq., 31, Cleveland-square, W.

M'Calmont, Robert, Esq., 87, Eaton-square.
Mackenzie, John W., Esq., Edinburgh.
McClintock, Commodore Sir Leopold, R.N., F.R.G.S, Port Royal, Jamaica.
Macready, W. C., Esq , Sherborne House, Dorset.
Madras Literary Society.
Maguire, Captain Rochfort, R.N.
Major, R. H., Esq., F.S.A., British Museum.
Malcolm, W. Elphinstone, Esq., Burnfoot, Langholm, Carlisle.
Mantell, Walter, Esq., New Zealand.
Markham, Clements R., Esq., 21, Eccleston-square, S.W.
Markham, Lieut. Albert H.
Massie, Admiral T. L., R.N., Chester.
Maxwell, Sir Wm. Stirling, of Keir, Bart., M.P., 7, Park-street.
Melbourne, Public Library of, per Mr. Guillaume.
Merewether, Lieut. Col. W. L., C.B.
Moore, Adolphus W., Esq., India Office.
Munich, Royal Library at.
Murchison, Sir Roderick Impey, Bart., K.C.B., F.R.S., &c., 16, Belgrave-square.
Murray, John, Esq., F.R.G.S., Albemarle-street.

Naval College (Royal), Portsmouth.
Newcastle-upon-Tyne Literary and Scientific Institute.
New York State Library.
Nicholson, Sir Charles, Bart., F.R.G.S., 26, Devonshire-place, W.
Northbrook, Lord, Stratton, Micheldever Station, Hants.

Ommanney, Rear Admiral Erasmus, C.B., 6, Talbot-square.
Oriental Club, Hanover-square.

Osborn, Captain Sherard, R.N., C.B., F.R.G.S., 119, Gloucester Terrace, Hyde-park.
Ouvry, F., Esq., F.S.A., 66, Lincoln's Inn Fields.

Paine, W. Dunkley, Esq., Cockshutt Hill, Reigate.
Palmer, John L., Esq., R.N., *H.M.S. Topaze*, Pacific Station.
Peabody Institute, Baltimore, U.S.
Peacock, Septimus, Esq., Alexandria.
Perry, Sir Erskine, 36, Eaton-place.
Petit, Rev. J. Louis, The Uplands, Shiffnal.
Petit, Miss, 9, New-square, Lincoln's Inn.
Phillimore, Charles B., Esq., F.R.G.S., 25, Upper Berkeley-street.
Plowden, W. H. Chicheley, Esq., F.R.S.
Porcher, Captain Edwin, R.N., F.R.G.S., *H.M.S. Sparrowhawk*, Pacific Station.
Portland, His Grace the Duke of.
Potts, Captain H. H., 1, Somerfield-terrace, Maidstone.
Powis, Earl of, 45, Berkeley-square.
Prescott, Admiral Sir Henry, K.C.B., Senior United Service Club.

Rawlinson, Major-General Sir H., K.C.B., M.P., 1, Hill-street, Berkeley-square.
Reed, F. J., Esq., 3, Gresham-street, City.
Richard, John E., Esq., Wandsworth, Surrey.
Royal Geographical Society, 15, Whitehall-place (*copies presented*)
Royal Society, Burlington House.
Rushout, The Hon. Miss, 26, Onslow-square, Brompton.
Ryder, Rear Admiral Alfred, R.N., 5, Victoria-street, Westminster.
Rye, W. B., Esq., British Museum.

Sassoon, David, Esq., F.R.S., 17, Cumberland-terrace, Regent's-park.
Sedgwick, Rev. Professor, Trinity College, Cambridge.
Sheffield, Earl of, 20, Portland-place.
Simpson, Lieutenant.
Smith, Edmund, Esq., Hull.
Smith, George, Esq., 21, Russell-square.
Smith, J., Esq. (Messrs. Smith and Elder.)
Somers, Earl, 33, Princes-gate, Hyde Park.
Somerville, Captain Phillip, R.N., 61, Belgrave-square, Brighton.

Sotheby, Mrs., Kingston.
Spottiswoode, William, Esq., F.R.S., 50, Grosvenor-place.
Stanford, Mr. E., Charing-cross.
Stanley, Lord, M.P., 23, St. James's-square, S.W.
Stanley, Hon. Henry, 40, Dover-street, W.
St. Andrew's University.
St. David's, the Right Rev. the Lord Bishop of, Abergwili, Carmarth
Stewart, M. J. Maxwell Shaw, Esq., Calcutta
Strangford, Viscount, 58, Cumberland-street.
Stuart, Alexander, Esq., New York.
Stuart, R. L., Esq., New York.
Stubbs, Commander Edward, R.N., Raleigh Cottage, Harrow.
Stockholm, Royal Library of.

Thomas, Edward, Esq., Athenæum Club.
Thomas, Luke, Esq., Carlton-villa, Blackheath Park.
Tolstoy, George, Esq., St. Petersburgh.
Toronto University.
Trade, the Board of, Whitehall.
Traveller's Club, 106, Pall Mall.
Trinity College, Cambridge.
Trinity Corporation, Tower Hill.

Union Society, Oxford
United Service Institution, Scotland Yard.

Van de Weyer, His Excellency M. Sylvain, 3, Grosvenor-square.
Vefyk, His Excellency Ahmed Effendi, Constantinople.
Victoria Library and Reading Rooms, Hong Kong.
Vienna Royal Imperial Library.
Vivian, Geo., Esq., 11, Upper Grosvenor-street.
Van Ryckevorsel, H., Consul de Venezuela, Conseiller à la Régence de Rotterdam.

Watkinson Library, Hertford, Connecticut, U.S.
Watts, Thomas, Esq., British Museum.
Webb, Captain John Sydney, The Trinity House.
Webb, William Frederick, Esq., Newstead Abbey.

Wellington, the Duke of, K.G., Apsley House, Piccadilly.
Whiteman, Mrs., Theydon Grove, Epping.
Wilkinson, John, Esq., 3, Wellington-street, Strand.
Williams, T., Esq., Northumberland-house, Strand.
Willis and Sotheran, Messrs., Strand
Wilson, Edward S., Esq., Hull.
Wolff, Sir H. Drummond, K.C.M.G., 15, Rutland-gate.
Woodd, Basil T., Esq., M.P., Conyngham Hall, Knaresborough.

Young, Allen, Esq., R.N.R., Riversdale, Twickenham.

WORKS ISSUED BY

The Hakluyt Society.

THE

THREE VOYAGES OF MARTIN FROBISHER.

M.DCCC.LXVII.

LONDON: T. RICHARDS, 37, GREAT QUEEN STREET, W.C.

MARTINUS FROBISHERUS, EQUES AURATUS.

THE

THREE VOYAGES

OF

MARTIN FROBISHER,

IN SEARCH OF A PASSAGE TO

CATHAIA AND INDIA BY THE NORTH-WEST,

A.D. 1576-8,

Reprinted from the First Edition of Hakluyt's Voyages,

WITH SELECTIONS FROM

MANUSCRIPT DOCUMENTS IN THE BRITISH MUSEUM
AND STATE PAPER OFFICE.

BY

REAR-ADMIRAL RICHARD COLLINSON,

C.B.

LONDON:

PRINTED FOR THE HAKLUYT SOCIETY.

M.DCCC.LXVII.

TO

HENRY GRINNELL, ESQ.,

OF NEW YORK,

THIS EDITION OF MARTIN FROBISHER'S THREE VOYAGES IN SEARCH OF A PASSAGE TO CATHAIA BY THE N.W.

IS DEDICATED,

AS A TRIBUTE OF RESPECT AND ADMIRATION,

NOT ONLY FOR HIS CORDIAL AND GENEROUS CO-OPERATION IN THE SEARCH FOR SIR JOHN FRANKLIN AND HIS COMPANIONS,

BUT ALSO FOR THE

INTEREST HE HAS SHOWN IN, AND THE AID HE HAS AFFORDED TO, POLAR EXPLORATION IN THE PRESENT DAY,

BY HIS OBEDIENT SERVANT,

RICHARD COLLINSON.

COUNCIL

OF

THE HAKLUYT SOCIETY.

INTRODUCTION.

Five years after the discovery of America by Columbus, the English, baffled in their attempts to reach Kathay by the N.E., turned their attention in another direction, and on the morning of the 24th of June, 1497, Newfoundland was discovered by John Cabot. Thus began those series of memorable voyages which have been continued, unto our day, with but short interruption, until the northern seaboard of the American continent has been perfectly discovered. The annals of these Arctic voyages have been read and re-read, published and re-published, evincing the deep interest which generation after generation has taken in these touching records of skill and daring, perseverance and long-suffering; and well may we turn to them with pride and pleasure, exhibiting as they do such proof of that spirit of maritime enterprise which always has been Great Britain's boast and glory.

In the year 1500 the discovery of the Cabots was followed up by Gaspar de Cortereal, in two ships from Lisbon, and attention was attracted to the value of the fisheries on the coast of Newfoundland, and in 1504 small vessels from Biscay, Bretagne, and Normandy resorted thither for this purpose. In 1506 Jean Denys drew a map of the Gulf of St. Lawrence; and in 1517 no less than fifty Spanish, French, and Portuguese ships were employed in this fishery. In 1527, R. Thorne of Bristol (who assisted the Cabots in the equipment of their vessels for the first voyage) sailed with two ships for the discovery of the N.W. passage, but was never after heard of.

In 1534 Jacques Cartier sailed from St. Malo with two ships, and explored the Gulf of St. Lawrence. In 1536 an attempt was made by one hundred and twenty Englishmen to form a settlement on Newfoundland, but they suffered the extremity of famine.

In 1548 the English fishery on the American coast had become an object of national importance and legislative encouragement.

The result of these discoveries was published to the world in Gerard Mercator's *Mappe Monde*, in 1569; and as this date will bring us to the period when we are told (see p. 70) "Captaine Frobisher began first with himself to devise and then with his friendes to conferre, and layd a plaine platte unto them that that voyage was not only possible by the Northweast, but also, as he coulde prove, easie to bee performed," this will be the place to describe the arrangement which has been adhered to in this edition of Frobisher's voyages. The text is taken from the first edition of Hakluyt's voyages (1578), in the Grenville library at the British Museum, an extremely rare book, with two maps, to be found in only one other copy. On the requisition of our President, and by the kindness of the gentlemen in charge of the manuscripts at the British Museum and at the public Record Office, access has been obtained to several important documents hitherto unpublished, and which have been arranged previous and subsequent to the several voyages to which they refer. George Beste, the author, served in the second and third voyages; and in his preface to the first voyage will be found a curious account of the knowledge of the world at that period, which will greatly interest those who are not already familiar with Hakluyt's volumes.

I am indebted to Mr. W. B. Rye, of the British Museum, for the following account of the expenses of the first voyage, which is abstracted from the report of the Commissioners on the Public Records, folio, 1837.

The amount of subscription to the first voyage amounted to £875.

Bill for Maps and Nautical Instruments.

	£	s.	d.
Paid for a book of cosmographie in French of Andreas Thevet	2	4	0
Paid to Humphry Cole and others—			
For a greate globe of metal in blanke in a case .	7	13	4
For a great instrument of brasse named Armilla Tolomei or Hemisperium . . .	4	6	8
For an instrument of brasse named Sphera Nautica	4	6	8
For a great instrument of brasse named Compassum Meridianum	4	6	8
For a great instrument of brasse named Holometrum Geometricum	4	0	0
For a great instrument of brasse named Horologium Universale	2	6	8
For a ringe of brasse named Annulus Astronomicus	1	10	0
For a little standing level of brasse . .	0	6	8
For an instrument of wood a stafe named Balestetta	0	13	4

	£	s.	d.
For a very great carte of navigation . .	5	0	0
For a great mappe universall of Mercator in prente	1	6	8
For three other small mappes prented . .	0	6	8
For 6 cartes of navigation written in blacke parchment whereof 4 ruled playne & 2 rounde .	2	0	0
For a Bible Englishe great volume . .	1	0	0
For a cosmographical glasse & castell knowlege .	0	10	0
For a new World of Andreas Thevett Englishe & French	0	6	8
For a Regiment of Medena (Spanishe) . .	0	3	4
For Sir John Mandevylle (Englishe) . .	0	1	0
For 20 compasses of divers sorts . .	3	3	0
For 18 hower glasses . . .	0	17	0
For a astrolabium	3	10	0

The following drugs shew the contents of a ship's medicine chest in Queen Elizabeth's reign:—Ambra Grisi oriental, Cibetti, Masche oriental, Agallorbi, Ligne Aloes, Rubarbi agarisi, Turpenti, Dragridii, Cipri India, Turmerick, Calam aromatica, Irios, Galanga, Myrrha fine, Mastichus, Argenti viti, Ladderi, Aumne Gomme, Oppoponax, Oppen, Alloes, Bellzonica, Styrax Calmuc, Myrobboralia chebue Bellerichi, Indioru citrini, Ledoria, Spica Nardi, Cardamomi, Ligne Rhode, Colucuthes, Magarite, Boli oriental, Lapis Lazuli, Cantatri Citemi, Corralina, Coralli Rubili, Borax, Camphora, Castorium.

Among the payments made by Michael Lok for the furniture of the first voyage the following occur—

For bote hyre of Mr. Furbisher following his bussyness alle this tyme . . .	10	10	0
Paid to Ducke upholster for beddinge for Mr. Captayne Frobiser	3	16	5
Paid for a bottell of aquavite for Mr. Frobiser paid it to his manne Borrowes . . .	0	10	0
Paid to Mr. Frobiser on accompte as followithe for beare and breade at launchinge of the *Gabriell* and for maryners dyners then . . .	0	19	0
Paid to Nicholas Cooke for aquavite 3 hogsheads paid to Anthonye Duffilde bruer . . .	13	18	0

	£	s.	d.
Paid for v tonne of beare at 42*s.* bought of my Lord Admiral by Arthur Pett . . .	10	10	0
Paid to Mr. Frobiser at divers tymes for his paynes takeing on this voyage & his endevor untill his retorne which was paid to clere him out of England one the voyage . . .	80	0	0
Paid for divers implements of houshold necessarye for the shippes furniture as followithe:—			
For a great kettle pan brasse with yron ball .	0	18	0
For a great bassone of brasse to bake one .	0	6	8
For a bakinge pan of yron with cover . .	0	2	8
For a chaffinge dish of brasse . . .	0	4	0
For a skimer of brasse . . .	0	1	4
For a greate potte of yron for meat . .	0	6	8
For a little pane brasse with handle yrone .	0	1	4
For a tryvet yrone	0	1	4
For ij fringe panes	0	5	0
For a drippinge pane yron . . .	0	2	0
For a grydyron	0	1	0
For ij spyttes	0	3	4
For a payre of potte hokes . . .	0	0	8
For a slyse of yron	0	0	8
For a fleshoke of yron . . .	0	0	8
For ij hokes yron flat . . .	0	0	8
For a clever great choppinge knyfe of yron .	0	1	6
For iij wooden platters Muskovia painted .	0	1	6
For a great bassone or ewar of pewtar . .	0	6	8
For iij pynte bottes of beare & wyne . .	0	5	4
For a saltesellar of pewtare . . .	0	1	0
Summe of all the said charges of furnyture of the said shippes outwardes coste as followithe:—			
For implements howshold . . .	8	11	0
For wages of men	213	17	0
For instrumentes of navigatione . .	50	14	0
For vyttelles	387	14	10
For ordonans munition . . .	100	8	4
For tackelinge of shippes . . .	172	5	6
For buyldinge the shippe *Gabriell* & the pynace (newe)	152	0	4
For the shipe *Mickael* with furnitur of her bought	120	0	0
Somme outwardes of shippinge...	£1205	11	8

In the State Papers subsequent to the first voyage will be found—Michael Loks account of his connection with Captain Frobiser, p. 87; Mr. Lockes discours touching the eure, p. 92; and an account of the cost provision, together with the names of the venturars in the second voyage, p. 103.

The subscriptions for the second voyage amounted to £5,150. The expedition consisted of 143 persons, viz., 36 officers and gentlemen, 14 mynars and fynars, 64 mariners on board the *Ayde*, 16 in the *Michael*, and 13 in the *Gabriel*.

The account of the second voyage will be found at p. 117. The collection of State Papers subsequent to the second voyage contains, among other things—The bryefe account of the expenses of the second voyage, and the names of the venturars, p. 164; the trials of the ore, p. 170.

The third voyage was undertaken upon a much larger scale, consisting of the ships *Ayde, Michael, Gabriel*, and *Judith*, belonging to the Company, together with nine other ships hired for the voyage, and arrangements were made for Captain E. Fenton, with one hundred men, to establish a fort at Meta Incognita. The ships brought home 1,296 tons of ore, which were deposited at Dartford, and considerable works seem to have been carried on there in smelting and refining the ore.

The State Papers relative to the outfit for the third voyage contain—A proportion of the charges for a thyrd voyage, p. 209; the inventorie of the ship *Ayde* (a curious document describing her rig and furniture), p. 218.

The third voyage commences at p. 225. The State Papers subsequent to the third voyage relate principally to the difficulty experienced in collecting the subscriptions, pp. 319-321; Mr. Lok's accounts and the answers thereto, pp. 325, 326, 332; all the stock of the venturers in all the three voyages, p. 358; the abuses of Captain Furbisher against the Companye, p. 359.

On the conclusion of the third voyage, when it was discovered that the ore would yield no return, Messrs. Neale and William Baynham were appointed, by letters dated August 12th, 1580, and May 6th, 1581, to audit the accounts. This report recapitulates the names of all the subscribers for the three voyages and the buildings at Dartford. The subscriptions for the three several voyages amounted to £20,345, of which the Queen advanced £4,000. In the account of the property of the Company it is mentioned that Thomas Allen received of Captain Frobisher two ingots of fine gold, weighing 9 pennyweights 8 grains, and two ingots of fine silver, weighing 7 ounces 18 pennyweights, which said gold and silver proceeded of the melting and working of four cwts. of the ore brought from Meta Incognita in the second voyage. That of the foresaid workes done at Dartford in the melting and rifining 16 tonnes of ore whereof proceeded 210 ounces of fine silver mixed with gold, which was delivered to Richard Young.

Amongst the assets of the Company is stated to be at Dartford 1,300 tons of ore remaining, valued at £13 : 6 : 8 per ton = £1,733 : 6 : 8. No further information can be collected respecting the ore, but it is to be presumed that it did not turn out so valuable, because we find it subsequently recited that the like ore may be obtained for £6 a ton, whereas this cost the Company £16.

In the appendix will be found a list of the relics of the Frobisher expedition brought home by Mr. C. F. Hall in 1863, which are now deposited at the Royal Geographical Society; and I am one of those who believe that his exertions in exploring King William's Land for the journals and records of the Franklin expedition will be attended with success. When this island was visited by Sir L. McClintock and Captain Hobson, the ground was covered with snow. Mr. Hall intends passing the summer upon it, and the knowledge he has obtained of the Esquimaux language and character during his two years' residence in Frobisher Sound will enable him to gain their confidence.

The two maps which accompany the narrative are facsimiles of those in the first edition of Hakluyt (1578). The island "Croc land," in the N.W. corner of the second map, is in all probability a misprint in the original, as in Mercator's "Mappe Monde" (1569) there appears an island called Groetland in this position.

The portrait is taken from the *Herwologia*, and has been engraved by Mr. Scott.

In Watts' *Bibliotheca Britannia*, ed. 1824, the following account is given of Frobisher's voyages:—

"A true report of Mr. Martin Frobisher his third and last voyage, 334 *o*, 1577.

"A true report of the last voyage into the west and

northern regions, etc., worthely atcheiued by Captaine Frobishor, of the said voyage the first Finder and Generall, 846 *r*, 1578.

"A Prayse & reporte of Maister Martin Frobishers Voyage to Meta Incognita, 225 *g*, 1579.

"A Welcome home to Mr. M. Frobisher & all those gentlemen and souldiers that have been with him this last iourney in the countrey called Meta Incognita, which welcome was written since this booke was put to the printing & ioyned to the same booke for a true testimony of Churchyardes good will for the furtherance of Maister F.'s fame, 225 *f*."

In the *Bibliotheca Grenvilliana*, under Frobisher, p. 259, vol. i, is the following:—

"A true discourse of the late voyages of discoverie for the finding of the passage to Cathaya by the Northweast, vnder the conduct of Martin Frobisher, Generall: Deuided into three bookes, London, by Henry Bynnyman, maps, 4to., 1578."

This is the first account of all the three voyages of Frobisher in 1576-77-78 by George Beste, who sailed with him: it is extremely rare; a separate and different narrative of the second voyage only by Settle, who likewise sailed with Frobisher, was printed in 1577, in 12mo., and is also extremely rare; but the peculiar value of this copy is in its possessing the two maps.*

"La Navigation du Cap. Martin Forbisher Anglois es regions de west et nordwest en l'année 1577. Pour Antoine Chuppen." 1578, woodcut, 8vo.

This French translation is of great rarity.

* These are the two maps which are given in this edition.

"De Martini Forbisseri Angli Navigatione in regione occidentis et septentrionis. Narratio Historia ex Gallico sermone in Latinum translata por Joan. Tho. Frugium Noribergæ in off. Catharine Gerlachen." 1580, 8vo.

This is the first Latin edition of Frobisher's second voyage.

"Historia navigationis Martini Forbisseri Angli Prætoris sive Capitanii A.C. 1577 ex Anglia in septentrionis et occidentis tractum suscepta ephemerides sive diarii more conscripta et stilo triennioque post ex Gallico in Latinum sermonem a J. T. Freigio translata Hamburgi sumptibus J. Naumanni." 1675, plate, 4to.*

It appears that the account of the voyage was also translated into Italian, as in Lowndes' *Biographical Manual* there is the following: "Scopumento dello Stretto Artico et de Meta Incognita dar Geo. Lor. Anania." Naples, 1582, 8vo.

In the *Restituta*, by Sir Egerton Brydges, vol. ii, will be found "A Rythm Decasybillical upon this last luckie voyage of worthie Captaine Frobisher," of which the following are the first and third verses:—

I.

"Through sundrie foming fretes and storming streightes,
That venturous knight of Ithac's soyle did sayle;
Against the force of Syren's caulmed heightes
His noble skill and courage did prevaile.
His hap was hard, his hope yet nothing fraile;
Not ragged rocks, not sinking sertes or sands,
His stoutness stayed from viewing foreign lands.

* I am indebted to our Vice-President, the Right Hon. Sir David Dundas, for the loan of a copy of this curious volume, as well as for references, which have assisted me greatly in drawing up this account. Ed.

III.

"A right heroical heart of Britanne blood,
Vlysses' match in skill and martial might,
For Princes fame and countries special good,
Through brackish seas where Neptune reignes by right,
Hath safely sailed in perils great despight.
The golden fleece like Jason hath he got,
And rich retourned saunce losse or lucklesse lot."

ABRAHAM FLEMING.

In conclusion, I have to acknowledge the assistance and ready help which I have received from Mr. R. H. Major of the British Museum, whose knowledge, not only of what was required, but where it was to be obtained, has been of the greatest service to me in preparing this edition.

SIR MARTIN FROBISHER, Knt.

"A most valourous man, and one that is to be reckoned amongst the famousest men of our age for counsell and glory gotten at sea." Camden, 3rd edition, 1635, p. 433.

The following account of the life of Sir Martin Frobisher has been derived from Fuller's *Worthies*, Camden, Campbell's *Lives of the Admirals*, Barrow's *Naval Worthies of Queen Elizabeth's Reign*, *Notes and Queries*, p. 478, June 11th, 1859, *History and Antiquities of Doncaster*, by Dr. Miller, and MS. papers in the British Museum.

Dr. Miller says—"It appears that Francis Frobisher* was mayor of Doncaster in 1535, and was probably the father of Martin. Unfortunately, the parish register does not commence the baptisms until 1558. However, I have found the baptism of several of his relations, viz.: 1561, May 30th, Christian, daughter of William Frobisher; 1564, March 2nd, Darcye, son of the same; 1566, March 18th, Matthew, son of the same; 1567, Jan. 18th, Elizabeth, daughter of the same. In Manerser's *Account of Yorkshire Families* it is stated that the father of Sir Martin Frobisher resided some time at Finningley; his mother was daughter to Mr. Rogers of Everton. His grandfather Wil-

* *History and Antiquities of Doncaster.*

liam married Margaret, daughter of William Boynton, of Burmston, Esq. His great great grandfather, Francis, was recorder of Doncaster, and married Christian, daughter of Sir Brian Hastings, Knt."

Campbell, in his *Lives of the Admirals*, tells us that his father bred him to the sea, but we have little account of his early years.

In the State Paper Office (*Domestic, Elizabeth*, vol. xl, June 11th, 1566) there is a paper entitled Examination of Martin Frobisher, of Normanton, co. of York, on suspicion of his having fitted out a vessel to go to sea as a pirate; and there is little doubt but that he was engaged on a voyage to Guinea about this time.

Campbell continues—"He distinguished himself first by undertaking the discovery of the north-west passage, wherein he had no success; yet it gained him great reputation, as he discovered a new promontory or cape, which he called the Queen's Foreland. In 1577 he undertook a second expedition, and in 1578 a third, in all which he gave the highest proof of his courage and conduct in providing for the safety of his men, and yet pushing the discovery he went upon as far as it was possible; so that, notwithstanding his disappointment, he still preserved his credit in spite of a little accident, which would certainly have overturned the good opinion entertained of a less esteemed commander."

Among the State Papers (*Domestic, Elizabeth*, cxlvi, 1580) there is a grant to M. F. (Martin Frobisher) of the office of clerk of H.M. ships; and in the same series, vol. cli, 17, 1581, is a petition of Isabel Furbusher, complaining that Capt. F. (whome God forgive) had spent all the money left her and her children by Thomas Ruggat, her first husband. It is however to be hoped that he was not long in getting over the difficulties occasioned by the failure of the north-west ore, for we find him in 1585 commanding the *Primrose* as vice-admiral, Sir F. Drake being admiral in the *Elizabeth Bonaventure*, in the fleet that was sent to the West Indies, when the booty brought home was £60,000 and two hundred pieces of brass cannon; and there is no doubt that in this expedition he must have added to the reputation which he had already gained, for when the country was threatened with invasion by the Spanish Armada, the Lord High Admiral, in writing to the Queen, says—"Sir F. Drake, Mr. Hawkins, Mr. Frobisher, and Mr. T. Fenner are those whom the world doth judge to be men of the greatest experience that this realm hath." Hoisting his flag on board the *Triumph*, one of the largest ships in the navy, he, in company with Sir F. Drake in the *Revenge*, and Sir J. Hawkins in the *Victory*, made the first attack on the Spaniards, and took an important part in each of the actions which led to the dispersion of the Armada, and therein did such excellent service, that he was among the number of the few knights made by the Lord High Admiral on that signal occasion.*

* A.D. 1588. Stowe's *Annals*, p. 1255. Upon Friday therefore, being the 26 of the moneth of July, ceasing from fighting, the Lord Admiral (as well for their good deserts and honorable service, as also to encourage others to the like valor) was desirous

He then appears to have remained in the *Triumph* to watch the Narrow Seas, as several documents in the State Paper Office prove the exertions which were made to provision the fleet under Sir M. Frobisher. In 1590 he commanded an expedition to the coast of Spain and the Islands, and in 1592 he took charge of the fleet fitted out by Sir Walter Raleigh; and though he had but three ships, yet he made a shift to burn one rich galleon and to bring home another. In 1591 the King of Spain sent 3,000 troops to the neighbourhood of Brest, where they had taken up a strong position. Queen Elizabeth being applied to for assistance, ordered a squadron to be prepared under the command of Sir Martin Frobisher, and in the course of the operations against Fort Crozon, addressed him the following characteristic letter :—

"Elizabeth R.

"Trustie and welbeloved, wee greet you well: wee have seen your letter to our Threasuror and our Admyrall, and thereby perceive your love of our service, also by others your owne good carriage, whereby you have wonne yourself reputation; whereof, for that wee

to advance certaine personages to the degree of knighthood, for that, behaving themselves manfully, as well with their ships as their good advice, they were worthie that degree of honor, and so much the more worthie in that, being farre separated from all courtly favour, which manie times imparteth the chiefest honours unto the least deserving men, they declared their valour in the eyes of either fleet.

Therefore the two Lords, viz., the Lord Howard and the Lord Sheffield, Roger Townesend, John Hawkins and Martin Frobisher were called foorth, and the order of knighthood given them by the Lord H. Admirall as their generall.

imagine it wil be comfort unto you to understand, wee have thought good to vouchsafe to take knowledge of it by our owne hande writinge.

"Wee know you are sufficiently instructed from our Admyrall, besides your owne circumspection, howe to prevent any soddaine mischeife by fire or otherwise upon our fleete under your charge; and yet do wee thinke it will worke in you the more impression to be by ourselfe againe remembred, who have observed by former experience that the Spaniards, for all their boaste, will truste more to their devices than they dare in deed with force look upon you. For the rest of my directions, we leave them to such letters as you shall receave from our Counsaile.

"Given under our privie signet at our mansion of Richmond the 14th of November, in the thirty-sixth yeare of our reigne, 1594.

"L. S.

"To our trustie and welbeloved

"Sir Martine Furbussher, knight."*

This letter can only have reached him on his return to Plymouth after the fort was taken, when Campbell tells us, "The garrrison defended themselves bravely till such time as Sir Martin landed his sailors, and desperately storming the place, carried it at once, but with the loss of several captains. Sir Martin himself received a shot in his side, and this, through want of skill in his surgeon, proved the cause of his death, which happened at Plymouth within a few days after his return."

* Cottonian MS., Otho, 2, 9.

The following letter to the Lord High Admiral must have been written fourteen days before his death.

"Englan[d].—My humble dutie my honorable good L[ord] the viith [of this] mounth by a batterie, underegininge and a verie dan[gerous] assault wee have taken this fort with the losse [of] of our people but non of any accoumpt. They [defended] it verie resolutlie. And never asked mercie. S[o] [they] were put all to the swoord savinge five or six th[at] hid themselfes in the rockes, many of them were slaine [with] our Cannonn and greatt ordenaunce in defendinge o[f the] breatch with there Captaine one Perithos:

"It was tyme for us to goa through with it for Don [John] is advanst within six leagges of our armie with a[n] intente to have succoured them. Sir John No[rris] doth rise this daie and doth martch towarde th[em] to a place called old Croydon :—

"Wee are about to gett in our ordenaunce as fast as w[e] can and so to make our repaire homewardes. Sir J[ohn] Norris would willinglie have some five hundred of [the] sayllers for his bettar streinght against the da[ie] of meetinge with don John w[hi]ch I would verie willinglie have don yf we had vittles to contin[ent all] our fleett heare for the tyme :—

"I was shoott in with a bullett at the battrie alongst [the] huckell bone. So as I was driven to have an ins[ision] made to take out the bullett. So as I am neither [able] to goa nor ride. And the mariners are verie unwi[llinge] to goa except I goa with them myselfe: yett [yf] I find it to come to an extremitie we will [try] what we are able: yf we had vittles it were [verie] easilie done but heare is non to be had. I ha[ve sente] accordinge to your honours derections tow shipp[es to] Plymouth and Dartmouth, we most presentlie s[aile] away yf they come not to us with vittles :—

"This bearer is able to certiffie your honours [with] all thinges at large. So with my humble p[rayers] to the Almyghtie for your increasse in hon[our].

"Croydon this viiith of Novembre, 1594.

"Your honours most h[umble]
"to comande

"Mr. Mondaie arived the xxviiith of Octobre at Breste and brought with him a thousand crownes for our vittlinge the which was distributed amongst the shippes.

"MARTIN FROOBISER."*

In the register of St. Andrew's parish, Plymouth, 1594, there appears the following :—

"Nov. 22nd. Sir Martin Frobisher, knight, being at the fort built against Brest by the Spaniards, deceased at Plymouth this day, whose entrails were here interred, but his corpse was carried hence to be buried in London."

"Thus fell," says Camden, "a man of undaunted courage, inferior to none of that age in experience and conduct, or the reputation of a brave commander."

Fuller, in his *Worthies of England*, says he was "verie valiant, but withal harsh and violent† (faults

* Caligula, E ix, Pars i, f. 206.

† In the State Paper Office, Domestic, Elizabeth, ccxix, August 10th, there is the following evidence of unbecoming words

which may be dispensed with in one of his profession), and our chronicles loudly resound the signal service in '88 for which he was knighted."

Camden, in the third edition, 1635, p. 433, thus speaks of him :—"Neither was this victory gotten by the English without bloud, very many valliant souldiers being slaine, and Sir Martine Fourbisher hurt with a small shot in the hip, who, when he had brought back the fleet to Plimmouth, dyed, a most valorous man, and one that is to be reckoned amongst the famousest men of our age for counsell and glory gotten at sea, as by the things which I have before spoken plainly appeareth."

Campbell concludes thus :—"He was one of the most able seamen of his time, of undaunted courage, great presence of mind, and equal to almost any undertaking, a true patriot, yet in his courage blunt, and a very strict observer of discipline, even to a degree of severity, which hindered his being beloved."

spoken by Sir Martin Frobisher against Sir Francis Drake, calling him a cowardly knave and traitor.

STATE PAPERS PREVIOUS TO THE FIRST VOYAGE.

I. NOTE OF NAVIGATIONS HERETOFORE ATTEMPTED.

II. A DISCOVERY OF LANDS BEYOND THE EQUINOCTIAL.

III. A DISCOURSE CONCERNING A STRAIGHTE TO BE DISCOVERED TOWARDE THE NORTHWESTE.

IV. A REPLY TO THE ABOVE DISCOURSE.

STATE PAPERS PREVIOUS TO THE FIRST VOYAGE.

[*Colonial*, No. 21. *Domestic*, cvi, No. 77. *Eliz.*, 1575 ?]

A NOTE OF CERTAYNE NAVIGATIONS HERTOFFORE ATTEMPTED FOR THE DISCOVERIE OF A PASSAGE THROUGHE THE STRAIGHTES OWT OF THE NORTHE SEA INTO THE SOUTH SEA.

In the countrey of America towardes the northe, aboute the sixtie degree, there is an elbowe of a land lying verie farre into the sea, which is called the head of Laborer. And on the southe side there is a verie broade bay lying towardes the weste, and of suche a breadth that it semeth, bothe in the verie entry and after, to be a greate sea, ffor yt lyeth oute aboute three or foure hundred myles, and hathe verie many ilandes, and all the yere throughe there are in the same huge heapes of ise, which bay is called Dusmendas.

Anno 1496. In the yere of our Lord 1496, in the reigne of kyng Henry the Seventh, Sebastian Cabotte, who afterward was chieffe pilot of Spayne, was sent oute of England by the said king, with two shippes, to fynd oute the passage oute of the Northe Sea unto the South, that the way into the countreys which are called Mangi Sepango and Cataya might be opened; which Sebastiane Cabotte, going furth on his voyage by the coastes of the ilandes, that so he might come into America about the sixtie degree, found greate mountains of ise and ilandes covered with snowe in the moneth of Julie when he was but under the sixtie degree onlie towardes the north, which countrey, finding contrary to his expectacion, he went round aboute, and beholding so greate abundance of ise, was in doubte that he should find any waye, and therfore retourned into England again, which hilles of ise there growe because dyvers rivers of sweete waters round downe from either side of the promontory which is not of the salte sea water; ffor this is to be noted, that the sea it self never freesethe. This daylie experience which we have by the shippes which yerelie go oute of England into Moscovia teacheth us whiche in the somer season retorne from thence into England in fyve monethes space. At which tyme of the yere oure countrey men fynd no suche ise or snowe there. Althoughe they passe under the 72 or 73 degree which is xij. or xiij. degrees nerer the Pole than Cabot was.

Anno 1500. Moreover, in the yere of our Lord 1500, one Gaspar Cortesreales, a pilot of Portingale from the northe parte of America was in these ilandes with two shippes, and brought with hym from thence threescore captyves or slaves.

But to find oute the passage oute of the North Sea into the Southe we must sayle to the 60 degree, that is, from 66 unto 68. And this passage is called the Narowe Sea or Streicte of the three Bretheren; in which passage, at no tyme in the yere, is ise wonte to be found. The cause is the swifte ronnyng downe of sea into sea. In the north side of this passage, John Scolus,[1] a pilot of Denmerke, was in anno 1476.

The southe side also of this passage was found of a Spanyard in anno 1541, who, travayling oute of Newe Spayne with a certain band of souldiers, was sent by the vice roy into this coaste; who, when he was come to this coaste, found certain shippes in a certain haven which came thither oute of Cataya laden with merchandise, having in theire ffagges hanging oute of the foreshippes certain burdes paynted called alcatrizæ. The mariners also declared by signes that they came oute of Cataya into that port in xxx. dayes.

[*Lansdowne MS.*, *C.*, fol. 142-6.]

A DISCOVERY OF LANDS BEYOND THE EQUINOCTIAL.

1. The matter hit selfe that is offred to be attempted.
2. That hit is feisible.
3. What meanes we haue commodiously to attchiue yt.
4. The Commodities to grow of hit.
5. An awnswere of suche difficulties and matters as maie be obiected.
6. That there is no injurie offred to any Prince or countreye, nor any offence of amitie.
7. The offer for performinge therof withoute her Majestie's chardge.
8. Matters thought vppon to be praied for her Majestie's good allowance of the Enterprise and direction of the procedinge, alwaies both referring the particularities therof to further consideration and to your Lordships' advice and judgement.

1. The matter hit self that is offred to be attempted.

The discouerie, traffique and enioyenge for the Quenes Majestie and her subiectes of all or anie landes, islandes and countries southewardes beyonde the æquinoctial, or where the Pole Antartik hathe anie elevation above the Horison, and which landes, islandes and countries be not alredie possessed or subdued by or to the vse of anie Christian Prince in Europe as by the charts and descriptions shall appere.

2. That hit is feisible.

The seas and passage, as farre as Bresill and Magellanes streight and the Portugal's navigations to the Moluccas, which all doe lie beyonde the zona torrida, beinge ofte and dailie passed bie theise nations and knowen to oure owen mariners doe shew hit possible. And the more for

[1] The person here referred to is the Polish pilot John Szkolny, whose name is misspelt Scolvus by Wytfliet (*Descript. Ptol. Augmentum*, Lovanii, 1597, p. 188); Pontanus (*Rerum. Danicarum Historia*, Amst., 1631, p. 763); and Horn (*Ulyssea*, Ludg. Bat., 1671, p. 335). He was, as here stated, in the service of Christian II, King of Denmark in 1476, and is said to have landed on the coasts of Labrador, after passing Norway, Greenland, and the Friesland of the Zeni.

that the landes which we seke lieng not onelie beyonde the said zone, but also beyonde the course of the Portugalls saylynge, and approchinge more to the Pole, from the æquinoctial draweth stylle more to the temper of Englonde and the knowen regions of Europe.

3. The meanes that we haue to attchiue hit.

Ships of our owen wel prepared.

The weste contrie lienge the apteste of all partes of Englonde for navigation southewarde.

Marriners and sailers to whome the passage as most thither is knowen.

The good and welkome commodities that from Englond shalbe caried to that people, who, lienge in the temper of Englond and other partes of Europe, cannot but lyke well of clothe wherin we most haboundе, and the transportation wherof is most necessarie for our people at home.

4. The commodities, etc.

The enlarginge of Christian faithe which those naked barbarous people are most apte to receiue, and especiallie when hit shal not carie with hit the unnaturall and incredible absurdities of papistrie.

The grete honor to her Majestie to have encresed the faïth and her d[ominions].

The aptnes and, as hit were, a fatall convenience that since the Portugall hathe atteined one parte of the newefounde worlde to the Este, the Spaniarde an other to the Weste, the Frenche the thirde to the Northe, nowe the fourthe to the southe is by God's providence lefte for Englonde, to whom the other in tymes paste haue bene fyrste offred.

The encrese of the nauigation of Englonde, of which commoditie, both for welthe and saffetie, enoughe can not be saide.

The lyklihoode of bringinge in grete tresure of gold, sylver and perle into this relme from those countries, as other Princes haue oute of the lyke regions.

The enrichinge of the relme with all other sortes of commodities that the same landes doe beare, which are lyke to be infynite and had with small price and for the onelie fetchinge; and accordinge to the diversyties of clymes, yt is moste lykelie that the manifolde diversytie of commodities wilbe fownde and muste nedes habunde, for that by traffique and exportance they haue not hitherto bene wasted.

The settinge of our idle and nedie people to worke and providinge for theim bothe in the travaile of the navigation and the worke of clothes and thinges to be caried thither.

The avoydinge of discommodities and perills that we be nowe subiecte vnto, when the welthe and worke of our lande and people dependethe partlie vpon the will of our skante trustie neighbours for ventinge our clothes and commodities.

The abatinge of the prices of spices and suche commodities that we now haue at the Portugals and Spaniardes handes, wherby they encrese their riches vppon our losse, when much spices and suche lyke here spente and bought deare of theim do with the lesse quantitie consume the vallewe of our clothes that they receiue.

The encrese of the quantitie of golde and sylver that shalbe brought oute of Spaine hit self into Englond when the commodities cominge oute of Spaine, becominge this waie cheper, and so lesse countervailing the vallewe of our clothes caried thyther, the ouerplus shal come more plentifullie hither in treasure.

That we shall receiue lesse of spices and suche commodities from Spaine havinge them from elswhere: and so the more of the retorne of our commodyties from theim in gold and sylver, which nedes muste be a grete commoditie when at this daie recevinge muche of our spices and southerne wares from Spaine and at dere prices: yet the sylver brought from thence is said to be the chief furniture of her Majesties mynte.

5. Answere to the difficulties, etc.

The passinge of the *whote* [*hot*] clyme or zona torrida. This hathe bene passed vi tymes by Magellans. The zona torrida is yerlie in everie voyage of the Portugalle to the Moluccæ passed iiij tymes, and everie voyage of the Spaniardes to Brasyle hit is passed twice. Sondrie of our owen nation and some suche as are to goe in these voyages haue passed hit to Guynie, Brasyle and other places.

The Portugals whole navigation to the Moluccæ, besydes his iiij tymes in everie voyage passinge vnder the æquinoctial, liethe whollie nigh the same lyne.

The contries that we seke soe lie that our course continuethe not nere the lyne, but crossinge the same, styll hastethe directlie to the temper of our owen regions.

5. The perils of the Portugals or Spaniards violence that shall envie our passage. Our strengthe shalbe suche as we feare hit not, besydes that we meane to kepe the Ocean and not to enter in or nere any their portes or places, kepte by their force.

The dispeopling of Englonde. It is no dispeoplinge. The people abonde as apperethe by the nomber greter then can welbe provided for: and the dailie losse by execution of lawe, and no evill pollicie to disburthen the land of some excesse of people.

The wastinge of marriners and furniture of shippinge. It is the encrese of marriners and the skylfulleste sorte and the provisyon of shippinge as by the ensample of Spaine and Portugall, and the Frenche is sene who haue by meanes of their traffique to the Indies and the Newfoundlande a grete nomber of grete ships more then ere that tyme they had or could set on work.

The absence of merriners and shippinge in farre voyages when we maie nede them at home. This reason is generall against all navigation to forren partes which yet is the verie true defense of the relme.

And in all theise reasons is to be noted that none are to passe withowt her Majestie's permission, and as to her heighnes and her counsell from tyme to tyme shall apere mete to be spared.

6. That there is no injurie, &c. ;

The Ffrenche have their portion to the northwarde directlie contrarie to that which we seke.

For the places alredie subdued and inhabited by the Spaniard or Portugall we seke no possession nor interest. But if occasion be free frendlie traffique with theim and their subiectes which is as lawfull as muche wythout iniurie as for the Quenes subiectes to traffiques as merchants in Portugall or Spaine hit self.

The passage by the same seas that they doe, offringe to take nothing from them that they haue or clayme to haue ; is not prohibited nor can be without iniurie or offense of amitie on their parte that shall forbyd hit.

The voyages to Guynea and traffikinge in Mexico and in the verie places of the Spaniards possession hathe in the president of Hawkyns voyage bene defended by her Majestie and counsell as frendlie and lawfull doenges ; much more this which is but passinge in the open sea by theim to places that they nether hold nor knowe. Besyde that not onelie trafyke but also possession, plantinge of people and habitation hathe bene alredie iudged lawfull for other nations in suche places as the Spaniardes or Portugals haue not alredie added to ther possession. As is proved by her Majesties most honorable and lawfull graunte to Thomas Stucle and his companie for terra Florida. Also the Ffrenche mens inhabitynge in Florida and Bresile, who albeit they acknowledge the Pope's authoritie in suche thinges as they grant to perteine to him, yet in this vniuersall and naturall right of traffique and temporall dominion they haue not holden them bounde by his power ; but do expounde his donation to the Spaniardes and Portugals either as a matter not perteyninge to the Pope's authoritie, or at leste not byndinge any other persons princes or nations but the Spaniards or Portugals onelie, who onelie submitted themselues, and were parties to the Pope's judgment in that behalf.

7. The offre for performinge, &c.

The gentlemen that offre this enterprise shall at their charge and adventure of them selves and suche as shall willinglie ioyne themselves to their companie performe the whole voyage at their owen chardges and toward the same shall set forward iiij good ships, wherin they will emploie v. Mll., viz., 2,000ll. in shippinge and furniture, 2,000ll. in victails and necessaries for the companie, and one 1,000ll. in clothe and merchandise fytte for the people ; wherwithe we truste hit wilbe atchived. And afterward as God shall prospere or sende occasion they will at their owen charge pursue the same.

8. Matters thought vpon, &c. :

That her Majestie wilbe plesed to give her letters patentes to the authors and fellowship of this voyage in nature of a Corporacion.

That hit will please her Majestye in the same letters patentes to [put] wordes of her good allowance and lykinge ef their good meaninge [and] add suche franchize and priveledge as in this case is requisyte [and] in the lyke hathe bene graunted.

That hit will plese her Maiestie by the same letters patentes to stablishe some forme of gouernance and authoritie in some persons of the companie of this adventure so as by some regimente, obedience, quiet vnitie and order maie be preserved.

That hit will also plese her Majestie to give her Highnes speciall letters bothe of testimoniall that these adventures be her h[ighnes] subiectes enterprisinge this voyage with her favore and also her letters of commendations to all princes and peoples for their lovinge and favorable enterteinement and traffique.

That some speciall rules and orders suche as the companie shall thincke mete to be kepte emongste theim maie be confirmed by her Maiesties authorytie, and further supplie of lyke ordinances to be made from tyme to tyme by the gouerners of her Highnes, to be appointed for the direction of the voyage ; for the agreement and obedience of the parties, for the contribution and charge, for the equallitie and partytion ; and severallie orders to be appointed by her Majestie for the stablishinge of her Majesties domynion and amitie in suche places as the shall arrive vnto, where the same shalbe to be donne, and for the rate and trew answering of her Majesties portion. Theise thinges brieflie at the fyrste we haue thought mete to exhibite to your honore, who are hable therof to judge muche better then we are hable to shewe. Howbeit yf your l[ordship] shall not be satisfied in anythinge concerninge this matter, hit maie plese you to assigne the same, that w[e] maie attende upon you wythe suche resolucion as we can give therin.

[*Lansdowne MS., C.*, No. 4.]

A DISCOURSE CONCERNINGE A STRAIGHTE TO BE DISCOVERED TOWARDE THE NORTHWESTE, PASSINGE TO CATHAIA AND THE ORIENTALL INDIANS, WITH A CONFUTACION OF THEIR ERROUR THAT THINKE THE DISCOVERYE THEROF TO BE MOSTE CONVENIENTLYE ATTEMPTED TO THE NORTHE OF BACCALAOS.

Consideringe Groynelande is well knowen to be an ilande, and that it is not conioyned to America in any parte, there is no cause of doubte but that upon the northe of Baccalaos the seas are open and no straighte to be there discovered, neither was it ever doubted but that America was an ilande if it were not ioyned with Cathaia. So that the straighte is there and not upon the Baccalaos to be fownde. And this is also by Sebastian Cabottos navigacion to be moste manifestly approved, who sailinge to the northweste of Noua Francia founde the seas open many daies sailinge, till by the mutynie of the mariners he was caused to retorne.

This straighte that disioynethe Asia and America of Gerardus Mercator and other moderne cosmographers is called the Straighte of Anian, and liethe by their descriptions at the leaste northweste. So that from Inglonde it is not lesse then 200 grades distaunte.

Now let vs consider which were the more conveniente waie to discover the said straighte, either passinge vnder the congeled Artike circle, for so highe the maine of America rechethe, or by passinge the straighte of Magilianus to ascende from the equinoctiall alonge the westerne course of that Atlanticall Ilande, as Plato semethe in his Timæo to terme it.

The which shall the better apeare if the comoditie and discomodities of the one and the other be compared.

Ffirst therfore of the southerne voiadge, the discomodities are only these:

The lengthe of the jorneye and the crossinge twise of Zona Torrida.

The lengthe of the jorneye is easilie examined, considering Magilianus Straite is not above 120 grades distaunte from the west of Inglonde, and from this straighte to Anian Straighte, as they are by cosmographers supposed, are not so many grades more, so that the vttermoste of that voiadge is not above 240 grades sailinge. By the other northerne passadge we shall, as is before shewed, be enforced to saile 200 grades in longitude and in latitude 10 grades at the leaste to ascende to the climate of the Baccalaos, northerne Cape, and then 10 degrees more descendinge to the supposed place of Anian Straighte. So that there differ not betwene these courses above 20 grades in true computacion. It wilbe obiected that the grades in the one are acompted in circle of position which are equall to grades equinoctiall, and in the other by grades of paralelle not 30 grades distaunte from the pole, so that althoughe in nomber of grades they smally differre, yet in quantitie the southerne voiadge is farre the greater. Heere I awnswere, true it is that the degrees of the equinoctiall differ in quantitie from the degrees of a paralelle in 60 grades of latitude, for so is the paralelle that is like in the northerne navigacion to be passed, and the difference is exactlie to be knowen; and by supputation the proportion is fownde dupla, every grade of the one being doble in quantitie to a degree in the other, so as the one voiadge maie be truly saide to be doble to the other at the leaste. But consideringe that in discoverye of newe unknowen seas I muste neither beare stiffe saile by nighte ne yet in the daie when fogges or mistes shall happen (which in these partes are almoste contynuallye) wheras contrarywise in the other, passinge altogether by seas knowen and alredy discovered, even till we come to the straighte soughte, I nede not refuse nighte or daie to packe on saile for my moste speede, being no lesse cleere in those whote and temperate zones then darke and mistie in the other. And therfore albeit in quantitie the grades differ, yet all circumstaunces dulye waied I may well affirme that in one naturall daie, and so consequently in one weeke or monethe, I will passe more grades of my southerne voidage then can be passed of the other.

But more particularly to examine the trothe, admitte (the soone being in the tropique of Cancer,) I hoise saile departinge Inglonde folowinge the soone before he come to the equinoctiall lyne, I maie easilie reache Magilianus Straightes and bestowe three weekes at the leaste in plattinge and discoveringe the ilandes and other commodities for fortification of the said straightes if neede were. And then before the soone aryve to the Brumale tropique I maie withe facilitie aryve to the Straightes of Anian. So haue I nowe one whole quarter of a yeare to discover the said straighte and to make plattes of every baie, roade, porte or chanell therein, and to sounde all suche places as in that passadge maie cause perill. In which tyme the soone wilbe arrived againe to the equinoctiall, aprochinge to the congeled Artike circle. And so haue I the whole Summer to retorne from the Northerne Seas, and the 3 firste monethes to employe in trafique with Cathaia or any other ilandes to the saide straighte adioyninge, which may sufficiently occupie the fleete till the seas be resolued. But contrariewise by the northe, it is vtterly inpossible or not without extreme perills of liefe and expence of victualles, without any advauntage in the meane, to discover the said straighte, as by the reasons ensuinge shalbe manifeste.

The distante of the Straightes of Anian to the northweste course beinge 200 grades in longitude maketh 6000 myles, alowinge 30 miles to a grade, for suche is the quantitie of a grade in 60 of latitude. Herto if we maye adioyne 1200 myles, which is the quantitie of 10 grades ascendinge and 10 descendinge tofore mencioned, there amountethe 7200 myles.

Nowe consideringe the seas and ayre vnder the Artike circle are so congeled that they are navigable only 3 monethes in the yeare, wherof it is requisite to reserve at the leaste one monethe to retorne, if the said passadge if the said passadge[1] sholde not be mette withall. Then examyne howe farre in the moyetie of that quarter a man maie passe, and the possibilitie of this voiadge will soone apeare.

It cannot be (consideringe the nighte muste not be navigate for daunger of the coaste, and many tymes in the daie we muste beare slacke saile by reson of mistes and ffogges) that in one daie we sholde saile above one grade or two at the vttermoste, and so in the meane tyme before lymyted not possible to reache the thirde parte of the waie to the desired straighte, the winde being alwaie favorable. I omitte infynite impedymentes that maie lette, as newe landes, ilandes, capes or other, also bayes entering into the contynente, which muste be thoroughly searched, or els the thinge we seeke mighte easilie be pretermitted. Seing therfore without thies impedimentes there is no tyme

[1] *Sic* duplicatur in MS.

sufficiente, howe impossible it is, all circumstances considered, to doe any good this waie, any man maie easilie judge.

Againe the discomodities by reason of the heate in the one are nothinge so manye nor so extreame in the Southe as those of the colde proceding in the Northe. The one beinge tempered by the coole of the nighte, which are alwaies nighe equall to the daie. And the dietinge of men so well knowen in those partes that no daunger is to be feared. But in the Northe bothe daie and nighte being freesing colde, not only men's bodies, but also the very lynes and tacklinge are so frosen, that with very greate difficultie maryners can handell their sailes, I omytte the rages of the seas and tempestuous wether, wherwith we shalbe farre more ofte endaungered in the Northe then in the Southe. Then seinge by this that hathe bene saide it manifestlye apearethe that by the Southe in one yeare, the straighte maie be discovered, and by the Northe it cannot be in a furre longer tyme, let vs also examyne, whether in the meane tyme, the one or the other voiadge, for any other accident, maye happen to be more serviceable or commodious. Wherin this is apparaunte, that whatsoever Northerne Ilande shalbe discovered, there is no other commoditie to be expected from it then only sutche as our Moscovian adventurers bring from Ruscia, seinge they are bothe subiecte to the artike cirkell. But from any lande that shall in the other voiadge be founde, we are assured to expecte, golde, siluer, pearle, spice, riche grayne, and suche moste precious marchaundize, besides countreis of moste excellente temperature to be inhabited, if we thinke it necessary, and if we aryve to tymely to enter the said straighte of Anian, yet haue we Cathaia, and all the Orientall Indians open vnto vs for trafique, besides the waste occeane to the Southe, which cannot but be replenished with numbers of Ilandes, the leaste wherof mighte aboundantly suffice to furnishe our navie with the forenamed comodities. If gemmes, turkesses, rubies, and other precious juells sholde not be there fownde, wherof there cannot but be greate aboundaunce in somme of them. Considering that in the ilande of Ormus and St. Laurence lyinge in the same temperature and clymate there was of olde tyme great plentie; and in this our age in these barbarous ilandes more likely to be founde, being not yet ever soughte and sifted by men of knowledge.

By this conference it maie apeare that as by the Southerne voyadge this Straighte of Anian may more sooner and withe farre lesse perill and exspence be discovered then by the Northerne; so dothe it also for comodities if this streighte were not founde, as farre excell the other as golde, siluer, and spice dothe waxe tarre and tallow, and in ease and safetie to the travailer as furre excedinge as the daie dothe the nighte, or the somer the winter; and yet I denie not that after the straighte shall once be founde, and all the chanels and roades sounded, the capes, fforlands, and bayes perfectly discovered, the enterchaungeable course of curante tried, perfecte plattes and cartes of every goolfe and passadge made, and every parte and harboroughe in his due longitude and latitude, situate in such sorte that both daie and nighte in the cleere and ffogge a man neede not feare to packe on saile with all celeritie to exploicte his voyadge without any doubte or scrupule, but that this waie he maie safelye comodiouslie and most spedelye passe into that 'reatche' riche and bountifull sea abounding with innumerable ilandes of incomperable ritches and unknowen treasure. But whosoever shall before suche exacte discouerye made that waies attempte the same I averre he shall proceade to the shame and dishonor of him selfe to the destruction and ruyne of his 'countrey' companye, and to the vtter discouradgmente of this nation ffurther to adventure in this gainfull honeste honorable enterprise. And reporte me to the judgment of the wise, these reasons before alleaged well weyed.

[*Otho E.*, viii, fol. 216 (225). *Colonial*, 23.]

REPLY TO THE DISCOURSE.

North passage or straighte of 67 degrees, and is not so daungerous as [the passage to] [Mus]covia is which is in 72 degrees and the[rfore] [moste] [da]ngerous for coulde and ise and notwithstandeing [that the] passage to Muscovia is traded v. monethes in the [yeare].

And this passage by the northwest at 67 degrees [oughte] to be searched, and the same may be sayled in xxx^tie^ da[ies from] England to the said passage of 67 degrees.

The which passage beinge knowne wolde make a grete tra[de in] those weste partes, where be manye riche merchandizes, and [the] passage lyeth farre from anye prince that might hinder y[t].

And I thinke verely that with the value of ccc^li^ of mon[nye] this passage might be knowne and truely certefied by mea[nes] of some of the shippes that trade yerely to Iselande for fyshe.

Ffor this passage is to be sayled from Iselande in viii. or x. dai[es], and they havinge ccc^li^ allowed them towards their charges wolde willingly searche the said passage, and ii. or iii. to be sent from hence in the said shippes to bringe true knowledge of the same.

And be yt remembered this passage at 67 degrees to Catayo is but 6,000 leagues, and to passe by the streight of Magilanus to the said Catayo is 15000 leagues. As also the passage of 67 degrees in the moneth of June ther is no darke nightes, but is brighte daye all the 24 howers.

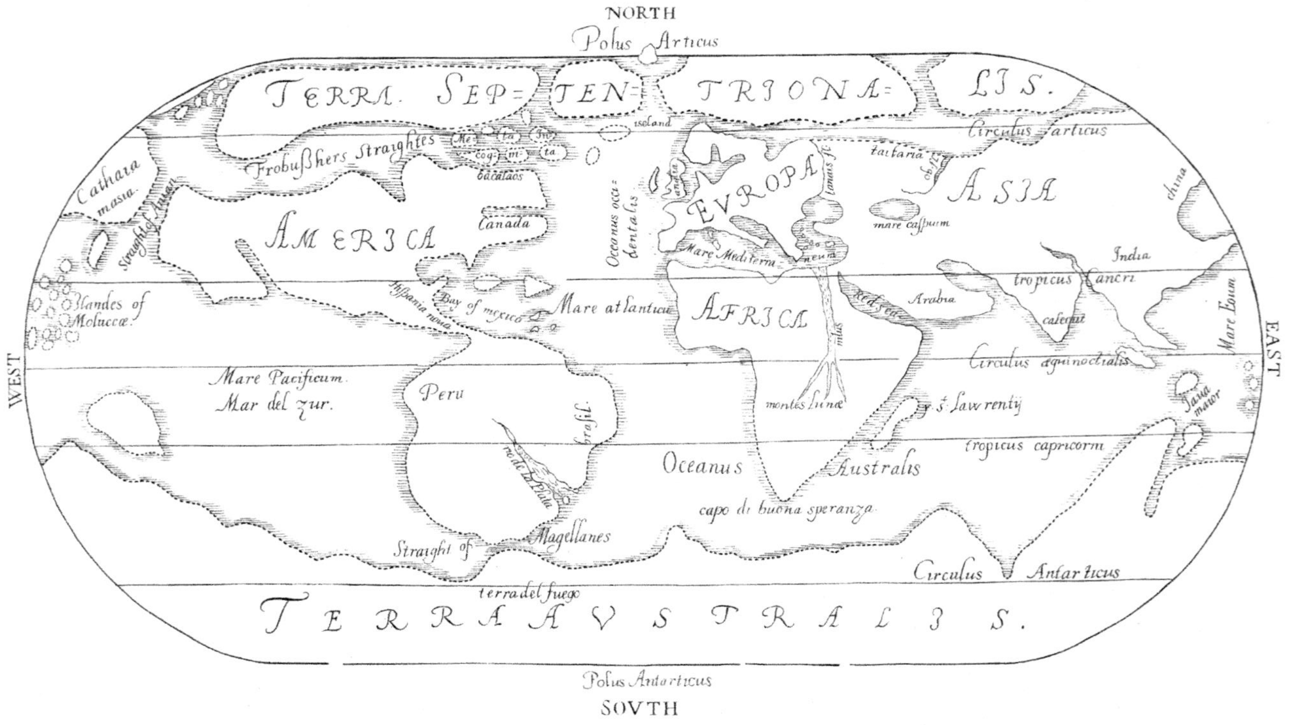
NORTH
Polus Articus
TERRA SEPTENTRIONALIS.
Circulus articus
isoland
Frobußhers Straightes
Meta incognita
bacalaos
tartaria
ASIA
Cathaia
masia
Straight of Anian
AMERICA
Canada
Oceanus occidentalis
EVROPA
tanais fl.
mare caspium
Mare Mediterra
India
china
tropicus Cancri
Ilandes of Moluccæ.
Hispania noua
Bay of mexico
Mare atlanticu
AFRICA
nilus
Red sea
Arabia
calecut
Mare Eoum
EAST
WEST
Circulus æquinoctialis
Mare Pacificum.
Mar del zur.
Peru
brasil
montes lunæ
St. Lawrentij
Iaua maior
tropicus capricorni
rio de la Plata
Oceanus
Australis
capo di buona speranza
Straight of
Magellanes
Circulus
Antarticus
terra del fuego
TERRA AVSTRALIS.
Polus Antarticus
SOVTH

A TRUE DISCOURSE

OF THE

LATE VOYAGES OF DISCOVERIE FOR FINDING OF A PASSAGE TO CATHAYA, BY THE NORTH-WEAST, UNDER THE CONDUCT OF *MARTIN FROBISHER* GENERAL.

DEVIDED INTO THREE BOOKES.

In the First whereof is shewed, his first voyage. Wherein also by the way is sette out a Geographicall description of the Worlde, and what partes thereof have bin discovered by the Navigations of the Englishmen. Also, there are annexed certayne reasons, to prove all partes of the Worlde habitable, with a generall Mappe adjoyned.

In the Second, is set out his second voyage, with the adventures and accidents thereof.

In the third, is declared the strange fortunes which hapned in the third, with a severall description of the countrey and the people there inhabiting. With a particular Card thereunto adjoined of *Meta Incognita*, so farre forth as the secretes of the voyage may permit.

AT LONDON,

Imprinted by HENRY BYNNYMAN, servant to the right Honourable Sir Christopher Hatton, Vizchamberlaine.

Anno Domini 1578.

WHAT COMMODITIES AND INSTRUCTIONS MAY BE REAPED BY DILIGENT READING THIS DISCOURSE.

---

1. FIRST, by example may be gathered, how a discoverer of new countries is to proceede in his first attempt of any discoverie.

2. Item, how he shoulde be provided of shipping, victuals, munition, and choice of men.

3. Howe to proceede and deale with straunge people, be they never so barbarous, cruell and fierce, eyther by lenitie or otherwise.

4. How trade of marchandize may be made withoute money.

5. How a pilot may deale, being environed wyth mountaines of ise in the frosen sea.

6. How lengths of dayes, chaunge of seasons, sommers and winters, do differ in sundry regions.

7. How dangerous it is to attempt new discoveries, either for the length of the voyage or the ignorance of the language, the want of interpretors, newe and unaccustomed elementes and ayres, straunge and unsavery meats, daunger of theeves and robbers, fiercenesse of wilde beasts and fishes, hugenesse of wooddes, daungerousnesse of seas, dreade of tempestes, feare of hidden rockes, steepenesse of mountaines, darknesse of sodaine falling fogges, continuall paines taking withoute anye reste, and infinite others.

8. How pleasaunt and profitable it is to attempt new discoveries, either for the sundry sights and shapes of strange beastes and fishes, the wonderful workes of nature, the different manners and fashions of diverse nations, the sundry sortes of gouernmente, the sight of straunge trees, fruite, foules, and beastes, the infinite treasure of pearle, gold and silver, the newes of new found landes, the sundry positions of the sphere, and many others.

9. How valiaunt captaines use to deale upon extremitie, and otherwise.

10. How trustie souldiers dutifully use to serue.

11. Also here may be seene a good example to be obserued of any priuate person, in taking notes, and making obseruations of al such things as are requisite for a discouerer of new countries.

12. Lastly, the reader here may see a good paterne of a well governed service, sundrye instructions of matters of cosmographie, geographie, and navigation, as in reading more at large may be seene.

TO

THE RIGHT HONOURABLE, MY SINGULAR GOOD MAYSTER, SIR CHRISTOPHER HATTON, KNIGHT, CAPTAINE OF THE QUEENES MAJESTIES GARDE, VIZCHAMBERLAINE TO HIR HIGHNESSE, AND ONE OF HIR MAJESTIES MOST HONOURABLE PRIVIE COUNSALE.

[The Epistle Dedicatory.]

RIGHT honorable, when I first entended the voyage of discoverie wyth *Mr. Frobisher*, for the finding of the passage to Cataya (beyng a matter in oure age above all other notable) I applyed myselfe wholy to the science of cosmographie, and secrets of navigation, to the ende, I might enable myselfe the better for the service of my countrie, not only to understande what I read and heard others speake, but also to execute in effect, and practise with my owne hands, the dutie and office appertayning to a marriner: and so thereby be better able to make a true reporte of al occurrents in the same voyage. And for that now the common reporte thereof is so vaine and uncertaine, by cause some men rather contendyng what they are able to say, than considering what in truth they should and ought to say, whereby, by sundrie men's fantasies, sundry untruths are spred abroad, to the gret slaunder of this so honest and honorable an action, I have thought good to lay open to your honorable judgement, the plain truth, and ful discourse of the whole service, which I have taken upon me (though altogether unable) to write, and to dedicate unto your Honor especially, for these speciall causes following. Firste, the world doth witnesse, and I myselfe by good proofe have tasted and found, being a man by your honorable goodnesse and good countenance, specially supported, and even (as it were) the handy worke of your owne hands, how honorable a regard you beare to vertue, howe readye to countenance the meanest man that truely serveth his countrie, howe willing to give unto suche both grace and opinion with hir Majestie, howe ready to procure rewarde there, for those that shall justlye merite the same. And there withall considering the sounde judgement you have to discern, as wel in this, as in al other causes of waight. And knowing wel what place you hold with hir highnesse, (who for the faithfull service you dayly doe hir, as wel in courte, as common weale, whyche nowe by the true tuchstone of time, and long experience, shee hath founde, and therefore confirmeth a faste and sure opinion in you wyth the chiefest) I have specially thought it necessarye, besides my dutie (whiche above all the worlde my alleageaunce reserved, I owe you moste) for these respects to make relation of this service unto your Honoure above others. And for that this action, both for the worthinesse of the attempt, for the good and quiet government, for the greate and marvelous daungers, for the straunge and unknowne accidents of the unknowne corners of the worlde, above all others, may appeare moste notable and famous : I have bene the rather desirous to take some pain therein, and what I have a ship-boorde rudely and unorderly framed or observed, to commend to your honourable construction the same: being willing rather to hazarde mine own shame, by shewing my selfe an insufficient writer which perchance maye seeme somewhat besides my profession) than that so honest and worthy attempts of our owne nation, with the example of so wel a governed service, should lye hidden from your Honour's sight. And for that I will be injurious to no man, whyche in this action hath borne place, and well discharged the same, and that those men with the maner of their dayly proceedings there, by name may be knowen unto you, I have in their place remembred them in order as becommeth: and have not onely named each principall, but everye private person (if by any speciall service hee hath merited the same) to the ende, that the wel deserving man, receyving the due commendation of his deserte, may be encouraged to continue, and take pleasure in wel doing after, and others being animated by like example, may for hope of like reward also, desire to deserve wel.

By this discourse, it may please your Honour to behold the greate industrie of oure present age, and the invincible mindes of our Englishe nation, who have never lefte anye worthy thing unattempted, nor anye parte almoste of the whole world unsearched, whome lately, neyther stormes of seas by long and tedious voyages, danger of darke fogs and hidden rockes in unknown coastes, congealed and frozen seas, with mountains of fleeting ise, nor yet present dayly before their face, coulde anye white dismay, or cause to desiste from intended enterprises; but rather preferring an honourable death before a shameful retourne, have (notwithstanding the former dangers,) after many perillous repulses, recovered their desired port. So that, if now the passage to CATAYA thereby be made open unto us, (which only matter hytherto hath occupied the finest heades of the world, and promiseth us a more riches by a nearer way than eyther *Spaine* or *Portugale* possesseth) whereof the hope (by the good industrie and great attemptes of these men is greatly augmented) or if the golde ore in these new discoveries founde out, doe in goodnesse as in greate plenty aunswere expectation, and the successe do followe as good, as the proofe thereof hitherto made, is great, we may truely infer, that the Englishman in these our dayes, in his notable discoveries, to the Spaniard and Portingale is nothing inferior: and for his hard adventures, and valiant resolutions, greatly superior. For what hath the Spaniarde or Portingale done by the southeast and southweast, that the Englishman by the northeast and northweast hath not countervailed the same? c 2

And albeit I confesse that the Englishe have not hytherto had so ful successe of profit and commoditie of pleasaunt place (considering that the former nations have happily chanced to travel by more temperate clymates, where they had not onlye good meates and drinkes, but all other things necessarie for the use of man) all whiche things, the English, travelling by more intemperate places, and as it were with mayne force, making waye throughe seas of ise, have wanted, which notwithstanding argueth a more resolution: for *Difficiliora pulchriora*, that is, the adventure the more hard the more honorable: yet concerning the perfecter knowledge of the world, and geographicall description, (wherein the present age and posteritie also, by a more universal understanding is much furthered, as appeareth by my universall mappe with pricked boundes here annexed) herein, the Englishman deserveth chiefe honour above any other. For neyther Spaniard nor Portugale, nor anye other besides the English, have bin found, by so great dangers of ise, so neare the Pole, to adventure any discoverie, whereby the obscure and unknowen partes of the world (which otherwise had laine hid) have bin made knowen unto us.

So that it may appeare, that by our Englishmen's industries, and these late voyages, the world is grown to a more fulnesse and perfection; many unknowen lands and ilands, (not so much as thought upon before) made knowen unto us: Christ's name spred: the Gospell preached; infidels like to be converted to Christianitie, in places where before the name of God had not once bin hearde of: shipping and seafaring men, have bin employed: navigation and the navie (which is the chief strength of our realm) maintayned: and gentlemen in the sea service, for the better service of their country, wel experienced. Al whiche things are (no doubt) of so gret importance, as being wel wayed, may seeme to countervayle the adventures charges; although the passage to CATAYA were not found out, neither yet the golde ore prove good, wher of both the hope is good and gret. But notwithstanding all these, even in this (if no otherwise) hyr most excellent Majestie hath reaped no small profit, that she may now stand assured, to have many more tried, able and sufficient men against time of need, that are (which without vaunt may be spoken) of valour gret, for any great adventure, and of governement good for any good place of service. For this may truly be spoken of these men, that there hath not bin seene in any nation, being so many in number, and so far from home, more civill order, better governement, or agreement. For even from the beginning of the service hitherto, there hath neither passed mutinie, quarrel, or notorious fact, either to the slaunder of the men, or daunger of the voyage, although the gentlemen, souldiers, and marriners (whiche seldome can agree) were by companies matched togither.

But I may perchance (right Honourable) seeme to discourse somewhat too largely, especially in a cause that (as a partie) somewhat concerneth my selfe; which I doe, not for that I doubt of your honorable opinion already conceived of the men, but for that I know, the ignorant multitude is rather ready to slander, than to give good encouragement by due commendation to good causes, who, respecting nothinge but a present gaine, and being more than needefully suspitious of the matter, do therwithall condemne the men, and that without any further respect, either of their honest intents, either of their wel performing the matter they dyd undertake (which according to their direction, was specially to bring home ore) either else of their painful travel (which for their Prince, and the publicke profite of their countries cause they have sustained.)

But by the way, it is not unknown to the world, that this our native country of England in al ages hath bred up (and specially at this present aboundeth with) many forward and valiant minds, fit to take in hand any notable enterprise; wherby appeareth, that if the Englishman had bin in times paste as fortunate and foreseeing to accept occasion offered, as he hath bin always forwarde in executing anye cause once taken in hand, he had bin worthily preferred before all nations of the worlde, and the Weast *Indies* had now bin in the possession of the Englishe.

For *Columbus*, the firste Discoverer of the Weast *Indies*, made firste offer thereof, with his service, to King *Henry* the seaventh, then Kyng of Englande, and was not accepted: Whereuppon, for want of entertainement here, hee was forced to go into *Spaine*, and offered there (as before) the same to *Ferdinando*, Kyng of *Castyle*, who presently acceptyng the occasion, did first himselfe, and now his successors, enjoy the benefite thereof.

Also *Sebastian Cabota*, being an Englishman, and born in *Bristowe*, after he had discovered sundrie parts of new found lande, and attempted the passage to Cataya by the North-west, for the King of England, for lacke of entertainment here, (notwithstanding his good desert) was forced to seeke to the Kinge of Spaine, to whose use hee discovered all that tract of *Brazil*, and about the famous river *Rio de la Plata*, and for the same, and other good services there, was afterwards renowmed, by title of *Piloto Maggiore*, that is Graunde Pylote, and constituted chiefe officer of the Contractation house of Sivilla: in whiche house are handled all matters concerning the Weast Indies, and the revenues therof; and further, that no Pylot shoulde be admitted for any discoverie but by his direction.

But there hath bin two speciall causes in former age, that have greatly hindered the English nation in their attempts. The one hath bin, lacke of liberalitie in the nobilitie, and the other want of skill in the cosmographie, and the arte of navigation. Whiche kinde of knowledge is verye necessary for all oure noblemen, for that wee being ilanders, our chiefest strength consisteth by sea. But these twoo causes are nowe in this present age (God be thanked) very well reformed; for not only hir majestie now, but all the nobilitie also, having perfect knowledge in Cosmographie, doe not only with good wordes countenance the forward minds of men, but also with their purses do liberally and bountifully contribute unto the same, whereby it cometh to passe, that navigation, whiche in the time of King Henry the 7th was very rawe, and toke (as it were) but beginning (and ever since hath had by little and little continuall increase) is now in hir Majestie's raign growen to his highest perfection.

Thus right Honorable, as I have in these my first travels in these late voyages, upon such occasions as passed there, nowe rendered your honour this bare and true accompte: so being further resolved to offer myself a continual sacrifice with the first, for hir Majestie and my country, in thys or any other like service, I intend (God willing) according to this beginning, if any thing hereafter fall out worth the memorie to present your honoure therewithall, and from time to time to advertise you of every particular. And in all these things which I deliver now, or shal hereafter advertise, I humbly praye, your honour would vouchsafe to give some credit thereunto, and rather to thinke, I may be deceived, than that I meane to deceive, colour, or conceale any thing, for I neither can, nor wil, use any flourish in the matter, but a bare truth in all: and thereupon I give my poore credite unto your honour in pawne. And herein I humbly pray pardon, for my rude order of writing, which proceedeth from the barren brayne of a souldier and one professing armes, who desireth rather to be wel thought of with your honour, for his well meaning, than for anye hys cunning writing at all.

And thus, having presumed to present these untimely and unripe fruites to your honoures beste and favourable construction, I humbly take my leave, beseeching God to blesse you, as I do faithfully serve, and will honor you ever.

The handie worke of your Honours handes and faithfully to serve you ever,

George Beste.

THE PRINTER TO THE READER.

Forasmuch as (gentle Reader) these three voyages lately made by our countrymen performed, do both for the matter of discoverie, for the strange and unknown accidentes, for the rare and hard adventures, and also for the good and discrete order of government, appeare above all others most notable and famous: I have bin specially desirous, by all meanes possible I could, to procure the publication thereof, thinking it too great an injurie to our common wealth, to burie in oblivion so worthy attemptes of our owne nation, and to hide the ensample of so good and so well a governed service. And for that (as I understand) many trifling Pamphlets have bin secretly thrust out, not only without the consent of the captaynes and executioners of the same, but also rather to the great disgrace of the worthy voyage, than otherwise, I having intelligence of a substantiall discourse whiche was diligently written thereof, and privately dedicated to my very Honourable Mayster, Sir Christopher Hatton Knight, by a gentleman of his own, who was personally present a captain in all the same service; I have, without making privie the authour, procured his coppie out of the handes of a friende of mine, who had the writing and perusing therof, and have presumed to publish and imprint the same, to the ende that thereby I mighte (gentle reader) as well satisfye thy greedy expectation, by unfolding these newe and unknowen matters, whereof the nature of man is most desirous, as also to performe that dutie whiche I owe unto my sayde Honourable Mayster, in publishing such things as are directed unto him. And for that the mater is worthy to passe under the protection of his honourable name, I have heerein bin willing, rather to beare the burthen of the authores private displeasure, if therewith he should afterwards be offended, than not by publishing the same, seeme not only to do a publicke injurie unto my native countrey, but also shew a lighte regarde of my duetie, in obscuring the doyngs and travels of him, or anye of his, whose honour (as I am chiefely bound) I tender more than my owne safetie. And albeit I have in a fewe places somewhat altered from my copie, and wronged therby the authoure, and have soughte to conceale upon good causes some secretes not fitte to be published or revealed to the world (as the degrees of longitude and latitude, the distance, and true position of places, and the variation of the compasse,) and whiche neverthelesse, by a generall, and particular mappe concerning the same, heereunto annexed, is so sufficiently explained, that easilie anything apperteyning unto the voyage, or in this discourse mentioned, may sensibly be understode; and though the matter be entirelie the authours owne, yet am I contented (for thy sake) rather than the same shoulde not be published, to beare the burthen of blame, and to abide the reprofe of the faultes escaped, taking upon me that reproche of presumption, and hazarding my name to the world, all which things the author peradventure taketh for so great disgraces, as willingly he would not adventure in his owne name the publishing thereof. But specially, for that the commendation of a historie consisteth in truthe and playnenesse, I have desired to bring forth, and prefer (before other pamphlets) the same, knowing that the authore thereof, in nothing more than in truth, desireth to maynteyne credite with this honourable personage, unto whome with his owne hand written, he hath privately dedicated the same, as by the epistle dedicatorie may appeare. If therefore thou shalte accept my well meaning in good parte, and yeelde but deserved prayse to the authoure for doing, and thankes to me for publishing the same, it shall suffice to make me thinke my travell well therein employed. And so fare you well.

THE FYRST BOOKE

OF THE FIRST VOYAGE OF *MARTIN FROBISHER*, ESQUIER, CAPTAYNE GENERALL FOR THE DISCOVERIE OF THE PASSAGE TO CATAYA AND THE EAST INDIA, BY THE NORTHWEAST,

FIRST ATTEMPTED IN *ANNO DOM.* 1576, THE 15. OF MAY.

Cicero, Offic., lib. i.

MAN is borne not only to serve his owne turne (as Tullie sayeth), but his kinsfolke, friends, and the common wealth especially, loke for some furtherance at hys handes, and some frutes of his laboure: where upon sundry men finding themselves as it were tyed by this bond and dutie of humane society, have willinglye endeavoured sundry wayes to shew themselves profitable members of their common weale. Some men by study of the minde, have employed themselves to give out good lawes and ordinances for governement, as *Solon*, *Lycurgus*, and others. Some have spente their time in devising artes and sciences, for the better sharpening of man's witte, and the easier expressing his conceytes, as in time past *Aristotle* for Logicke and Philosophie, *Cicero* and *Demosthenes* for Rethoricke, *Euclide* and others for Arithmeticke and Geometrie. Others againe by long and diligent observation, have found out the motion and courses of the celestiall Orbes, that thereby man might have the distinction of times and seasons, the better to direct his doings both for taking paynes and rest, as occasion and circumstances doth require. Some delight in feates of armes, thereby to be better able to defend their countreys from the force of the enimie, and rightfully (when occasion is) to enlarge their

Astronomie.

dominions. And many others in sundrie faculties and sciences, have both heretofore, and especially now in these later dayes do so bestow and employ their time, that rightly they may be sayd to have deserved the name of profitable members in the common wealth; so that now by continuall practise, and exercising of good wittes, the world is waxed finer, and growen to more perfection, not only in all the speculative artes and sciences, but also in the practicall application of the same, to man's use, whereof as the one doth exceedingly delighte the inward mind, in seeing the sequele of things by arte and reason, so the other in the mechanicall and practicall application (whiche of late yeares, more than ever heeretofore hath bin used) dothe so pleasure and profite the world, that this time only may rightely be called the liberall and flourishing age. For when was there ever heard of such abundance of gold and silver (whiche no doubt being well used, is the great benefite and good blessing of God to mankind) as in these our dayes. No, *Solomon* himselfe, with all the pretious mettall of *Ophir*, which he (one only king) had in that only place, can not be comparable to the great store of golde, and all other mettals, which dayly are digged out of the bowels of the earth, almost in all parts of the world, and now lately in the supposed hard and congealed frozen lands, almost under the Poles. Yea, now every private man can witnesse this with me, that he is no more contented with the wealthe and riches that his auncesters hadde, but thinkes himself base minded, if by his industrie he encrease not his privat wealth proportionallie, as the whole world increaseth in common wealth, and not only of gold and silver is such great encrease, but also of all other things, serving as well for pleasure and delightes of the mind, as for the necessarie uses of man's life. For, as we are placed in these lower elementes firste to know and acknowledge the high Creator, and then thankefully to take the fruition of things for oure mayntenance, which are especially two, that

This is the flourishing age.

Abundance of all things.

To what end man is created.

is, meate and drinke to susteyne the body, and coverture to defend the same from the rigor of heate and cold, and so thereby to glorifie God in his workes: what age hath bin ever heeretofore, that hath so abounded with store, not only of necessarie meates, but also of pleasant and delectable confections, to delight man withal: for whatsoever sundry sorte of corne, grayne, and meates former yeares have had, we not only have all the same in farre greater abundance, but thereunto are added thousandes of new things simple and compound, never heretofore seene or heard of. And as for coverture to defende the bodye, the matter is growen to such excellencie of architecture and building, to such finenesse of cloth and silkes of all sortes and colours; that man studieth no more to multiplye the encrease thereof; so much as to devise fashions, to make it serve more for ornament, than for necessarie uses. And the chiefest cause of all these effects (next after ye divine Providence) is the searching wit of man, whiche being more curious and inquisitive of new and strange devises than heeretofore, bringeth out dayly more strange inventions, and causeth others, through emulation, to do the like—not only in providing ye necessary things aforesaid, but also a continual care and constancie to find out other new arts, occupations, and faculties. For to remember one or two inventions for al, found out of late

Printing of bookes.

yeares. The use and benefite of printing bookes, a devise so commodious and necessarie, saving within these few yeares in respect, hath layne utterly hid and unknowen. The arte

The arte of warre.

of war is nowe growen to that excellencie, that if *Achilles*, *Alexander* the Great, Julius Cæsar, and other, should come in these later dayes, they themselves would more admire and wonder at the courages of our men, their engines, and their policies in warre, than the ignorant and barbarous multitude in their dayes did to them in celebrating their solemnities with all the honor that might be. But to drawe neere to my purposed scope, that is to discourse of inventions by way of discoveries, I say, that one of the excellentest artes that ever hath bin devised is the arte of navigation, which in times past was so raw and unknowen, that no man durst travel by sea, saving only alongst the shore: and if by wind, currant, or tempest, he were driven against his will so far from the land that he lost the sight thereof, he made no other accompte but to be cast away, his vessell was so rude and his skill so little. Navigation.

In those dayes they knew not the singular use and benefite of the loadestone, called in Latin *Magnes*, whiche, besides the property of drawing iron unto it, it directeth, and with opposite poyntes sheweth two principall partes of the worlde, the north and the south, and that more distinctly than the rising of the sunne doth shew east and west (excepte it be onely in the dayes of Æquinoctium which is but twice a yeare), whiche rare propertie of the loadestone, if any man desire at large to see, let him put the sayd stone into a round dish, and they both so together in some vessel of water, wherein they might swim at pleasure voluntarily, which dish when it standeth still then doe the two principall and opposite poyntes of the stone firmely and constantly poynt out north and south; and if, before the quarters of north and south were knowen, by this experience he may find out the two principal poyntes of the stone; so that the one being knowen, the other can not be wanting. And that a man may be the better persuaded of this effect, let him remove or turn round the dish after it hath once stoode still, and he shall ever finde it to returne constantly to the same poynt againe. Also a pillier or piece of steele being but touched with the foresaid *Magnes*, playing Æquilibra upon some piramid or point, receyveth such virtue that it produceth like effect. Whereunto, if wood or paper in circular forme devided into 32 equall parts be handsomely compacted, it will distinguishe and poynte out all parts of the horizon, and direct into all coasts of the worlde, and that onely by the influent spirite

The stone called magnes.

Two and thirty poynts of the compasse.

of the two principall poyntes respecting ever north and south.

This excellent propertie and benefite of the lodestone I the rather remember at large, because some seamen whiche knowe this rare and miraculous effecte as well as I, doe not sufficiently admire the same, bycause it is now so commonly knowen, and yet indeede is to be preferred before all pretious stones in the worlde, whiche only tend to ornament, and have no other vertue, whereas this serveth to so necessarie use. The vertue of this stone, as it is not long since it was first found, so in these dayes it is like to receive his perfection concerning his north-easting and north-westing to be brought in rule, and particularly in this noble voyage of our worthy Captaine *Martine Frobisher*, who, as you shall after understande in the discourse, hath diligentlye observed the variation of the needle. And such observations of skylfull pylotts is the onlye waye to bring it in rule; for it passeth the reach of naturall philosophy. The making and pricking of cardes, the shifting of sunne and moone, the use of the compasse, the houre-glasse for observing time, instrumentes of astronomie to take longitudes and latitudes of countreys, and many other helps, are so commonly knowen of every mariner now adayes, that he that hathe bin twice at sea, is ashamed to come home if he be not able to render accounte of all these particularities. By whiche skill in navigation is brought to passe that the people of Europe can as easilye and far more easilier make long voyages by sea than by lande, whereby hathe come to passe that within the memorie of man within these foure-score yeares, there hath beene more newe countries and regions discovered than in five thousande yeares before; yea, more than halfe the worlde hath beene discovered by men that are yet (or might very well for their age be) alive. When I name the world in this sense, I meane the uppermoste face and *superficies* of the earth and sea, which, unite together, make one globe or

The variation of the needle.

New discoveries.

sphere. And this face of the earth whiche Almightie God hath given man as most convenient place to inhabite in, thorowe the negligence of man hathe, until of late dayes, layne so hidde and unknowen that he hathe loste the fruition and benefit of more than halfe the earth.

A marvellous thing, that man, who hath always abhorred so muche thraldome and restrainte, and so greedily desired liberty, coulde be contented so many thousande yeares, to be shut up in so narrow bounds. For it is to be thought that only such countries in times paste have bin known as either did bounde and hang togither, or else were separated by very narrow seas, as are Europa, Affrica, and Asia, out of which from either to other a man may travaile by lande, or else shall finde in some places very narrow seas separating them, and so mighte saile from the one to the other onelye by lande-markes wythoute the arte of navigation, bycause the one was wythin a ken of the other.

For even the greate strength and stoutnesse of Hercules himselfe, when out of *Græcia* westward he had travelled and conquered all the regions and countries comming to the straight betweene Spaine and Barbarie, made accompte to have beene at the west ende of the worlde, and therefore there created two pillers as a perpetual monument of his fame, whiche to this day are called *Herculæ Columnæ*, the pillers of Hercules, the one standing in *Spaine* of *Europe*, the other in *Affrica*, and called the straight *Fretum Herculeum*: and nowe commonlye is named the straightes of Malega or Gibraltar. And having come so farre westwarde, contented himselfe, and said, *Non plus ultra*, no further.

The west end of the old worlde.

Likewise, Alexander Magnus, out of *Macedonia* in *Greece*, passing throw *Armenia*, *Persia*, and *India*, comming to the great River Ganges, and conquering all these countries (althoughe he was persuaded that *Asia* extended somewhat further into the east and north-east) yet knowing them not to be very great countries, and thinking them to be of small moment, erected there certaine aultars, whiche are yet called *Aræ Alexandrinæ*, as beyond which no man else in those dayes had passed, or neede to passe more eastwarde, and this was accompted as it were a bounder of the easte side of the worlde, althoughe indeede Asia doeth extende further, twenty degrees, and is environed with *Mare Eoum*, and the straight Anian, which our Captaine Frobisher pretendeth to finde out.

The east end of the old worlde.

Touching the south parts of the world towards *Affrica*, Ptolomeus King of *Ægipt*, a famous cosmographer, who was more sollicite and curious in describing al the face of the earth than any king before him or after (excepte of late dayes), delivered in plat described and knowen only sixteen degrees beyond ye equinoctiall to the southwardes or pole Antartique, and that bounder was called Montes Lunæ, out of whiche the greate river *Nilus* is supposed to have his beginning and spring. And, as for the knowen land on the north partes of the world, *Thyle* being one of ye ilands of Orcades (more probably than *Iseland*), was so long pronounced and continued *Ultima*, that it was esteemed a great erroure for anye man to imagine anye lande more north than that.

The end of the old world southward.

The end of the old world northward.

Thus have I briefly named the foure principall bounders of the worlde, which was onely known from the beginning of the worlde (as some thinke) untill within these eighty last yeares. That is, the Straights of *Gibraltar* or *Malaga* westward. The east part of *Asia* beyond the Aræ Alexandrinæ eastwarde. *Vltima* Thyle by Scotlande northward, and sixteen grades beyond the equinoctiall southward. But these sixteen degrees of south latitude are to be understoode only in the continent of Africa, whiche extendeth not passyng seventy degrees in longitude. Therefore, whatsoever countries or regions have since been discovered and knowne beyonde 180 degrees in longitude, 60 degrees in north latitude, and 16 degrees in south latitude, all the commendation, honour, renoume, glorie, and fame thereof, must be attributed to the

The greate discoveries of late yeres.

Englishmen, Spaniardes, Portingales, Frenchmenne, and Italians, whose valiaunt courage and high mindes be suche that either they alreadye have, or shortly will discover and searche out every narrowe corner of the world. By these means valours and industries, the knowne regions of the worlde, whiche before were divided into three partes; that is, Europa, Affrica, and Asia, are now made sixe, by addition of other three. For like as the whole massie frame of the world being firste divided into two principall regions, the one elementiall, the other heavenly, the elementiall containeth foure partes; that is, the four elements, the earth, the water, the ayre, and the fyre: the heavenly region, although one yet for diversitie of motion, may be compted two; that is, *Primum Mobile*, moving onelye uppon the poles, Articke and Antarticke, and all the reste of orbes and planets, moving uppon the poales of the zodiacke are by this difference of motion imagined two, whereby ariseth the number of sixe substances partes of the world; that is, the foure elements and the two varieties of orbes. So likewise the inferiour world, I meane the *superfices* of the earth, is also divided into sixe partes; that is, *Europa*, *Affrica*, *Asia*, *Terra Septentrionalis*, *America*, and *Terra Australis*, whose bounders bycause this division seemeth somewhat strange, I thought good for the more particularitie, here briefly to repeate.

The earth devided into six partes.

THE CHIEFE BOUNDERS OF THE PRINCIPALL PARTES OF THE WORLD.

Europa.

Europa is bounded on the weast side with our Weaste Ocean; on the south side wyth *Mare Mediterraneum*; on the east with *Mare Ægeum*, *Pontus Euxinus*, and the river *Taniæs*, folowing the meridian thereof northwarde; on the north side it was thoughte sometime to be bounded with islandes—*Hebrides*, *Orcades*, and *Hyperborei montes* in *Sarmatia* of Europe. But nowe, by the navigation of the Englishemen, the boundes are extended unto that sea which compasseth Norway, Laplande, and Moscovia.

Africa. Affrica is bounded westwarde with the sea *Atlanticum*; southward with the South Ocean, passing by Cape d'Buona Speranza; eastwarde with the Red Sea; and northwarde by the sea *Mediterraneum*.

Asia. Asia is bounded on the south side with the South Ocean; on the easte side with *Mare Eoum*, and the Straighte *Anian*; on the north side with the Scithian Sea; on the weaste side with the meridian of the river *Tanais*, and parte of the sea *Mediterraneum*, as *Pontus Euxinus*, *Mare Egeum*, *Sinus Issicus*, and the Red Sea.

Terra Septentrionalis. *Terra Septentrionalis* is divided from *Asia* by the Scythian Sea, from *Europe* by the North Sea aboute *Iselande*, called in times past *Mare Congelatum*, the Frosen Sea, and from *America* is divided by Frobisher's Straights. It lyeth rounde about the Pole Artike, and is included by a paralell passing about 70 degrees in north latitude, as it is also more at large described in Mercators and Ortelius Universall Mappes.

This parte of the world hath beene most or onely made knowen by the Englishmen's industrie. For, as Mercato mentioneth out of a probable author, there was a frier of *Oxforde*, a greate mathmatician,[1] who himselfe went verye farre north above 200 yeares agoe, and, with an astrolabe, described almoste all the lande aboute the Pole, finding it divided into foure partes or ilandes by foure greate guttes, indrafts, or channels, running violently and delivering themselves into a monstrous receptacle and swallowing sincke, with suche a violent force and currant, that a shippe beyng entred never so little within one of these foure indraftes, cannot be holden backe by the force of any great winde, but runneth in headlong by that deepe swallowing sinke into

[1] Nicholas de Linna, *i.e.*, of Lynn in Norfolk, whose voyage to the Arctic regions in 1360 is quoted by Mercator in his map of the wôrld dated 1569, from the Itinerary of Jacob Croyen of Bois le Duc, and also referred to by Dr. John Dee. See *Hakluyt*, vol. i, pp. 121, 122.

the bowels of the earth. Hee reporteth that the south-weast parte of that lande is a fruitfull and a holesome soyle. The north-east part (in respect of England) is inhabited with a people called Pygmœi, whiche are not at the uttermoste above foure foote highe. One of these foure greate monstrous gulfes wyth hys violent raging course followeth the meridian of the fortunate ilandes, and receiveth the ocean with three mouths, and is frozen over three moneths in the yeare, and is 37 leagues in breadth. The next eastwarde beyonde the iland *Vagats* is at 110 degrees in longitude and receyveth the East Ocean with five mouths, and, being narrowe and swifte, is never frozen. The third is at 190 degrees in longitude, and receiveth the East Ocean with nineteen receits. The fourth is at 280 degrees in longitude. All these indraftes and raging channels runne directly towards a point under the Pole, where is also said to be a monstrous gret mountain of wonderful gret height and about 35 leagues in compasse at the foot.

Pigmei.

A great rock under the pole.

Guilielmus Postellus saith, that here under and aboute the Pole is best habitation for man, and that they ever have continuall daye, and know not what night and darknesse meaneth. But this seemeth contrary to the principles of the sphere, whyche alloweth well that they shoulde see the sunne halfe a yeare togither without any night. During the time of his being in the north signes from the one *Equinoctium* to the other, yet, that in ye other halfe they shold have continuall night without any day. But I thinke Postellus (being a good astronomer) doubted nothing of ye reason of ye sphere, but meaneth y^{t} for their great twilights, and ye high swelling of ye erth, and ye high mountaine under the Pole, they have continual light; but hereof you shall heare more at large hereafter in this treatise, when I speake of the temperature of ye north regions. This so particular a description of ye land and countries lying about the Pole, argueth that this *Oxford* frier tooke great pains therein, and induceth great

Continual day for ever.

A frier of Oxford.

probabilitie and likelihood of yᵉ truth thereof, bicause he observed so diligently by measure, the bredth of the indrafts, what time, and how long they continued frosen, and with how manye mouths or receipts every one of them received the ocean.

Upon yᵉ bounds and description of this part of the erth, I have yᵉ longer staid, because I find it discovered only by the English nation. And although yᵉ greatest part herof was made knowen 200 and odde yeres past, yet some bounders thereof were described and set out by yᵉ travel of Sr. Hugh Willoby Knight, an Englishman, who ventured and lost his life in yᵉ cause, and so died an honorable death, and with him Ric. Chancelor, chiefe Pilot in that voyage, in an. 1554, who discovered and founde out, yᵗ Norway and Sweden &c., conjoined not to *Groneland*, or any part of yᵉ Northern regions, as one firme and continent, but yᵗ by sea a man might travel to yᵉ country of *Moscovia*, and a gret way more estward, as far as the gret river Obby. Also oure worthy General Ca. Frobisher in his three last voyages, wherof we are briefly to entreat in these three books, hath discovered and described a gret part of yᵉ Southwest bounds thereof. and meneth (God willing) not only to describe the one halfe therof in going to *Cataia* by yᵉ Northwest, but also to put in triall, whether he may return into England by the Northeast, and so also to describe yᵉ other part, which to do, is one of yᵉ waightiest matters of the world, and a thing that will cause other Princes to admire yᵉ fortunate state, and yᵉ gret valor of yᵉ English nation. But to retourne againe to the bounding of the other parts of the world.

Sir Hughe Willoughby and Rich. Chancelor.

Frobisher's 3 voyages.

America.

America an ilande is included on the east side with the sea *Antartique*; on the weast side with *Mare del Sur*, or *Mare Pacificum*; on the south side it is bounded wyth the straight of Magellanus; and on the north with Frobisher's straights.

Terra Australis.

Terra Australis seemeth to be a great firme land, lying

under and aboute the south pole, being in many places a fruitefull soyle, and is not yet thorowly discovered, but onlye seene and touched on the north edge therof, by the travaile of the Portingales and Spaniards, in their voyages to their East and Weast *Indies*.

It is included almost by a paralell, passing at 40 degrees in south latitude, yet in some places it reacheth into the sea with greate promontories, even into the tropicke Capricornus. Onely these partes of it are beste knowen, as over against *Capo d' buona Speranza* (where the Portingales see popingayes commonly of a wonderfull greatnesse,) and againe it is knowen at the south side of the straight of Magellanus, and is called Terra del Fuego.

It is thoughte this southlande, about the pole Antartike, is farre bigger than the north land aboute the pole Artike; but whether it be so or not, we have no certaine knowledge, for we have no particular description hereof, as we have of the lande under and aboute the north pole.

Thus I have briefly butted and bounded out all the parts of the earth, according unto thys latter division into sixe parts. Which, that it might be more apparent and sensible to every man's understanding, I have hereunto adjoyned an universall map, wherein my minde was to make knowne to the eye what countries have been discovered of late yeares, and what before of olde time. The olde knowen partes have their boundes traced and drawen with whole lines, the newe discovered countries have theyr bounds drawen wyth points or broken lines, whereby the reader shall at the firste sight see both the shape and fashion of the whole universall face of the earth, compared all togyther, and also all the severall partes thereof, whether they were of old time discovered, or of late yeares, the which mappe, though it be roughly framed, withoute degrees of longitude or latitude, yet is it sufficient for the purpose it was ordeyned, for heerein, as in all the rest of this discourse, of the three voyages of our worthy Generall Captayne Frobisher, my intente is, more to sette out simply the true and playne proceeding and handling of the whole matter, than to use circumstance of more words, or fyne eloquent phrases, wherein if I shoulde once goe about to entangle myselfe, it would doe nothing else, but bewray my owne ignorance, and lack of schole skyll. Therefore, of me there is nothing else to be loked for, but such playne talke and writing, as souldiers and marriners doe use in theyr dayly meetings and voyages, and this of necessity must anye man use that will deale with suche a matter as thys is, although he were curious to the contrarie.

By this discourse and mappe, is to be seene, the valiante courages of men in this later age within these eighty yeares, that have so muche enlarged the boundes of the worlde, that now we have twice and thrice so muche scope for oure earthlie peregrination, as we have hadde in times past, so that nowe men neede no more contentiously to strive for roume to build an house on, or for a little turffe of ground, of one acre or two, when greate countreys, and whole worldes, offer and reache out themselves, to them that will first voutsafe to possesse, inhabite, and till them. Yea, there are countreys yet remayning withoute maysters and possessors, whiche are fertile to bring forth all manner of corne and grayne, infinite sortes of land, cattell, as horse, elephantes, kine, sheepe, great varietie of flying fowles of the ayre, as phesants, partridge, quayle, popingeys, ostridges, etc., infinite kinde of fruits, as almonds, dates, quinces, pomgranats, oringes, etc., holesome, medicinable, and delectable. Greate varietie of floures continuallie springing, winter and sommer, beautifull for coloure, odififerous, and comfortable. Abundance of faire hilles and valleys, furnished with all maner woddes and pleasante rivers. Millions of newe fashions, and strange beastes and fishes, both in sea and fresh waters. Mountaines bringing forth all maner of mettals, as gold, silver, yron, etc. All sorts of pretious stones and spices, in all which land wanteth nothing that may be desired, eyther for pleasure, profit, or necessarie uses. Which sundrey countreys to possess and obteyne, as it is an easie thing, so would I not have our Englishe nation to be slacke therein, least perhaps agayne they overshoote themselves, in refusing occasion offered, as it was in the time of King Henry the seventh, when all the West Indies were firste profered to the Englishmen, to be given into their handes, whiche they little regarding, was afterwards offered to the Spaniards, who presently accepted the occasion, and now enjoy the infinite treasure and commoditie thereof. I would not wishe Englishmen to be now unlike themselves, for in all the later discoveries the English nation hath bin as forward as any other. As firste, by their navigations northeastward, the bounds of Europe were made perfect on the north syde, for Ptolomie, Strabo, and al other geographers lefte it described but onley to the ilandes Orcades in Irelande, and Hyperboreas Montes in Sarmatia, and finding the land on the north side of Germany, Poland, Moscovia, and Asia, to extend northward, they left it confusedly, and knew not whether it reached to the pole as one firme lande, or whether it were devided by some sea they knew not. But this doubt hath long since bin dissolved, by the valiant attempt of Sr. Hugh Willoughby Knight, who (as I said before) in this noble discoverie, died an honourable death, and the voyage hath since been perfected by ye two brethren the Borowes, and other valiant yong men of our time eastward, beyonde ye great River Obij, as farre as ye empire of ye great Cam or Cane of Tartaria, as appeereth in my general mapp by ye pricked bounds thereof. That voyage was then taken in hand, of ye valiant Knight, with pretence to have gone eastward to the rich countrey of Cataya, and was grounded briefely upon these reasons. First, bicause there was a unicornes horne found upon the coast of Tartaria by the River Obij, which (said he) was like by no other ways to come thither, but from India or Cataya, where the saide

West Indies profered to the English nation.

The bounds of Europe perfected by Englishmen.

The two Barowes.

Reasons to prove the passage by the north-east.

unicornes are only found, and that by some sea bringing it thither. Also a fisherman of Tartaria reporteth, y[t] he sailed verye farre south-eastward, and found no end of sea, or likelyhoode therof. Lastly, a Tartarian, inhabiting neere y[e] Scithian Sea, reported such a streame and currant to runne there continually, towards the west, that if you cast anything therein, it would presently be caried out of your sight towards the west, whereby necessarily foloweth, ther should be some passage to some larger sea, wherein this continual streame might emptie itself. And by y[e] experience of this voyage, it was found, y[t] the frozen zones were not frozen, but habitable and navigable, a thing y[t] almost all the old philosophers did deny, and went about with sundry reasons to impugne, for in this voyage to Moscovia, our men passed beyond seventy-two degrees in north latitude, wheras y[e] frosen zone beginneth at sixty-six degrees and a halfe. This enterprise, although it toke not effect, to finde y[e] passage to Cataya eastward, because y[e] worthy knight, the chief author therof, dyed in y[e] way thither, yet hath it bin very beneficiall to England, in finding out y[e] trace to S. Nicholas, both for y[e] maintenance of y[e] navie, and the yerely profit is reaped therby, the which voyage is known to be more dangerous and painful, than any y[e] Spanyards or Portugals have ever dealt in, for they being borne in a somewhat hote countrey, hapned to deale with easie voyages, although they were long out, not much differing from their own temperature.

Commodities by Moscovie voyage.

And I thinke, a man mighte be bolde to saye, that in all their long voyages, to the East and West Indies, they were never so muche distressed and oppressed with so infinite numbers, and sundrie kindes of dangers, as oure valiante Generall Captayne Frobisher, and his companye were in every one of these his three voyages, as readyng it, you shall understand more at large. And yet they courageously persist and continue on their purposed enterprise, and will not surceasse untill they have (God willing) found oute that long wished passage to Cataya, to the everlasting renoune, glorie, and fame of the English nation.

Also, the valor of the Englishmen, did first of all discover and finde out all that part of America, whiche nowe is called Baccalaos: for Sebastian Cabot, an Englishman, borne in Bristow, was by commandment of Kyng Henry the seaventh in anno 1508, furnished with shipping, munition, and men, and sayled along all that tract, pretending to discover the passage to Cataya, and went alande in many places, and brought home sundry of the people, and manye other things of that countrey, in token of possession, beeing (I say) the firste Christians that ever there sette foote on land.

Also, the sayde Englishman Cabot, did first discover, at the procurement of the Kyng of Spayne, all that other porte of America, adjoyning next beyond Brasill, lying aboute the famous river called Rio de la Plata.

Also, the Englishmen have made sundrye voyages to Guinea and Binny, although the Spanyardes and Portugalles, bycause of their neare dwelling thereunto, got thyther the firste starte of them, and there prevented them in building townes and castels, whereby appeareth, that the English nation, by their long and dangerous navigations, have diligently and paynefullie searched out by sea the temperature of all the zones, whether they were burning, frosen, hot, colde, or indifferent, even from the pole Artike to the equinoctiall, and crossing it also passed beyonde the tropicke of Capricorne, and returned agayne. And therefore, as we are inferioure to no other nation, in making greate and long voyages by sea, so knowe I no nation comparable unto us in taking in hande long travels and voyages by lande. For what nation is it that hath ever had such a long trade by land as is the Englishmans into Persia, which, besides two monethes sayling by sea along the weast and northerne coastes of Norway and Lapland, by Wardhouse unto the Bay of Saint Nicholas, it remayneth more in

Englishmene great travellers by land.

voyage by land and fresh rivers, aboute three thousande Englishe myles: for from the merchantes house at Saint Nicholas, by the river Duina and Lughana, to the citie Volugda, is compted seaven hundred English miles; from thence to the citie Yearuslaue, standing upon the great river Volga, travelling by only land, is reckned about one hundred and forty miles, where the merchants making new shipping for the freshe river Volga, goe eastwarde aboute seven hundred miles: then the sayde river turning agayne south by many windings, at the last by the greate citie Astracan, delivereth it selfe into the south side of the Sea Caspium, that tract being above nine hundred miles: then after in two or three dayes, with a good winde crossing the Caspium Sea, they arrive at a port named Bilbill, where after by lande journeying with camels in one and twenty dayes, Voyage to Persia. being almost six hundred miles, they come to the famous city of Tauris or Teuris, being the greatest citie of Persia for trade of merchandise. This long and paynefull voyage by land was taken in hand by a worthy gentleman, Mayster Anthony Jenkenson, who made thereof a plat, with the first particular description that I have seene of the whole countrey of Moscovia whiche is yet extant, and therefore the Englishmen are to be preferred before all other nations in making long voyages by lande. The Spanyards and Portugalles undoubtedly are worthye immortal fame and glorie, for their greate enterprises and good successes they have therein: yet have they never seene nor hard such straunge and extraordinarie accidents of the sphere as hath happened unto the Englishmen. For neyther Spaniarde nor Portugal ever sawe in all their long voyages, the sun and the moone to make whole and perfect revolutions above the horizon, as our men yearely do see in their voyage to Moscovia, where when they abide any time at Wardhouse they see the sunne goe continually above ground the space of above two moneths togither, where if they take not great heed, they shall not know what day of the moneth it is, after the order of our calender, for that they have no nights. But yet bycause once everye twenty-four houres the sunne draweth neare to the horizon in the north parts, it is there commonly shadowed with vapours and thicke fogges, whiche usually rise from the earth, and seeme a little to shadowe the bodye of the sunne; and that lowest approaching of the sun to the earthward, they counte night, and so make good enough reckning of the days of the moneth, according to our usuall fashion. But one inconvenience there is that dismayeth and deterreth moste men (though they be of valiant courage) from taking in hande long voyages, eyther by sea or by lande, and that is the newe and uncustomed elements, and the extreme ayres of hot and cold, whereby (as some think) if they travel far northward, they shall be frozen to death in the harde congealed and frozen sea: and again, if they travel far toward the south, they fear they should be parched and broyled to death with the extreme heat of ye middle burnyng zone: or else if perhaps they escape alive, yet at least they shold be burned as black as a cole, as the Indians or black Moors there are; and this to believe they are partlye perswaded by the sight of those Indians, and partlye by the perswasions of certaine philosophers, who went with reasons to prove that between the two tropicks was no dwelling or being for the extreme heate, the sunne beating on them continuallye: neyther neare eyther pole, for the extreme frostes, colde and snow whiche continuallye hath there (from the beginning of the world as some thinke) increased, the sunne being so farre distante from them. Which opinion of some, bycause it importeth very much, I thought good here to do my indevour to refell, both bycause I know the contrarie by my owne experience, and also for that I finde the course of the sunne in zodiacke (which God hath ordayned to give light and life to all things) can induce no such kinde of extremitie: and so, lastly, to confirme all partes of the worlde to be habitable.

Day of two months.

EXPERIENCES AND REASONS OF THE SPHERE, TO PROVE AL PARTES OF THE WORLD HABITABLE, AND THEREBY TO CONFUTE THE POSITION OF THE FIVE ZONES.

Experience to prove that *Torrida Zono* is habitable.

First it may be gathered by experience of our Englishmen in *an.* 1553; for Captain Windam made a voyage with merchandise to *Guinea*, and entred so far within the *Torrida Zona*, that he was within three or four degrees of the Equinoctial, and abiding there certain moneths, returned with gain. Also the Englishmen made another voyage verye prosperous and gainfull, *an.* 1554, to y[e] Coasts of *Binin*, lying east from *Guinea*, being within three degrees of y[e] Equinoctial. And yet it is reporteth of a truth, that al y[e] tract from *Cape de las Palmas*, trending by *C. de tres Puntas*, alongst *Benin*, until the Ile of Saint Thomas (which is perpendicular under the Equinoctial): al y[e] whole bay is more subject to many bloming and smothering heates, with infectious and contagious ayres than any other place in al *Torrida Zona*, and the cause thereof is some accidents in y[e] land. For it is most certain that mountains, seas, woods, and lakes, etc., may cause through their sundrie kind of situation sundry straunge and extraordinarie effects whych the reason of the clyme otherwise woulde not give. I mention these voyages of oure Englishemenne, not so much to prove that *Torrida Zona* may be, and is inhabited, as to shew their readynesse in attempting long and dangerous navigations. We also among us in England have blacke Moores, Ethiopians, out of all partes of Torrida Zona, whiche after a small continuance can wel endure the colde of our countrey, and why should not we as well abide the heate of their countrey. But what shoulde I name anye more experiences, seeying that all the coastes of Guynea and Bynnin are inhabited of Portugals, Spanyards, French, and some Englishmen, and there have built castels and townes. Onely this I will say to the merchants of London that trade yeerely to Marochus, it is very certayne that the greatest part of the burning zone is far more temperate and coole in June than the country of Marochus, as shall appeere by these reasons and experiences following. For let us first consider the breadth and bignesse of this burning zone (which, as every man knoweth, is forty-seven degrees eache tropicke, whiche are the bounders thereof) being twenty-three degrees and a halfe distant from the Equinoctiall. Imagine againe two other paralels on each side the Equinoctiall one, eyther of them distant from the Equinoctiall about twentie degrees, whiche paralels maye be described eyther of them twice a yeare by the sunne beinge in the firste degrees of Gemini the eleventh of May, and in Leo the thirtenth of July, having north latitude. And agayne, the sunne beeyng in the first degrees of Sagittarius the twelfth of November, and in Aquarius the ninth of January, havyng south latitude, I am to prove by experience and reason that all that distance included between these two parralels last named (conteyning fortye degrees in latitude, goyng rounde aboute the earthe, according to longitude) is not only habitable, but the same most frutefull and delectable, and that if anye extremitie of heate bee the same not to bee within the space of twentye degrees of the Equinoctiall on eyther side, but onely under and about the two tropicks, and so proportionally the nearer you do approache to either tropicke the more you are subject to extremitie of heate (if any suche be), and so Marochus being scituate but six or seven degrees from the Tropicke of Cancer, shall be more subject to heate than any place under the Equinoctiall line.

Marochus more hote than about the equinoctiall.

And first by the experience of sundrie men, yea thousands travailers and merchaunts to the East and Weast *Indies* in many places, both directly under, and harde by the Equinoctiall, they with one consent affirme that it aboundeth in the middest of *Torrida Zona* with all maner of grain, hearbes, grasse, fruite, wood, and cattell, that we have here, and thousands other sortes farre more holesome, delectable, and pretious than anye wee have in these northerne climates, as very well shall appeare to him that wil reade the Histories and Navigations of such as have travelled *Arabia, India intra* and *extra, Gangem,* the Ilandes Moluccæ, America, etc., which all lye about ye middle of ye burning zone, where it is truly reported that the great hearbes, as are radishe, lettuce, colewortes, borage, and suche like, doe waxe ripe, greater, more saverie and delectable in taste than ours within sixteene dayes after the seed is sowen. Wheat being sowed the first of Februarie, was found ripe the firste of May, and generally, where it is lesse fruitefull, the wheate will be ripe the fourth moneth after the seede is sowne, and in some places will bring forth an eare as bigge as the wriste of a man's arme, containing a thousand graines. Beanes, pease, etc., are there ripe twice a yeare. Also grasse being cutte downe will growe up in sixe dayes above one foot highe. If our cattell be transported thither within a small time their yong ones become of bigger stature and more fatte than ever they would have been in these countries. There are found in everie wood in great numbers such timber trees as twelve men holding handes togither are not able to fadome. And to be short, all they that have bene there, with one consent affirme that there are the goodlyest greene meddowes and playnes, the faireste mountaines, covered with all sorts of trees and fruits, the fairest vallies, the goodliest pleasaunt fresh rivers, stoared with infinite kinde of fishes, the thickest woods, greene and bearing fruite al the whole yeare, that are in al the worlde. And as for gold, silver, and al other kinde of metals, al kind of spices, and delectable fruites, both for delicacie and health, are there in such abundance as hitherto they have bene thought to have bene bred no where else but there. And in conclusion, it is now thought that no where else but under the Equinoctiall, or not far from thence, is the earthlye Paradise and the only place of per-

Marvellous fruitful soile under the equinoctiall.

Great trees.

Commodities and pictures under the equinoctiall.

fection in the world. And that these things may seeme the lesse strange bycause it hath bin accompted of the old philosophers that there could nothing prosper for the extreme heate of the sunne continually going over their heades in the zodiacke, I thoughte goode here to alleage suche naturall causes as to mee seeme verie substanciall and sure reasons. First, you are to understande that the sunne doeth worke his more or lesse heate in these lower parts by two meanes, the one is by the kind of angle that the sun-beames doeth make with the earth, as in all *Torrida Zona*, it maketh perpendicularly righte angles in some place or other at noone and towardes the two Poles very oblique and uneven angles. And the other meane is the longer or shorter continuance of the sunne above the horizon. So that wheresover these two causes do most concurre there is moste excesse of heate; and when this one is wanting the rigor of the heate is lesse. For though the sunne-beames do beate perpendicularly upon any region subject to it, if it hath no continuance or abode above the horizon to worke his operation in, there can no hote effect proceede. For nothing can be don in a momente. And this seconde cause, *mora supra horizontem*, the time of the sunne's abiding above the horizon, the old philosophers hence remembered, but regarded only the manner of angles y[t] the sun-beames made with the horizon, which if they were equall and right, the heate was the greater, as in *Torrida Zona;* if they were unequall and oblique, the heat was the lesse, as towardes both Poles, which reason is very good and substancial; for the perpendicular beames reflect and reverberate in themselves, so that the heate is doubled, every beame striking twice, and by uniting are multiplied and continue strong in forme of a columne. But in our latitude of fifty and sixty degrees the sun-beams descend oblique and slanting wise, and so striketh but once and departeth, and therefore oure heate is the lesse for any effect that the angle of the sun's beames make. Yet, because we have a

Heat is caused by two meanes, that is. by his maner of angle and by his continuance.

Note thys reason.

longer continuance of the sunne's presence above the horizon than they have under the Equinoctiall, by whiche continuance the heate is increased, for it shineth to us xvj or xviij houres sometime, when it continueth with them but twelve houres alwayes.

And againe, oure night is very shorte, wherein colde vapors use to abound, being but six or eight houres long, whereas theirs is alwayes twelve houres long, by which two advantages of long dayes and shorte nights, thoughe we wante the equalitie of angle, it commeth to passe that in sommer oure heate here is as greate as theirs is there, as hath bin proved by experience, and is nothing dissonant from good reason.

Therefore, whosoever wil rightly way the force of colde and heate in any region, muste not onelye consider the angle that the sunne beames make, but also the continuance of the same above the horizon. As firste to them under the equinoctiall the sunne is twice a yeare at noone in their zenith perpendicular over their heades, and therfore during the ii houres of those two dayes the heat is very urgent, and so perhaps it will be in four or five days more, an houre everye daye, untill the sunne in his proper motion, have crossed the equinoctiall, so that this extreame heate caused by the perpendicular angle of the sunne beames, endureth but two houres of two dayes, in a yeare.

But if any man say the sunne maye scalde a good while, before and after it come to the meridian, so farre forthe as reason leadeth, I am content to allow it, and therefore I will measure and proportion the sunne's heate, by comparing the angles there, with the angles made here in England, bicause this temperature is best knowen to us. As for example, the 11th day of March, when under the equinoctiall it is halfe houre past eight of the clocke in the morning, the sun will be in the east about thirty-eight degrees above the horizon, bycause there it riseth always at six of the clock, and moveth every hour fifteen degrees, and so high very neare will it be with us at *London* the saide 11th of March at noone. And therefore looke what force the sunne hath with us at noone, the 11th of March, the same force it seemeth to have under the equinoctial at half houre past eight in the morning, or rather lesse force under the equinoctiall. For with us the sunne had beene alreadye sixe houres above the horizon, and so had purified and clensed all the vapours, and thereby his force encreased at noone, but under the equinoctiall, the sunne having been uppe but two and half houres hadde sufficient to doe, to purge and consume the cold and moyst vapors of the long night past, and as yet had wrought no effect of heate. And therefore I may boldely pronounce, that there is much less heate at halfe houre past eight under the equinoctiall, than is with us at noone (*à fortiori*). But in Marche, we are not onlye contented to have the sunne shining, but we greately desire the same. Likewise the 11th of June, the sunne in our meridian is sixty-two degrees highe at London; and under the equinoctiall it is so high after ten of the clocke, and seeing then it is beneficiall with us, *à fortiori*, it is beneficial to them after ten of the clocke.

And thus have we measured the force of the sun's greatest heate, the hottest dayes in the yeare, under the equinoctiall, that is, in March and September, from sixe tyll after tenne of the clocke in the morning, and from two untill sunne set. And this is concluded, by respecting only the first cause of heate, which is the consideration of the angle of y[e] sunnes beames, by a certaine similitude, that whereas the sunne shineth never above twelve houres, more than eight of them would be coole and pleasant even to us, much more to them that are acquainted alwayes with suche warme places. So there remayneth lesse than foure houres of any excessive heate, and that only in the two sommer dayes of the yare, that is, the 11th of March, and the 14th of September, for under the equinoctiall, they have two sommers the one in

March, and the other in September, which are our spring and autumne; and likewise two winters, in June and December, which are our sommer and winter, as may well appeare to him that hath onelye tasted the principles of the sphere. But if the sunne be in eyther tropicke, or approaching neare thereunto, then may we more easilye measure the force of his meridian altitude, that it striketh upon the equinoctial. As for example, the twelfth of June, the sunne will be in the first degree of cancer. Then loke what force the heate of the sunne hath under the equinoctiall, ye same force and greater, it hath in all that paralel, where the pole is elevated betweene forty-seven and forty-eight degrees. And therefore *Paris* in *France*, the 12th daye of June, sustayneth more heate of the sunne, than Saint Thomas Ilande, lying neere the same meridian, doth likewise at noone, or the ilandes *Taprobana, Molucco,* or the firme lande of *Peru* in *America,* which all lye underneath the equinoctial. For upon the 12th day of June aforesaid, the sunne beames at noone doe make an Isocheles triangle, whose *vertex* is the center of the sunne, the *basis* a lyne extended from Saint Thomas Ilande, under the equinoctiall, unto *Paris* in *France,* neare the same meridian; therefore the two angles of the base, must needes be equall ꝑ 5, *primi, ergo* the force of the heate equall, if there were no other cause, than the reason of the angle, as the olde philosophers have appointed. But bycause at *Paris* the sun riseth two houres before it riseth to them under the equinoctiall, and setteth likewise two houres after them, by means of the obliquity of the horizon, in which time of the sunnes presence 4 houres in one place more than the other, it worketh some effect more in one place than in the other, and being of equall height at noone, it muste then needes follow to be more hote in the paralell of *Paris* than it is under the equinoctiall. Also this is an other reason, that when the sun setteth to them under the equinoctiall, it goeth very deep and lowe under their hori-

Paris in France is as hote as under the equinoctiall in June.

zon, almost even to their *antipodes,* whereby their twylights are very shorte, and their nights are made verye extreame darke and long, and so the moysture and coldenesse of the long nightes wonderfully encreaseth, so that at length the sun rising can hardly in many houres consume and drive away the colde humoures and moyst vapours of the nighte paste, whiche is cleane contrarye in ye paralel of *Paris.* For ye sun goeth under their horizon but verye little, after a sloping sorte, whereby their nights are not verye darke, but lightsome; as looking into the north in a cleare night withoute cloudes, it doeth manifestlye appeare their twylightes are long, for the paralel Cancer cutteth not the horizon of *Paris* at right angles (but at angles very uneven and unlike), as it doth the horizon of the equinoctiall. Also the sommer day at *Paris* is sixteene houres long and the night but eight; where contrarie wise, under the equinoctial, the day is but twelve houres long, and so long is also the nighte, in what soever paralel the sun be; and therefore looke what oddes and difference of proportion there is betweene the sunnes abode above the horizon in *Paris* and the abode it hath under the equinoctiall (it being in Cancer), the same proportion would seeme to be between the heate of the one place and heate of the other: for other things (as the angle of the whole arcke of the sunnes progresse that day in both places) are equall.

The twylights are shorter and the nights darker under the equinoctial than at Paris.

But under the equinoctiall, the presence and abode of the sunne above the horizon is equall to his absence and abode under the horison, eache being twelve houres. And at *Paris* the continuance and abode of the sunne is above the horizon sixteene houres long and but eight hours absence, whiche proportion is double, from whiche, if the proportion of the equalitie be subtracted to find the difference, there will remaine stil a double proportion; whereby it seemeth to followe, that in June the heate at *Paris* were double to the heate under the equinoctiall. For (as I have saide) the angle of the sunne beames are in all points equall, and the cause of difference is *Mora solis supra horizontem*, the staye of the sunne in the one horizon more than in the other. Therefore, whosoever could finde out in what proportion the angle of the sunne beames heateth, and what encrease the sunnes continuance doeth adde thereunto, it might expresly be sette downe what force of heate and cold is in all regions.

Thus you partely see, by comparing a clymate to us well knowe and familiarlye acquainted by lyke height of the sunne in bothe places, that under the equinoctiall in June is no excessive heate, but a temperate ayre, rather tendinge to colde. For as they have there for the moste parte a continuall moderate heate, so yet sometime they are a little pintched wyth colde, and use the benefite of fyre as well as wee, especiallye in the evening when they goe to bedde; for as they lye in hanging beddes tyed faste in the upper parte of the house, so wyll they have fyres made on both sides their bed, of which two fires, y[e] one they devise superstitiouslye to drive awaye spyrites, and the other to keep away from them the coldnesse of the nights.

They use and have neede of fire under the equinoctiall.

Also in many places of *Torrida Zona*, especially in the higher landes somewhat mountainous, the people a little shrinke at the colde, and are often forced to provide themselves clothing, so that y[e] Spaniards have found in the West *Indies* many people clothed, especially in winter; whereby appeareth, that with their heate there is colde intermingled, else would they never provide this remedy of clothing, which to them is rather a griefe and trouble than otherwise. For when they go to warres they wil putte off al their apparell, thinking it to be cumbersome, and wil alwayes goe naked, y[t] they thereby might be more nimble in their flight. Some there be that thinke y[e] middle zone extreme hot, bycause y[e] people of y[e] countrie can and do live withoute clothing, wherein they childishly are deceived, for oure clime

Colde intermingled with heate under the equinoctiall.

rather tendeth to extreamitie of colde, bicause we cannot live without clothing: for this our dubble lining, furring and wearing so many cloths, is a remedy against extremitie, and argueth not ye goodnesse of ye habitation, but inconvenience and injury of cold: and that is rather ye moderate, temperate, and delectable habitation, where none of these troublesome things are required, but that we may live naked and bare, as nature bringeth us forth. Others again imagine ye midle zone to be extreme hote, bycause the people of *Affrica*, especially ye Ethiopians, are so cole blacke, and their haire, like wooll, curled short, which blacknesse and curled haire they suppose to come only by ye parching heate of ye sun, which how it should be possible I cannot see. For even under the equinoctiall in *America*, and in ye East *Indies*, and in the Ilands *Moluccæ* ye people are not blacke but white, with long haire uncurled as we have; so that if the Ethiopians blacknesse came by the heate of the sune, why shoulde not those *Americans* and *Indians* also be as blacke as they, seeyng the sunne is equally distant from them both, they abiding in one paralel: for the concave and convex *superfices* of the orbe of the sun is concentrike and equidistant to ye earth, except any man should imagine somewhat of *Aux* [Arx] *Solis* and *Oppositum*, whiche indifferently may be applied as wel to ye one place as to the other. But ye sunne is thought to give no otherwise heate but by way of angle in reflection, and not by his neerenes to ye earth; for throughout al Africa, yea in ye middest of ye middle zone, and in all other places, upon ye tops of mountains, there lyeth continuall snow, which is nerer to the orbe of the sunne than ye people in the valley by so muche as the height of these mountaynes amount unto; and yet the sunne, notwithstanding his nerenesse, can not melt the snow, for want of convenient place of reflexions. Also the middle region of the ayre, where all the hayle, frost and snowe is engendered, is neerer unto the sunne than the earth is, and

Ethiopians blacke with curled haire.

The sunne heateth not by his nearnesse, but only by reflexion.

yet there continueth perpetuall colde, bycause there is nothing that the sunnes beames may reflect against, whereby appeareth the neerenesse of the body of ye sunne worketh nothing. Therefore, to returne again to the blacke Moores, I my selfe have seene an Ethiopian as blacke as a cole brought into Englande, who taking a faire Englishe woman to wife, begatte a sonne in all respects as blacke as the father was, although England were his native countrey and an English woman his mother: whereby it seemeth this blacknesse proceedeth rather of some natural infection of that man, whiche was so strong, that neyther ye nature of ye clime neyther the good complexion of the mother concurring coulde any thing alter, and therefore we can not impute it to ye nature of the clime. And for a more fresh example our people of Meta Incognita (of whome and for whome thys discourse is taken in hande) that were broughte this last yeare into Englande, were all generallie of the same colour that many nations be lying in the middest of the middle zone. And this their coloure was not only in the face, whiche was subjecte to sunne and ayre, but also in their bodies, which were still covered with garments as oures are, yea the verye suckinge childe of twelve moneths age hadde his skinne of the very same coloure that most have under the equinoctiall; which thing can not proceed by reason of the clime, for that they are at least tenne degrees more towards the north than we in Englande are; no, the sunne never commeth neere their zenith by 40 degrees, for in effect they are within three or four degrees of that which they call the frosen zone, whereby it followeth that there is some other cause than the clymate, or the sunnes perpendicular reflection, that shoulde cause the Ethiopians great blacknesse. And the most probable cause to my judgemente is, that this blacknesse proceedeth of some naturall infection of the first inhabitants of that countrey, and so all the whole progenie of them descended are still poluted

A blacke Moores son borne in England.

The coloure of the people in Meta Incognita.

with the same blot of infection. Therefore it shall not be farre from our purpose to examine the first originall of these blacke men, and how by lineall discente they have hitherto continued this blacke. It manifestly and plainely appeareth by holy Scripture that after the generall inundation and overflowing of the earth, there remained no more men alive but Noe and his three sons, Sem, Cham, and Japeth, who only were lefte to possesse and inhabit the whole face of the earth: therefore all the land that untill this daye hath bin inhabited by sundry discents, must needes come of the offspring eyther of Sem, Cham, or Japhet, as the onely sonnes of Noe, who all three being white and their wives also, by course of nature should have begotten and brought forth white children. But the envie of our great and continuall enemie the wicked spirit is such, that as he could not suffer our old father Adam to live in the felicitie and angelike state wherein he was first created, but tempting him sought and procured his ruine and fal: so againe, finding at this floud none but a father and three sonnes living, he so caused one of them to transgresse and disobey his fathers commandement, that after him all his posteritie shoulde be accursed. The fact of disobedience was this. When Noe at the commandement of God had made and entered the Arke, and the floud gates of heaven were opened, so that the whole face of the earth, every tree and mountaine was covered with abundance of water, he straightly commanded his sonnes and their wives, that they should with reverance and feare behold the justice and mighty power of God, and that during the time of the floud, while they remained in the Arke, they should use continence and absteine from carnall copulation with their wives: and many other preceptes he gave unto them, and admonitions touching the justice of God in revenging sinne and his mercie in delivering them who nothing deserved it. Which good instructions and exhortations notwithstanding, his wicked sonne Cham

The cause of the Ethiopians blacknesse.

The Arke of Noe.

Chus ye sonne of Cham accursed.

disobeyed, and being persuaded that the first child borne after the flood (by right and law of nature) should inherit and possesse all the dominion of the earth, he, contrarie to his fathers commandement, while they were yet in the Arke used company with his wife, and craftily went about thereby to disinherit the offspring of his other two brethren; for the which wicked and detestable fact, as an example for contempte of Almightie God and disobedience of parents, God would a sonne shuld be borne whose name was Chus, who not only itselfe, but all his posteritie after him, should be so black and lothsome that it might remaine a spectacle of disobedience to all the world. And of this blacke and cursed Chus came all these blacke Moores which are in Africa, for after the water was vanished from off the face of the earth and that the land was drie, Sem chose that part of the land to inhabit in which now is called Asia, and Japhet had that which now is called Europa wherein we dwell, and Africa remained for Cham and his blacke sonne Chus, and was called Chamesis after ye fathers name, being perhaps a cursed, dry, sandy, and unfruiteful ground, fit for such a generation to inhabit in. Thus you see y^t the cause of ye Ethiopians blacknesse is the curse and natural infection of blood and not the distemperature of the clymate, which also may be proved by this example, that these black men are found in all partes of Africa, as well withoute the tropicks as within, even unto Capo d' buona Speranza southward, where by reason of the sphere should be the same temperature that is in Spayne, Sardigna, and Sicilia,[1] where all be of very good complexions. Wherefore I conclude that the blacknesse proceedeth, not of the hotenesse of the clime, but as I sayd of the infection of bloud, and therfore this their argumente gathered of the Africans blacknesse is not able to destroy the temperature of the middle zone. We may therefore very well be assertayned, that under the equinoctiall is

Africa was called Chamesis.

[1] In *Hakluyt*, it stands thus "Sicilia, Morea, and Candie."

the most pleasant and delectable place of the world to dwell in, where, although the sunne for two houres in a yeare be directe over their heads, and therefore the heate at that time somewhat of force, yet bycause it commeth so seldome and continueth so small a time, when it commeth it is not to be wayed, but rather the moderate heate of other times is all the yeare to be remembered. And if the heate at any time should in the shorte day waxe somewhat urgent, the coldnesse of the long night there would easilie refreshe it, according as Honterus[1] sayth, speaking of the temperature under the equinoctiall.

Greatest temperature under the equinoctiall.

Quodq. die solis violento incanduit æstu
Humida nox reficit, paribusq. refrigerat horis.

If the heate of the sunne in the day time doe burne or parch any thing, the moysture of the night dothe coole and refresh the same againe, the sunne being as long absente in the night as it was present in the day. Also our author of the *Sphere*, Joannes d' Sacrobosco,[2] in the chapter of the Zodiacke, deriveth the etimologie of *Zodiacus*, of the Greeke word *Zoe*, which in Latin signifyeth *Vita*, life, for out of Aristotle he alledgeth, that *secundum accessum et recessum solis in Zodiaco, fiunt generationes et corruptiones in rebus inferioribus:* according to the sunnes going to and fro in the zodiake, the inferiour bodies take their causes of generation and corruption. Then it followeth, that where there is most going too and fro there is most generation and corruption: whiche must needes be betweene the two tropikes, for there the sunne goeth too and fro most, and no where else but there. Therefore betweene the two tropikes, that is, in the midle zone, is greatest increase, multiplication, genera-

Under the equinoctiall is greatest generation.

[1] Joannes Honterus or Honter of Cronstadt in Transilvania, in the 3rd Book of his *Rudimentorum Cosmographicorum libri quatuor*, 1573.

[2] John Holywood, better known as Joannes de Sacrobosco, the famous author of the *Sphæra Mundi*, an astronomical work which maintained its reputation for four centuries. He was a native of Yorkshire, and died in 1256.

tion and corruption of things; which also we find by experience, for there is sommer twice in the yeare and twice winter, so that they have two harvests in the yeare and continuall spring. Seing then the middle zone falleth out so temperate, it resteth to declare where the hottest part of the world should be, for we fynde some places more hote than others. To answere this doubt, reason persuadeth the hotest place in the worlde to be under and aboute the two tropickes, for there more than in anye other place doe both the causes of heate concurre, that is, the perpendicular falling of the sun beames at right angles, and a greater continuance of the sunne above the horizon (the pole there being elevated three or foure and twentie degrees). And as before I concluded, that though the sunne were perpendicular to them under the equinoctiall, yet bycause the same continued but a small tyme (theyr dayes being short and theyr nightes long), and theyr speedie departure of the sunne from their zenith, bycause of the suddayne crossing of the zodiacke with the equinoctiall, and that by such continuall course and recourse of hote and colde the temperature grew moderate.

Greatest heate under the tropiks.

So now to them under y[e] two tropikes, the sun having once by his proper permotion declined 20 degrees from the equinoctiall, beginneth to draw neere theyr zenith, which may bee (as before) aboute the eleventh day of May, and then beginneth to send his beames almost at right angles, about which tyme the sunne entreth into the first degree of Gemini, and with this almost right angle the sunne beames will continue untill it be past Cancer, that is, the space of two moneths every day at noone, almost perpendicular over their heads, being then the time of *Solstitium Æstiuale*, whiche so long continuance of the sun aboute their zenith maye cause an extreame heate (if anye be in the world) but of necessitie farre more heate than can bee under the equinoctiall, where the sunne hathe no suche long abode in the zenith, but passeth away therehence very quickly. Also under the tropikes the day is longer by an houre and a halfe than it is under the equinoctiall, wherefore the heate of the sunne, having a longer tyme of operation, must needes be encreased, especially seeing the nighte, wherein colde and moysture doe abounde under the tropikes is lesse than it is under the equinoctiall. Therefore I gather that under the tropikes is the hottest place, not onely of *Torrida Zona*, but of any other parte of the worlde, especially bycause there both causes of heate doe concurre, that is, the perpendicular falling of the sunne beames two moneths togyther, and the longer abode of the sunnes presence above the horizon. And by this meanes more at large is proved that Marochus in sommer is farre more hote than at any tyme under the equinoctiall, bycause it is situate so neere the tropike Cancer, and also for the length of their dayes. Neyther yet doe I thinke that the regions scituate under the tropikes are not habitable, for they are founde to be verye frutefull also, although Marochus and some other partes of Africa neere the tropike, for the drynesse of the native sandie soile and some accidentes, maye seeme to some to be intemperate for overmuch heate. For Ferdinandus Ouiedus,[1] speaking of Cuba and Hispaniola, ilands of America, lying hard under Cuba. Hispaniola. or by the tropike Cancer, sayeth that these ilandes have as good pasture for cattell as any other countrey in the world.

Also they have most holesome and cleere water and temperate ayre, by reason whereof the heards of beastes are much bigger, fatter, and of better tast than any in Spayne, bycause of the rank pasture, whose moysture is better digested in the hearbe or grasse by continuall and temperate heate of the sunne, whereby being made more fatte and unctious, it is of better and more stedfast nourishment. For continuall and temperate heate dothe not only draw much moysture out of the earth, to the nourishmente of suche things as growe and are engendred in that clyme, but dothe

[1] Oviedo, the Spanish traveller and historian.

also by moderation preserve the same from putrifying, digesting also and condensating or thickening the sayd moyst nourishmente into a gummie and unctious substance, whereby appeareth also that under the tropikes is both holesome, frutefull, and pleasant habitation; whereby, lastly, it followeth that al the middle zone, whiche untyll of late dayes hathe bin compted and called the burning, broyling, and parched zone, is now found to bee the most delicate, temperate, commodious, pleasaunte, and delectable part of the world, and especially under the equinoctiall.

Under the tropics is a moderate temperature.

Havyng nowe sufficiently at large declared the temperature of the middle zone, it remayneth to speake somewhat also of the moderate and continuall heate in colde regions, as well in the nighte as in the daye all the sommer long, and also how these regions are habitable to the inhabitantes of the same, contrarie to the opinion of olde writers.

OF THE TEMPERATURE OF COLDE REGIONS ALL THE SOMMER LONG, AND ALSO HOWE IN WINTER THE SAME IS HABITABLE, ESPECIALLY TO THE INHABITANTS THEREOF.

The colde regions of the worlde are those whiche, tending towarde the pole artike and antartike, are withoute the circuit or bounds of the seaven climates, which, agreeable to the opinion of the olde writers, is founde and sette out in our authore of the *Sphere*, Joannes de Sacrobosco, where he playnely sayeth, that without the seaventh climate, which is bounded by a parallel passing at fiftie degrees in latitude, all the habitation beyonde that to be discommodious and intollerable: but Gemma Phrisius, a late writer, finding England and Scotland to be withoute the compasse of those climates wherein he knew to be very temperate and good habitation, added thereunto two other climates, the utter-

Nine climates.

most paralell whereof passeth by 56 degrees in latitude, and therein comprehendeth over and above the first computation, England, Scotland, Denmarke, Moscovia, etc., which all are rich and mightie kingdomes.

The old writers, perswaded by bare conjecture, went aboute to determine of those places, by comparing them to their own complexions, bycause they felt them to be hardly tolerable to themselves, and so toke thereby an argument of the whole habitable earth, as if a man borne in Morochus, or other part of Barbarie, should at the latter end of sommer, upon the suddayne, eyther naked, or wyth hys thinne vesture, be broughte into England, he would judge this region presently not to be habitable, bycause he being broughte up in so warme a countrey, is not able heere to live, for so sodaine an alteration of the cold ayre; but if the same man hadde come at the beginning of sommer, and so afterwarde by little and little by certaine degrees, had felt and acquainted himselfe with the frost of autumne, it would have seemed by degrees to harden him, and so to make it far more tollerable, and by use after one yeere or two, the ayre woulde seeme to hym more temperate. It was compted a greate matter in the olde time, that there was a brasse pot broken in sunder with frozen water in *Ponthus*, which after was broughte and shewed in *Delphis*, in token of a miraculous cold region and winter, and therefore consecrated to the Temple of Apollo.

A comparison betweene Marochus and England.

This effect being wroghte in the paralell of 48 degrees in latitude, it was presentlye compted a place verye hardlye and uneasily to be inhabited for the greate colde. And howe then can suche men define uppon other regions very farre without that paralell, wh'er they were inhabited or not, seeing that in so neare a place they so grossely mistooke the matter, and others their followers being contented with the inventions of the olde authors, have persisted willingly in the same opinion, with more confidence than consideration of the cause, so lightly was that opinion received, as touching the unhabitable clime neare and under the Poles.

All the north regions are habitable.

Therefore I am at this present to prove y^t all the land lying betweene the laste climate even unto the point directly under either Poles, is or maye be inhabited, especially of suche creatures as are ingendred and bredde therein. For indeed it is to be confessed, that some particular living creature cannot live in every particular place or region, especially wyth the same joy and felicitie, as it did where it was first bredde, for the certaine agreement of nature that is betweene the place, and the thing bredde in that place, as appeareth by the elephant, which being translated and brought out of the second or third climate, though they may live, yet will they never ingender or bring forth yong. Also wee see the like in many kinds of plants and hearbs: for example, the orange tree, although in Naples they bring forth fruit abundantly, in Rome and Florence they will beare onlye faire greene leaves, but not any fruite: and translated into England, they will hardly beare either flowers, fruite, or leaves, but are the next winter pinched and withered with colde: yet it followeth not for this, that England, Rome, and Florence should not be habitable.

Elephant.

Orange tree.

In the proving of these colde regions habitable, I shall be verye shorte, bicause the same reasons serve for this purpose, which were alleaged before in the proving the middle zone to be temperate, especially seeing all heate and colde proceede from the sunne, by the meanes eyther of the angle his beames doeth make with the horizon, or else by ye long or shorte continuance of the sun's presence above ground: so that if the sunnes beames do heate perpendicularlye at righte angles, then there is one cause of heate, and if the sunne doe also long continue above the horizon, then the heate thereby is muche encreased by accesse of this other cause, and so groweth to a kind of extremitie. And these ii causes, as I said before, doe moste concurre under the two tropickes, and

Two causes of heate.

therefore there is the greatest heate of ye worlde. And likewise, where both these causes are most absent, there is greatest want of heate, and encrease of colde (seeing that colde is nothing but the privation and absence of heat), and if one cause be wanting and the other present the effect will grow indifferent. Therefore this is to be understanded, that the nearer anye region is to the equinoctiall the higher the sunne doeth rise over their heades at noone, and so maketh either righte or neare righte angles, but the sun tarryeth with them so much the shorter time, and causeth shorter dayes, with longer and colder nights, to restore the domage of the daye paste, by reason of the moisture consumed by vapour. But in such reasons, over the which the sun riseth lower (as in regions extended towardes eyther pole) it maketh there unequall angles, but the sunne continueth longer, and maketh longer dayes, and causeth so much shorter and warmer nights, as retayning warme vapoures of the daye paste. For there are found by experience sommer nights in Scotland and Gothland very hot, when under the equinoctiall they are found very colde. This benefit of the sunnes long continuance and encrease of the day, doth augment so muche the more in colde regions, as they are nearer the poles, and ceaseth not encreasing, until it come directly under the point of the pole articke, where the sunne continueth above grounde the space of sixe moneths or halfe a yeare togither, and so the daye is halfe a yere longe, that is the time of ye suns being in the north signes, from the first degree of Aries until the last of Virgo, that is all the time from our 10 day of March, untill the 14th of September. The sun therefore during the time of these 6 moneths without any offence or hinderaunce of the nighte, gyveth his influence upon those landes with heate that never ceaseth during that time, which maketh to the great increase of sommer, by reason of the sunnes continuance. Therefore it followeth, that though the sunne be not there very high over their heads to cause right

Hot nights near ye pole. Cold nights under the equinoctiall.

One daye of sixe moneths.

angle beams and to give great heate, yet the sun being there sometime almost 24 degrees high, doth caste a convenient and meane heate which there continueth without hinderance of the night the space of six moneths (as is before saide) during whiche time there foloweth to be a convenient, moderate and temperate heat, or else rather it is to be suspected the heat there to be very great, both for continuance and also *quia virtus unita crescit,* the vertue and strength of heat united in one encreaseth. If then there be suche a moderate heat under the poles, and the same to continue so long time, what shoulde move the olde writers to saye there cannot be place for habitation. And that the certaintie of this temperate heat under both the poles might more manifestlye appeare, lette us consider the position and qualitie of the sphere, the length of the day, and so to gather the heighte of the sunne at all times, and by consequent the quantitie of his angle, and so lastely the strength of his heate.

Those landes and regions lying under the pole and having the pole for their zenith, muste needes have the equinoctiall circle for their horizon, therefore the sunne entring into the north signes, and describing every 24 houres a paralell to the equinoctiall by the diurnall motion of *Primum Mobile,* the same paralels must needes be wholely above the horizon, and so looke how many degrees there are from the fyrst of Aries to the last of Virgo, so many whole revolutions there are above theyr horizon y[t] dwell under the pole, whiche amounteth to 182, and so manye of oure dayes the sunne continueth with them. During whych tyme they have there continuall daye and lighte withoute anye hinderaunce of moiste nightes. Yet it is to be noted that the sunne being in the fyrst degree of Aries, and laste degree of Virgo, maketh his revolution in the very horizon, so that in these 24 houres halfe the body of the sunne is above the horizon and the other halfe is under this only center, describing both the horizon and the equinoctiall circle.

The sun never setteth in a 182 dayes.

Horizon and equinoctiall one under the pole.

And therefore seeing the greatest declination of the sun is almost 24 degrees, it followeth his greatest height in those countries to be almost 24 degrees. And so high is the sun at noone to us in London about y[e] 29 of October, being in the 15 degree of Scorpio, and likewise the 21 of January being in the 15 of Aquarius. Therefore looke what force the sun at noone hath in London the 29th of October, the same force of heate it hathe, to them that dwell under the pole, the space almost of two moneths, during the time of the sommer *solstitium*, and that without intermingling of any colde night: so that if the heate of the sunne at noone coulde be well measured in London (which is verye harde to do, bycause of the long nights, whiche engender greate moysture and colde), then woulde manifestlye appeare by expresse numbers the maner of the heate under the poles, which certainly must needes be to the inhabitants verye commodious and profitable, if it inclyne not to over much heate, and if moysture do not want. London.

For as in October in England we find temperate aire, and have in our gardens hearbes and floures notwithstanding our colde nights, how much more shoulde they have y[e] same good ayre, being continual without night. This heate of ours continueth but one houre while the sunne is in y[e] meridian, but theirs continueth a long time in one height. This our heate is weake, and by the coolenesse of the night vanisheth; that heate is strong, and by continual accesse is still increased and strengthened. And thus by a similitude of the equal height of the sunne in both places, appeareth the commodious and moderate heate of the regions under the poles. Commodious under the poles.

And surely I can not thinke that the divine providence hath made any thing uncommunicable, but to have given such order to all things that one way or other the same shoulde be emploied, and that every thing and place should be tollerable to the next. But especiallye all things in this lower world be given to man to have dominion and use thereof. Therefore wee neede no longer to doubt of the temperate and commodious habitation under the poles during the tyme of sommer.

And al the controversie consisteth in the winter, for then the sun leaveth those regions, and is no more seene for the space of other sixe months, in the which time al the sunnes course is under their horizon for the space of halfe a yeare, and then those regions (saye some) muste needs be deformed with horible darkenesse and continuall nyghte, whiche maye be the cause that beastes can not seeke theyr foode, and that also the cold should then be intollerable. By which double evils al living creatures should be constrayned to die, and were not able to indure the extremitie and injurie of winter and famine ensuing thereof, but that all things shoulde perish before the sommer folowing, when they should bring forth their broode and yong, and that for these causes yᵉ said clime about the pole shold be desolate and not habitable. To al which objections may be answered in this manner: first, that thoughe the sun be absent from them those five months, yet it followeth not there should be such extreame darkenesse, for as the sunne is departed under their horizon, so is it not farre from them. And not so soone as the sunne falleth, so sodainely commeth the darke night, but the evening doth substitute and prolong the daye a good while after by twilight. After which time the residue of yᵉ night receiveth light of the moone and starres untill the breake of the day, which giveth also a certaine light before the sunnes rising, so that by these means the nights are seldom dark, which is verified in all parts of the world, but least in the middle zone under the equinoctiall, where the twylights are short and the nights darker than in any other place, bycause the sun goeth under their horizon so deepe, even to their *antipodes*. Wee see in Englande in the sommer nights, when the sun goeth not far

The nightes under the pole.

The twylights give light under the pole almoste at the winter.

under the horizon, that by the light of the moone and stars wee may travel al night, and if occasion were do some other labour also. And there is no man that doubteth whether our cattel can see to feede in ye nights, seeing wee are so well certified thereof by our experience: and by reason of the sphere, our nights should be darker than any time under the poles. The ending of twylight.

The astronomers consent that the sun, descending from our upper hemisphere at the 18 paralell under the horizon, maketh an end of twylight, so that at length the darke night ensueth, and that afterward in the morning, the sun approaching againe within as many paralels, doth drive away ye night by accesse of ye twylight. Againe, by the position of the sphere under ye pole, the horizon and the equinoctiall are al one. These revolutions therefore that are paralell to the equinoctiall are also paralel to the horizon, so that the sun descending under ye horizon, and there describing certain paralels not farre distant, doeth not bringe darke nights to those regions until it come to the paralels distant 18 degrees from ye equinoctiall, that is, about ye 21 degree of Scorpio, which wil be about ye 4 day of our November and after the winter *solstitium*, ye sun retourning backe againe to ye 9 degree of Aquarius, whiche wil be aboute ye 19 of January, during which time only, that is from ye 4 of November untill the xix day of January, which is about six weeks space, those regions do want ye commoditie of twylights. Therefore, during ye time of these said six moneths of darknesse under ye poles, ye night is destitute of ye benefit of ye sun, and ye said twilights, only for ye space of six weeks or thereabout. And yet neither this time of six weeks is without remedy fro heaven. For ye moone with hir encreased light hathe accesse at that time and illuminateth the moneths, lacking light every one of themselves severally halfe the course of ye moneth, by whose benefite it commeth to passe y^t^ ye night named extreame dark possesseth those But six weeks dark under the pols.

regions no longer than one moneth, neither that continually or al at one time, but this also divided into two sorts of shorter nights, of y[e] which either of them endureth for y[e] space of 15 dayes, and are illuminate of y[e] moone accordingly. And this reason is gathered out of the sphere, whereby we may testifie y[t] the sommers are warme and fruitful, and the winters nights under the pole are tollerable to living creatures. And if it be so that the winter and time of darknes there be very cold, yet hath not nature left them unprovided therefore. For there y[e] beasts are covered with haire so muche the thicker in how much the vehemencie of cold is greater, by reason wherof the best and richest furres are broughte out of the coldest regions. Also the foules of these cold countries have thicker skins, thicker feathers, and more stored of down than in other hot places. Our Englishmen that travel to S. Nicholas, and go a fishing to Wardhouse, enter far within the circle artike, and so are in the frozen zone; and yet there, as well as in Iseland, and all along those northern seas, they finde the greatest store of the greatest fishes that are, as whales, etc., and also abundance of meane fishes, as herings, coddes, haddockes, brettes, etc., whiche argueth, that the sea as well as the land, maye bee and is well frequented and inhabited in the colde countries.

The creatures of that countrie are provided for the cold.

But some, perhaps, will marvel there should be such temperate places in y[e] regions aboute y[e] poles, when at under degrees in latitude, our Captaine Frobisher and his companye were troubled wyth so manye and so great mountaines of fleeting ise, with so great stormes of colde, with such continuall snow on toppes of mountaines, and with such barren soyle, there being neither woodde or trees, but lowe shrubbes, and suche like. To al which objections may be answered thus:—First, those infinite ilandes of ise were engendered and congealed in time of winter, and now by the gret heate of sommer were thawed, and then by

An objection of Meta Incognita.

ebbes, floudes, windes, and currants, were driven to and fro, and troubled the fleete, so that this is an argument to prove the heat in sommer there to be great, that was able to thaw so monstrous mountaines of ise. As for continuall snow on tops of mountains, it is ther no otherwise than is in the hottest parte of the middle zone, where also lyeth great snowe al the sommer long uppon toppes of mountaines, bycause there is not sufficient space for the sunnes reflection wherby the snowe should be molten. Touching the colde stormy windes, and the barrennesse of the country, it is there, as it is in *Cornwall* and *Devonshire* in England, which parts, thoughe we know to be fruitful and fertile, yet on the north side therof al alongst the coast within seaven or eight myles off the sea, there can neither hedge nor tree grow, althoughe they be diligently by art husbanded and seene unto; and the cause therof are the northerne driving windes, whiche, coming from the sea, are so bitter and sharp, that they kill al y[e] yong and tender plants, and suffer scarce anything to grow, and so is it in y[e] ilands of *Meta Incognita*, which are subject most to east and northerne winds, which y[e] last were choked up y[e] passage so with ise, that the fleet could hardly recover their port; yet, notwithstanding all the objections that may be, the countrey is habitable, for there are men, women, children, and sundrie kind of beastes in great plentie, as beares, dere, hares, foxes, and dogges: all kind of flying fowles, as duckes, seamews, wilmots, partriches, larkes, crowes, hawkes, and such like, as in the thirde booke you shall understand more at large. Then it appeareth, that not only the middle zone, but also the zones about the poles are habitable, which thing being well considered, and familiarly knowen to our generall Captaine Frobisher, as well for that he is thoroughly furnished of the knowledge of the sphere, and all other skilles apperteyning to the art of navigation, as also for the confirmation he hath of the same by many yeares experience, both by sea and

Meta Incognita inhabited.

Captaine Frobisher

land, and being persuaded of a new and neerer passage to Cataya, than by Capo d'buona Speranza, which the Portugalles yeerly use. He began first with himselfe to devise, and then with his friendes to conferre, and layde a playne platte unto them, that that voyage was not onely possible by the north-weast, but also, as he coulde prove, easie to bee performed. And further, he determined and resolved wyth himselfe, to go make full proofe thereof, and to accomplishe, or bring true certificate of the truth, or else never to retourne againe, knowing this to be the onely thing of the worlde that was left yet undone, whereby a notable mind mighte be made famous and fortunate. But although his will were great to performe this notable voyage, whereof hee had conceyved in his mind a great hope, by sundry sure reasons and secret intelligence, whiche heere, for sundry causes, I leave untouched—yet he wanted altogither meanes and abilitie to set forward and performe the same. Long tyme he conferred with his private friendes of these secretes, and made also manye offers for the performing of the same in effect unto sundry merchants of our countrey, above fifteen yeares before he attempted the same, as by good witnesse shall well appeare (albeit some evill willers whiche challenge to themselves the frutes of other mens laboures, have greately injured him in the reportes of the same, saying that they have bin the first authors of that action, and that they have learned him the way, which themselves, as yet, have never gone). But perceyving that hardly he was hearkened unto of the merchants, whiche never regarde vertue withoute sure, certayne, and present gaynes, hee repayred to the courte (from whence, as from the fountaine of oure commonwealth, all good causes have theyr chiefe encrease and mayntenance), and there layde open to manye great estates and learned men, the plot and summe of hys devise. And amongst manye honourable myndes whyche favoured hys honest and commendable enterprise, he was specially bounde and beholdyng to the

Frobisher's first voyage.

Captaine Frobisher pretended this discoverie above xv yeares agoe.

ryghte honourable Ambrose Dudley, Earle of Warwicke, whose favourable mynde and good disposition, hath alwayes bin readye to countenance and advance all honest actions wyth the authors and executors of the same ; and so by meanes of my lorde hys honourable countenance, hee recyved some comforte of hys cause, and by little and little, with no small expense and payne, brought hys cause to some perfection, and hadde drawen togither so many adventurers and suche summes of money as myghte well defray a reasonable charge, to furnishe hymselfe to sea withall.

He prepared two small barkes of twentie and fyve and twentie tunne a peece, wherein hee intended to accomplish his pretended voyage. Wherefore, beeying furnished wyth the foresayde two barkes and one small pinnesse of tenne tunne burthen, havyng therein victuals and other neccessaries for twelve monethes provision, he departed uppon the sayde voyage from Blackewall the fiftenth of June, *Anno Domini* 1576. Furniture for the first voyage.

One of the barkes wherein he wente, was named the Gabriell and the other the Michaell, and sayling northweast from Englande uppon the firste of July, at length he hadde sighte of a highe and ragged lande, whiche he judged Freeselande (whereof some authours have made mention), but durst not approche the same by reason of the greate store of ise that lay alongst the coast, and the greate mistes that troubled them not a little. Gabriell and Michaell.

Not farre from thence hee lost companye of his small pinnesse, whiche, by meanes of the greate storme, he supposed to bee swallowed uppe of the sea, wherein he lost onely foure men. The pinnasse lost.

Also the other barke named the Michaell mistrusting the matter, conveyed themselves privilie away from him, and retourned home wyth greate reporte that he was cast awaye. The Michaell returned home.

The worthye captayne, notwithstanding these discomfortes, although his mast was sprung, and his toppe mast blowen overboorde wyth extreame foule weather, continued hys course towardes the north-weast, knowing that the sea at length must needes have an endyng, and that some lande shoulde have a beginning that way; and determined, therefore, at the least, to bryng true proofe what lande and sea the same myght bee, so farre to the northweastwardes, beyonde anye man that hathe heeretofore discovered. And the twentieth of July hee hadde sighte of a highe lande, whyche hee called Queene Elizabeth's Forlande, after hyr Majesties name, and sayling more northerlie alongst the coast he descried another forlande with a greate gutte, bay, or passage, deviding as it were, two maynelands or continents asunder. There he met with store of exceeding great ise al this coast along, and coveting still to continue his course to the northwardes, was alwayes by contrarie winde deteyned overthwarte these straytes, and could not get beyonde. Within few days after he perceyved the ise to be well consumed and gone, eyther there engulfed in by some swifte currants or in draftes caried more to the southwardes of the same straytes, or else conveyed some other way; wherefore he determined to make profe of this place to see how far that gutte had continuance, and whether he mighte carrie himselfe through the same into some open sea on the backe syde, whereof he conceyved no small hope, and so entred the same the one-and-twentieth of July, and passed above fyftie leagues therein, as hee reported, having upon eyther hande a greate mayne or continent; and that land uppon hys right hande as hee sayled westward, he judged to be the continente of Asia, and there to bee devided from the firme of America, whiche lyeth uppon the lefte hande over against the same.

Queene Elizabeths Forlande.

Frobishers first entrance within the streightes.

Frobishers streytes.

This place he named after his name Frobisher's Streytes, lyke as Magellanus at the south-weast ende of the worlde having discovered the passage to the South Sea (where America is devided from the continente of that lande whiche lyeth under the south pole), and called the same straites Magellanes streightes. After he hadde passed 60 leagues into this foresayde strayte hee wente ashore, and founde signe where fire had bin made.

He saw mightie deere y[t] seemed to be mankind, which ranne at him, and hardly he escaped with his life in a narrow way, where he was faine to use defence and policie to save his life. Deere.

In this place he saw and perceyved sundry tokens of the peoples resorting thither, and being ashore upon the toppe of a hill, he perceived a number of small things fleeting in the sea afarre off, whyche hee supposed to be porposes or seales, or some kinde of strange fishe; but coming nearer, he discovered them to be men in small boates made of leather. And before he could descende downe from the hyll certain of those people had almost cut off his boate from him, having stollen secretly behinde the rocks for that purpose, where he speedily hasted to his boate and bente himselfe to his holberte, and narrowly escaped the daunger and saved his bote. Afterwards he had sundry conferences with them, and they came aborde his ship, and brought him salmon and raw fleshe and fishe, and greedily devoured the same before our mens faces. And to shewe their agilitie, they tryed many maisteries upon the ropes of the ship after our mariners fashion, and appeared to be very strong of theyr armes and nimble of their bodies. They exchaunged coates of seale and beares skinnes, and suche like, with oure men, and received belles, looking-glasses, and other toyes in recompence thereof againe. After great curtesie and many meetings, our mariners, contrarie to theyr captaines dyrection, began more easily to trust them, and five of oure men going ashoare, were by them intercepted with theyr boate, and were never since hearde of to this daye againe. So that the captaine being destitute of boate, barke, and al company, had scarcely sufficient number to conduct back his bark againe. He coulde nowe neither

The firste sight of the salvage.

Salmon.

Five Englishmen intercepted and taken.

convey himselfe ashore to rescue his men (if he had bin able), for want of a boate; and again, the subtile traytours were so warie as they would after that never come within our mens danger. The captaine, notwithstanding, desirous to bring some token from thence of his being there, was greatly discontented that he had not before apprehended some of them. And therefore to deceive the deceivers he wrought a prettie pollicie, for knowing well how they greatly delighted in our toyes, and specially in belles, he rang a pretie lowbel, making wise that he would give him the same that would come and fetch it. And bycause they would not come within his daunger for feare, he flung one bell unto them, which of purpose he threw short that it might fal into the sea and be lost. And to make them more greedie of the matter he rang a lowder bell, so that in the ende one of them came neare the ship side to receive the bell, which, when he thought to take at the captaine's hand he was thereby taken himself; for the captain being redily provided, let the bel fal and cought the man fast, and plucked him with maine force boate and al into his bark out of the sea. Wherupon, when he founde himself in captivitie, for very choller and disdain, he bit his tong in twayne within his mouth: notwithstanding, he died not therof, but lived untill he came in Englande, and then he died of colde which he had taken at sea.

Taking of the first savage.

Nowe with this newe pray (whiche was a sufficient witnesse of the captaines farre and tedious travell towards the unknowne partes of the worlde, as did well appeare by this strange Infidel, whose like was never seen, red, nor harde of before, and whose language was neyther knowne nor understoode of anye) the saide Captaine Frobisher retourned homeward, and arrived in England in August folowing, an. 1576, where he was highly commended of all men for his great and notable attempt, but specially famous for the great hope he brought of the passage to Cataya, which he

Frobishers returne.

doubted nothing at all to find and passe through in those parts, as he reporteth.

And it is especially to be remembred at the first arrivall in those partes, there laye so great store of ise all the coaste along so thicke togither, that hardely his boate coulde passe unto the shoare. At lengthe, after diverse attempts, he commaunded his company if by anye possible meanes they could get ashore, to bring him whatsoever thing they could first find, whether it were living or dead, stocke or stone, in token of Christian possession, which thereby he toke in behalfe of the Queenes most excellent Majestie, thinking that therby he might justify the having and enjoying of ye same things that grew in these unknowne partes.

The taking possession of Meta Incognita.

Some of his companye broughte floures, some greene grasse, and one brought a peece of a blacke stone, much lyke to a seacole in colloure, whiche by the waight seemed to be some kinde of mettall or mynerall. This was a thing of no accompt in the judgement of the captain at the first sight. And yet for novelty it was kept, in respect of the place from whence it came.

How the ore was found by chance.

After his arrival in London, being demanded of sundrie his friendes what thing he had brought them home of that country, he had nothing left to present them withall but a peece of this black stone. And it fortuned a gentlewoman, one of ye adventurers wives, to have a peece thereof, which by chance she threw and burned in the fire, so long, that at the length being taken forth and quenched in a little vinegre, it glistered with a bright Marquesset of golde. Whereupon the matter being called in some question, it was brought to certain goldfinders in London to make assay therof, who indeed found it to hold gold, and that very ritchly for the quantity. Afterwards, the same goldfinders promised great matters thereof if there were anye store to be found, and offred themselves to adventure for the serching of those partes from whence the same was brought. Some,

Many adventures.

that had great hope of the matter, sought secretly to have a lease at hir Majesties hands of those places, whereby to enjoy the masse of so great a publike profit unto their owne private gaines.

In conclusion, the hope of the same golde ore to be founde, kindled a greater opinion in the heartes of many to advaunce the voyage againe. Whereupon preparation was made for a newe voyage against the yeare following, and the captaine more specially directed by commission for the searching more of this golde ore than for the searching any further of the passage. And being wel accompanied with diverse resolute and forward gentlemen, hir Majestie then lying at the right honourable the Lord of Warwicks house in Essex, came to take theyr leaves, and kissing hir highnesse hands, with gracious countenance and comfortable words departed towardes their charge.

In the second voyage commission was given only for ye bringing of ore.

STATE PAPERS SUBSEQUENT TO THE FIRST VOYAGE.

STATE PAPERS SUBSEQUENT TO THE FIRST VOYAGE.

[*Otho, E.*, viii, fol. 46 (47) ; *Colonial*, 27.]

EAST INDIA BY THE NORTHWESTW[ARD].

Captayn hath now allready and hath put that matter owt of [all dowbt] that he there hath found the same seas [passing from] the one into the other. Neither nede I [say anythi]ng touching the naturall riches and infinit t[reas]or and the great traffik of rich merchandise th[at] is in those cuntries of Kathay, China, India, and [other] cuntries therabouts, for that every boke of histo[ry or] cosmography of those parts of the world, which a[re] to be had in euery prynters shop, doo declare [the] same at large: and the cuntries of Spayn and Portingale doo fynde and feele the same to their great ioy. But of the matters that chiefly moved me to enterprise and avance this new voyage, and to venture my mony therein so largely: I will say briefly that three things chiefly moved me thereto. First: The great hope to fy[nde] our English seas open into the seas of East India by that way, which I concevcd by the great likelyhood therof which I found in reading the histories of many mens travailes toward that parte of the world, whereby we might have passage by sea to those rich cuntries for traffik of merchandize, which was the thing I chiefly desyred. Secondly: I was assured by manifolde good proofs of dyvers travailers and histories, that the cuntries of Baccaleaw, Canada, and the new fownd lands thereto adioining, were full of people and full of such commodities and merchandize, as are in the cuntries of Lappia, Russia, Moscovia, Permia, Pechora, Samoietza, and the cuntries thereto adioyning, which are furres, hydes, wax, tallow, oyle, and other. Whereby yf yt should happen those new lands to stretch to the north pole so that we could not have passage by sea that way which we sought to the northwestward to pas into East India, yet in those same new lands to the northwestward might be established the like trade of merchandize, as is now, in the other sayd cuntries of the that on the sea coasts to the northwestwards [abunda]nce of fish of many kyndes, and of wha[les and other gre]at fisshes wherof the trane oyle is made [and the best] place for fisshing therof that is in any pla[ce in the w]orld whereby would allso grow to the realm, and [to all the] followers therof great riches and benefit. And now, to speake of the good mynde and sufficient [courage] of this rare and valiant Captayn Martyn Froby[sher], who hath thus put his lyfe in so great hazard and endured such great labours for the benefit of his cuntry, as the like is not to be read of in any history, yf his dooings and theirs be duly consydered and compared. My eloquence and wit are unsufficient duly to declare the same. Neverthelcs, according to my small talent therein, I will briefly show my good will towards him in declaring the truthe of him and his dooings according to my knowledge and true information had thereof, referring his due commendations therefore unto other whom God shall move in due tyme to doo the same, according to his great good deserts. He was borne at of honest parentage—jentlemen of a good house and antiquity, who, in his youth for lak of good scholes therabout, sent him to London, being of the age of yeres, where he was put to Sir John York, knight, now deceased, being his kinsman ; who, perceiving him to be of great spirit and bould courage, and naturall hardnes of body, sent him in the ship named to the hote cuntry of Guinea in company of other ships which were set owt by dyvers merchants of London, Anno Domini in which voyage &c., [t]hus being furnished with ship[s and all other things] necessary in as ample manner as the [funds supplied] would reach: in the name of God he set [sail and depa]rted withall on his voyage from Ratclyfe the vii day of June, anno domini 1576, and [Gr]avesend the xii day of June aforesayd. And bei[ng ath]wart of Harwith the xiii day with a contrary wynd he put into Harwich, and departed agayn from thence [on the] xiiii day, and passing along the coast northwards with skant wyndes was put agayn to Harwich three tym[es], and arived at Yarmouth Road the xviiith day, and set sail from thence the xixth day, and with fayr weather arived at the Ile of Shotland the xxvi day of June, wher[e] they ankered one tyde to refresh their water. And the same day at night set sayle agayne with a large wynd and fayr weather on their way northwestward untill th[e] xxxth day, wheron the weather grew to very great storm, which continued untill the viii day of July, in which tyme they could beare no manner of sayle. And in the sayd night they lost the sight of their small pynnes having three men therein, which they could never syns here of, though they used all possible diligence and means that the weather would suffer to seke and save the same. And when the storm ceased they set sayle and passed along on their way agayn, and on the xith day of July they had sight of land vnknown to them, for they could not come to set fote theron for the marveilous haboundance of monstrous great ilands of ise which lay dryving all alongst the coast therof. But by coniecture had owt of histories and cartes of cosmography yt should seeme to be the great Ilond of Friseland, which they saw all along by the eastern syde therof. And bearing in nerer to discover the same, they found yt marveilous high, and full of high ragged roks all along by the coast, and some of the ilands of ise were nere yt of such heigth as the clowds hanged about the tops of them, and the byrds that flew about them were owt [of sight] they lost [shi]p Michael, to their great discomfor[ture] [u]ntill their return to London for that [com]pany of that ship Michael being to make discovery of newe lands nor cou[rage] that he possessed at his departure from [Being] now (rather willingly then by force), separated from their captayn, and put to their own shift [and gove]rnment, toke counsaile with his mariners and com[rades] in the ship what they were best to doo. And among [them]selves concluded (as they say), that having yet a good [ti]me of the yere, and iudging themselves to be not far from [the] new land named Labrador, they wolde procede accor[din]g to their commission, at the least, to see that land and proceeding within iiij dayes, they saw that land. and found [it] so compassed with monstrous high ilands of ise fleting [b]y the sea shore, that they durst not approche with [th]eir ship, nor land theron with their bote. And [so] in great discomfort cast about with the ship the [n]ext day: and set their course bak agayn homward to [L]ondon, where they arived the first day of September. And in this mean tyme the sayd captayn with his ship Gabriel (as is sayd before) being overset with a sea which they shipped on the xiij day of July in the rage [of] an extreme storm which cast the ship flat on her syde. And being open in the wayste, fylled her with water so as she lay still for sunk: and would not weare nor steare with any help of the helm: and could never have rysen agayn but by the marveilous work of God's great mercy to help them all. In this distres when all the men in the ship had lost their courage, and did dispayre of lyfe: the captayne like himselfe with valiant courage stood up and passed alongst the ship's side in the chayn wales lying on her flat syde, and caught holde on the weather leche of the foresaile and in the weather coyling of the ship, the foreyard brake. And to lighten the ship, they cut over the misn maste. And the mariners allso would have cut over the mayn maste, which the capit[ayn] [u]pright agayn being full of w[ater]. And so with the rolling of the both sydes, the water yssued and withall [m]any things fleting over the ships sydes. [An]d so they put the ship before the sea all [that day] and next night in that storm: wherin allso they their mayn maste afterward, and mending it ag[ayn]. And the storm being ceased, and being now owt of [hope] any more to mete with his other ship: yet sti[ll de]termined alone to follow his enterprise and voyage a[ccor]ding to commission to the uttermost of his power. A[nd] rather to make a sacrifice onto God of his lyfe than to return home withowt the discovery of K[athay] except by compulsion of extreme force and necessity. And so returned to the course of his way toward t[he] Land of Labrador, according to commission. And by fay[re] and by fowle on the xxixth day of July the capitayn himself first had sight of a new land of a marveilous great heith : which by the account of the course and way they iudged to be the Land of Labr[ador] as in dede upon good proof afterward they iudge yt s[till] so to be. The hed land wherof he named Elizabeth Foreland in memory of the Quene's Majestie. And drawing still nere thereto in great comfort, when they approched nere they fownd the sea-shore full of monstrous high ilands and mountayns of ise fleting and driving with the wyndes and tydes and streams so as they durst not yet approche with their ship to land theron. Nevertheles remayning still with hope by some means of serch to fynde a safe place to enter with the ship ; and passing still to and fro along the coste, still in the sight of land as occasion required to avoyd dangers : within the tyme of xvj dayes the yse being well consumed and gone : they did land in three or four places upon ilands, and the master of the ship did land upon the first iland and named it Halls Iland after his own name, and there repayred his ship of her laks and leaks they fo[un]d y[t] betwene two great mayn [wh]ich they named Frobysher's streict, by [reason of his] name who discovered yt : and many ilands good harboroughs for ships which they dis[covered as] they passed. And on the xixth day of August [they f]ownd an iland which liked them, and named yt by the naem of one of the mariners which first [espyed] the same. And thereon they landed. And the capi[tayn] and six of his men went to the top of an high moun[tain] therof to discover about them : and there theie espyed [vij] strange botes with men rowing toward that iland. [Wh]erupon in great haste they ran down agayn to recover [the]ir own bote which hardly they recovered before the ari[va]ll of those vij botes. But so he returned to his ship [wi]th his bote to put all in readynes for defence yf nede [sh]ould be. And sent aland his skyf with men, [to] vewe the men and have speache with them as they could. They made of friendship to our men for desyre to have [th]em on land to take their rest. And by signes yt was agreed that one of their men came in the skyf aborde the ship, and in pledge for him one of our men went on land to them. And this man being in the ship made great wondering at all things : and the capitayn gave him to taste of the ship's meat and drink and wyne : but he made no countenance of liking any. And he gave him and other tryfles which he liked well and toke them with him to land where he was delyvered and our man received bak agayn. And hereby the captayn perceiving these strange people to be of countenance and conversation proceding of a nature geven to fyersnes and rapyne, and he being not yet well prepared in his ship for defence, he set sayle presently, and departed thens to take more tyme to prepare for defence to an other iland which they named by name of being very nere to the mayn land (on the northern syde) which they named on land. But be[in]g no [w] [w]ynde contrary they passed from [th]e western land : and there ankered [and] prepared them selves to defend the best they [could as] nede should be ; which was no force having so [small a] ship now armed with so faynt and weake men [who had] so great labours and disseases suffered at the sea [the] captayn only excepted, whose force and courag[e never] fayled for all his labours passed. And on this [wes]tern shore the capitayn with of his men went on [shore] on an iland mynding to have gone to the top of an high [moun]tayn to discover what he could of the straiets of the sea [and] land about, and there he saw far the two hed lands at [the] furdest end of the straiets and no likelyhood of land to th[e] northwards of them and the great open betwene th[em] which by reason of the great tydes of flood which they found comming owt of the same, and for many other good reasons they iudged to be the West Sea, whereby to pas to Cathay and to the East India. And on this syde the sa[yd] hedlands they saw many ilands not far asonder. A[nd] there allso they found the walls of xij olde houses of the cuntry-like cottages but no people in them. Which cottages seemed rather to have byn woork 'of' houses, th[an] dwelling houses where they perchance used to dres leather, trane oyle of some whales, or seales, or other great fisshes, of whose bones they saw there great store. And withall they allso espyed in a valley right under them iij houses covered with leather of seales skyns like tents, and allso two dogs. And presently to avoyd danger the captayn with his men repayred to the bote at the sea shore and assone as they were entred they espyed a great bote of that cuntry with men therin hard by them behynde a rok, who made signes of freendship (by laying their head in their hands) to them: to come on land and take their rest. But the captayn would not trust them, but made signes to them to come into his ship. And the master of our ship being one in the bote was his ship before. An[d we]nt on land to the people being who received him and led him by the [hand into their] houses, or rather cottages. And they thus [having got the] master among them, some of them made secret [signs to the] man for pledge in our bote that he should es[cape out] of the bote into the water, which signes th[e cap]itayn perceived. And, therefore, having in his han[d a f]ayer long partezan gylt, he held the poynt therof [to] the strange mans brest, threatening by signes to [ki]ll him yf he did ones stir. And thus the master [w]as led into their houses and there saw their manner of [f]oode and lyfe, which is very strange and beastly, as hereafter shall be shewed. And he being returned to the bote [and] entred therein the stranger kept for pledge was delyvered on land, and presently an other of those strange men went willingly in the capitayns bote aboord the ship to see the same: to whom was given many tryfles of haberdash, which pleased him well. And he being in the ship the capitayn had talke with him by signes in a bargayn which they made that he should be their pylot through the Streiets into the West Sea: to pas in his little bote rowing before the ship thither: which he agreed onto, and made signes that in two dayes rowing he should be there: and for his labour he should have many tryfling things which there were shewed him, and layd owt for him.

Yet yf they had thus passed no trust could have byn geven to such a pylot then the capitayn had sene good cause. And hervpon this strange pylot was caryed back agayn to land in our ship bote to the end to prepare himself in his own bote. And bycause the capitayn did wisely forsee that these strange people are not to be trusted for any cause nor shew of freendship that they would make, and allso did see the foolish mynde of the mariners that should row the bote with him to be desyrous at the land to have some traffik with the people for their danger of los of all of his men to go with them, and he had greater confidence than them that they should set that strange [pylot on lan]d at the poynt of a certayn rok of that i[land which he] assigned them: which was within his own sight and [so he might ha]ve rescued them yf any force should have byn off[ered a]gainst them. And that so done they should furth[with] return back to the ship. But these foolish men, be[ing] five of them in all in the bote, having set on land thi[s] stranger at the place appointed: the capitayn being [in] the ship saw them quietly put of their bote, and immedia[tely contrary to his commandment and charge geven th[ey] rowed furder beyond that poynt of the land owt of hi[s] sight, and there landed iij of them, and the other twayn reste[d] in the bote a little from the land so as he saw them agayn, to whom owt of the ship they made signes and noyse as well as they could to call them to the ship. And immediat[e]ly these two men with the bote rowed into the land agayn to their fellowes owt of his sight, and after that hower he never saw them, nor could here anything of them. And thus the capitayn having lost his bote and five of his best men, to his great discomfort he still remayned with the ship there at anker all that day and next night hoping to here of them agayn. But he could not here or know anything of them: and thereby he iudged they were taken and kept by force. Wherefore the next morning, which was the xx day of August, he set sayle with the ship and passed along by their houses, as nere as he could, and caused his trumpet to sound and shot a pece of ordonance over the houses but not to touche them. But with all this he could see nothing nor here of his bote or men. And therfere passed from thens to a bay not far of, where he ankered all that day and next night: and from thens returned bak agayn the next morning to the same place where he lost his bote, and there they of the land had sight of his ship. And he hard them of the land laugh or rather not lightly to make peace agayn [with them but rath]er depart from thens to other places ther[e to try and f]ynde some other people of that land to whom [their late doi]ngs were unknown, and of them to take some pry[soners for] his own men, which he did for the space of thr[ee days], and fynding none other people, he then determined of all his men in the ship (except the master) to re[tu]rn agayn to the same place where he lost his bote and men. And being there come to anker vnder that iland, he perceived that all the men were gone from thens and their houses allso, which was to his greatest discomfort: for [t]hat he remayned then in dispayre of the recovery of his bote and men any more. Whereby allso being thus maymed and disarmed he uttered dispayred how to procede furder on his voyage toward Cathay. And most of all other was oppressed with sorrow that he should return back agayn to his cuntry bringing any evidens or token of any place whereby to certify the world where he had byn. And so remayning in this great perplexity and sorrow more willing to dy than to lyve in that state: suddenly he espyed a number of the botes and men of that cuntry comming towards the ship. Wherat he was revived though he weak state being duly consydered, he had the more cause to be affrayd. But he with courage (more than a man) presented, armed, and prepared his ship with all things within necessary for defence; and allso without he covered the chayn wales and shrowds, and all other places (where the enemy might take any hold on 'places' any ropes to clamber into the ship) with canvas fast nayled to the ships sydes. So as they could take no hold thereby to enter into the ship being so low and so nere to the water. And in the waste of the ship he placed a pece of ordonance mynding to shote to synk one of their great botes having xx men therein. And so with the ship under sayle to have recovered some of them for prisoners, yf otherwise he could not come by any of them to redeme his own men. And when their botes being appr[oa]ch[ed] and perceiving the defens made men mustering in the ships waste fearfullnes of the men which stood before pece of ordonance they parting from yt pece appered to them: therwith they all themselves with their botes, and would approche no n[erer] but drew together in counsaile. And therupon we w[ent] to the contrary syde of the ship from the pece of ordonance, and so stayed far of, and onely one bote with one man therin, which was he that first of all came into the ship, approched very nere to 'to' the ships syde making signes of freendship that we should on land and ta[ke] our rest. Wherat the capitayn likewise made him signes of freendship as though that he would so doo, and thus entertayned him with signes of freendship, and placed him self at the waste of the ship at the syde alone having at his fete in secret his weapons, and caused all his men to withdraw from him, whereby he might appere to them open as though without any malice. And made offer of small things to geve him at the ship's syde, but the man a while stood in susspition and wolde not approche. Wheropon the captain cast into the sea a shirt and other things that would swym which the streame caryed from the ship, and he toke them up. And likewise made offer of a bell in his hand, which he toke of him hard at the ships syde.

Wherwithall one of the mariners mynded with a botehoke to have taken holde of his bote, which the man espyed and so suddenly put of his bote far from the ship, and in a long tyme would no more approche, which was no small grief to the capitayn and the rest. Yet at the last with the fayr offers and entisements with gifts of the capitayn he approached agayn with his bote to the ships syde, but stood upon garde with his ore in one hand next to the ship ready to put of his bote agayn suddenly yf nede should have byn through any cause of suspition that he might have perceived. And in this order of dealing in the presence of the rest of all his company he toke on bell more at the cap[tayn's hands] self whereby the ship into the [r]est of our men. But the capi[tayn] mischief as might happen offred him freendly countenance and made a short arme [and let the] bell fall into the sea to move the man the to approche more nere within him. Whereat the [man seemed] to be greatly sory for the los thereof and therevpon sudde[nly the] capitayn called for an other bell which allso he [rea]ched to him with a short arme, and in that reache [he] caught holde on the man's hand, and with his other hand [he] caught holde on his wrest; and suddenly by mayn force of strength plucked both the man and his light bote owt of the sea into the ship in a tryse and so kept him withowt any shew of enmity, and made signes to him presently that yf he would bring his v men he should go againe at liberty, but he would not seem to vnderstand his meaning, and therefore he was still kept in the ship with sure garde. This was done in the presence of all the rest of his fellows being within an arrow shote of the ship, whereat they were all marveilously amased and thereopon presently cast them selves into counsell and so departed in great haste toward the land with great hallowing or howling showts after their manner; like the howling of wolves or other beasts in the woods. And the capitayn with his ship remayned still there all that day, and ankered iij leges from thens all the night and the next day but could here no newes of his men nor bote nor could perceive by the prysoner that wold come agayn. Whereupon having this strange man prisoner in his ship he toke counsaile with the master and other in the ship what were best to be done. And they all 'all' agreed that consydering their evell and weak state by the los of their bote and five of their best men, and the weaknes and little hability of the rest of the men that were left in the ship, being but xiij men and boyes so tyred and sik with laboure of their hard voyage, passed as they were neither hable well to procede in any long voyage toward the tyme of winter nor yet of return home already passed than to pro[ceed] any further with so great danger of the vtt[er loss of the] enterprise for ever after if they should way. And therefore on the xxv day of Aug[ust they set] sayle with their ship keping their course ba[ck to]ward England, and in their way they had sight of the Iland Friseland, which they discovered round about but did not land thereon through the diffi[culty] of the monstrous great yse which lay fleeting still by [the] land. And after they departed from thens they endu[red] contynually extreme storms of weather but the w[ynde] still in their favour homewards. So as by the xxvth da[y] of September they were on the coast of Scotland in sight of the Iles of Orkney, and passed from thens so as they aryved at Harwich on the ij day of October in safety where they taryed to refresh their sick and weake men, and so came to London with their ship Gabriel the ixth day of October and there were ioyfully received with the great admiration of the people, bringing with them their strange man and his bote, which was such a wonder onto the whole city and to the rest of the realm that heard of yt as seemed never to have happened the like great matter to any man's knowledge. Wherefore I being not hable to geve to this capitayn his due commendation for this great and strange attempt so well accomplished I leave the dooing thereof to other which are better hable to enter therinto, that his good renown may lyve for ever according to the woorthines of his well dooings in this matter so greatly appertayning to the benefit of this whole realme of England which he is determined still to follow with the travaile of his body and spending of his lyfe ontill he have brought the same to such perfection as is desyred.

And because that I have heard report of many strange tales and fayned fables touching the personage and manners of this strange man, I have thought good therefore to declare the very truthe thereof to satisfy the world and allso to expres his picture as well as may be done with ink and paper. He was a very good shape and strongly pight made his head, his nek, his brest a very brode face and very fat and fu[ll] his body. But his legs shorter and smaller [than the pro]portion of his body required, and his hands h]is heare cole blak and long hanging and 'tyer' tyed [in a knot] above his forehead. His eyes little and a little [cole] blak beard. His cullor of skyn all over his bo[dy and fa]ce of a dark sallow, much like to the tawny Mores, [or ra]ther to the Tartar nation, whereof I think he was. [His] countenance sullen or churlish and sharp withall.

Colonial 27. *Otho E.* viii, fol. 41 (42.)

PASSED ANNO 1577.

[The] gracious favor of Allmighty God hath byn [alwaies my Pro]tector these xlv yeres in manner following [and I trust the] same will still protect me allso the rest of my [life to] his glory, to others benefit, and to the cum[fort of] me and myne.

My late father Sir William Lok, knight, alderman of [Lon]don, kept me at scholes of grammer in England [un]till I was xiij yeres olde, which was A.D. 1545, [and] he being sworn servant to King Henry VIIJth [as] his mercer ; and allso his agent beyond the seas [in] dyvers affayres, he then sent me over seas to Flan[d]ers and France to learn those languages and to know the world. Synce which tyme I have contynned these xxxij yeres in travaile of body and study of mynde, following my vocation in the trade of merchandise, whereoft I have spent the first xv yeres in contynuall [t]ravaile of body, passing through almost all the cun[t]ries of Christianity, namely owt of England [i]nto Scotland, Ireland, Flanders, Germany, France, Spayne, Italy, and Grece, both by land and by sea, not without great labours, cares, dangers and expenses of mony incident; having had the charge (as capitayn) of a great ship of burden 1000 tuns by the space of more then iij yeres in dyvers voyages in the Levant Seas wherewithall I returned into England. In which travailes besides the knowlege of all those famous 'common' languages of those cuntries I sought also for the knowledge of the state of all their commonwealths chiefly in all matters apperteining to the traffique of merchants. And the rest of my tyme I have spent in England under the happy raigne of the Quenes Majestie now being. Where by a certayn forcible inclination of mynde I have byn drawn contynually as my vocation and care for my family wolde. Cosmo[graphy] arts appertening as in voiages I could get for my mony. And [also] acquyring by dyvers conferences with many [foreign] nations, travailers and merchants fa[miliar knowledge] of the state of the whole worlde as might [appert]ayn to the benefit of myn aturall cuntry w[ith the] maintenance of myselfe and my family by the tr[ade in] merchandise according to my vocation. And as [Horatius][1] sayth: Impiger extremos currit mercator ad Indos: Pauperiem fugiens per saxa, per mare, per ignes. The diligent merchant runneth to the furdest Indians flying poverty by roks, by seas, by fyers; as by m[a]nifold notes thereof in writing and remaining still by m[e], which being put together wolde not be conteined in an[y] hundred shetes of paper that I have made for my own pryvate satisfaction yt may appere. Whereby I am perswaded of great matters. And of late by God's good providence renuyng myne old acquayntance with Martyn Frobisher gentleman; and fynding him sufficient and ready to execute the attemp[t] of so great matters, I ioyned with him, and to my power advanced him to the world with credit and above myne own power for my parte furnisshed him with things necessary for his fust voyage lately made to the northwestward for the discovery of Cathay and other new cuntries, to thintent the whole world might be opened unto England which hitherto hath byn hydden from yt by the slowthfulnes of some and policy of other. In the which voyage allready made by that way are discovered such new lands as the world now doth talk of which very shortly by God's grace the world shall playnly see to yelde to the Quenes Majestie great honor, and to the whole realm infinit treasor and benefit, which God graunt and make us thankfull.

And bycause that of late dayes syns the return home of Martyn Frobisher, dyvers men speake dyversly of his dooings.

[1] Epis. I, i, 45. Should be—

Per mare pauperiem fugiens, per saxa, per ignes.

MONETH OF ANNO 1574.

[Mart]yn Frobisher brought a letter under the [c]ertayn of the Quenes Majesties most honora[ble Privy] Cownsaile directed to the Cumpany of Mosco[via] conteining this effect: That, forasmuch as [the discov]ery of the cuntry of Cathay by sea wold be t[o En]gland, a matter of great commodity, and they being a [cum]pany priveleged and encorporated for the discovery of [n]ewe trades. Against whose privelege they would not [at]tempt any matter without their licence. Therefore [I ex]horted and perswaded them to attempt that matter now [o]nce agayn, themselves, after xx yeres allready past, syns their first enterprise thereof. Or els, to grant their licence to others which are desyrous now to 'attept' at[t]empt the same. Uppon the recept of which letter the said Cumpany assembled themselves at their Court, to consyder the same: And thereunto made answer by their letter, requiring to have conference with the parties that were desyrous to attempt that matter that thereby they might determyn what were mete to be done therein. Wherupon the sayd Martyn Frobisher agayn repayred to the sayd Cumpany with order for himselfe and others not then named to have conference with them; and theruppon the Cumpany appointed certayn of them selves, namely, Mr. George Barn, now Shrief of London, William Jowerson and Steven Borough, mariner; and me, as their agent, having the charge of all their busynes to understand the ground of this case. And in the conference of the matter, we perceiving the purpose to be to the northwestward, and no good evidence shewed by the parties for the proof of the matter: upon one relation therof made to they Company, they suspected some other matter to be meant by the parties. And forasmuch as they themselves with their very great charges allready had discovered more than half the way to Cathay by the northeastward, and purposed to doo the rest so sone as they might have good ad[vice] [a]ny good [gr]eatly hurtful to them to to any others. And therefore appo[inted] Heyward their Governour, and man and me to certify the right honora[ble Lord] [Bur]ghley, Lord High Treasorer of England of [the s]tate of the matter; which they did in the p[resence of] Martyn Frobisher aforesayd. Yet neverth[eless] very shortly afterward by the sute of the say[d] Martyn Frobysher, an other letter was brought [to the] Cumpany, requiring them either to attempt the matt[er] them selves or to grant licence to other to doo yt by the northwestwards, wherupon for dyvers consyderations then moving the Cumpany they did grant licence and privilege therof to me and Martyn Frobisher and such other as would be venturers with us in the sa[me] as appereth by the writings under their common sea[l], dated in the moneth of February, Anno Domini 1574.

Wherupon presently we made such preparation for ship[s], and all other necessaries as we could. But for lak of sufficient mony thereto in due tyme the enterprise was stayed that yere. Nevertheles, by the good assistance of the mony and favour of dyvers persons of honour and worship, and others hereunder named the matter toke such effect the yere following that we furnished two small barks of xxv tuns the pece: the one named the Gabriell, wherof was Master Christofer Hall of Lymehous, mariner. And the other named the Michaell wherof was Master Owen Gryffyn of , mariner. And a small pinnes of x tun with a close dek to sayle with them. And with them passed the sayd Martyn Frobisher for capitayn and pylot; and with them in all 34 persons, whose names be hereunder: who departed together from Gravesend, on their voyage, the xijth day of June, Anno Domini 1576. And, allthough the world in all the yere, and other tyme before hering talke of this purposed enterprise, did not beleve that yt wolde take any good effect; but rather the most of the others which were of [wi]sdom and dignity in the common the enterprise and assist the same as tyme. Wherin I will now speake p[recisely and s]ay the truthe that every mans good dede [may have] his iust commendation. The learned man, Mr. John Dee, hering the common [report] of this new enterprise and understanding of the prepa[rations] for furniture of the ships being thereby perswa[ded] that it would now procede, and having not byn acquain[ted] with our 'new enterprise', purpose in any parte before, [abo]ut the xxth day of May, Anno 1576, of his own good na[tu]re favoring this enterprise in respect of the service and commodity of his naturall cuntry came unto me, desy[r]ing to know of me the reasons 'and' of my foundation and purpose in this enterprise, and offering his furderance thereof with such instructions and advise, as by his learning he could geve therin. Wherupon I conceved a great good opinion of him: and therefore apointed a tyme of meeting in my house, wherat were present Martyn Frobysher, Steven Burrough, Christofer Hall, with other. Where freely and playnly I layd open to him at large my whole purpose in the traffike of merchandise by those new partes of the world for the benefit of the realm by many meanes as well in the cuntries of East India, yf the sea this way be open as allso otherwise, though that this 'ne' new land should chance to bar us from the sea of India. And allso declared such coniectures and probabilities as I had conceved of a passage by sea into the same sea of East India by that way of the northwest from England. And for the proof of these two matters I layd before him my bokes and authors, my cardes and instruments, and my notes therof made in writing, as I had made them of many yeres study before. Which matters, when he had thus hard and sene, he answered that he was right glad to know of me thus much of this matter, and that he was greatly satisfyed in his desyre about his expectation, and that I was so well grounded in this [pur]pose he sh[e]wed me all[so] his own. And allso shewed me I did very well like. And afterw[ards] [the while] the ships remayned here, he toke pay[ns to learn the] rules of geometry and cosmography for [the informat]ion of the masters and mariners in the use of [the in]struments for navigation in their voyage and fo[r cas]uallties happening at sea which did them service whereby he deserveth iust commendation. Allso [Sir] Humfrey Gilbert, knight, hath byn of many yeres (as I am enformed) a great good willer to this like enterpr[ise]. And syns I came acquainted with him which was abo[ut] Easter last, Anno Domini 1575, I have hard him make dyvers good discourses in the favour therof, and allso his go[od] will and study therein doth well appere in the boke which he made and put in prynt in the monthe of May, Anno 157, for the mayntenance of the good hope and likelyhood in this enterprise of new discovery. Whereby men may see many good causes to move them to like well thereof. Allthough to say the very truthe without geving any offence: neither that boke comming out so late nor yet his former discourses, being none others than were wel[l] known to us long before, were any manner of causes o[r] instructions to the chief enterprises of this new voyage of discovery to attempt the same or to direct us therin. And William Burrough, allthough he was not so well perswaded of this enterprise, that he would venter his money therein: yet, in respect of the service of his cuntry, he did take paynes to procure a master and many mariners for the ships. And gave his good advise in the furniture of the ships: and did consent unto the opinion and mynde of the capitayn in the direction of the ships course in the voyage which was to very good purpose. And besides these men, I know none other worthy of name for any thing done by them to the help of this enterprise, but onely the venturers which did help the same.

[*Otho, E.*, viii, fol. 45. b. (47); *Colonial*, 35.]

I crave pardon with the reading of this writ[ing] xiij day of October last, Mr. Fro[bysher gave me a] stone aboord his ship: Saying, that acco[rding to his promi]se he did geve me the fyrst thinge that he founde [in the new l]and, which he gave me openly in the presence of two [other] men, whome I know not. But Rowland York and many [others] were then in the ship; and they for the strangers the[rof brake off a] pece which they caryed away with them. Within the space of one month after, I gave a small pece to [Mr.] Williams, saymaster of the Towr, not telling what nor wh[ence]. He made proof and aunswered that it was but a marquesite s[tone]. And theruppon, I gave an other small pece to one Wheler g[old] fyner by Mr. Williams order. He aunswered allso tha[t] he made proof and founde it but a marquesite stone. And allso an other small pece to George Nedam: he aunswered allso that he made proof and colde fynde no mettall therin.

Herewithall I stayed, making small account of the stone, and at more leysure musing more thereon. In the begynning of January I delivered a small pece thereof to John Baptista Agnello, not telling what nor from whence. But prayed him to prove what mettall was therein. And within three dayes I came to hym for aunswer. He shewed me a very little powder of gold: Saying, it came therowt, and willed me to give him an other pece to make a better proof. I did so, and within three dayes agayne, he shewed me more powder of golde. I tolde hym I wold not beleve it, without better proof. He asked an other pece to make a better proof: Saying, that he wold make anatomy thereof, I gave it him: Saying, that I marveyled much of his doings, sith I had given peces to other iij to make proof who could fynde no such thinge therin: he aunswered me, 'Bisogna sapere adulare la natura', and so I departed.

The xviij day of January he sent me by his mayde this little scrap of paper written, No. 1, hereinclosed; and thereinclosed the grayne of gold, which afterward I delivered to your majesty, &c., 1577.

[*Colonial*, 34. *Domestic Eliz.*, cxii, No. 25.]

MR. LOCKES DISCOORS TOUCHING THE EWRE, 1577.

To the Quenes Moste Excellent Matie,

Moste humbly I crave pardon, in troublinge yor ma^tie^ wth the readynge of this wrytynge.

In the xiij day of Octobar laste, Mr. Furbosher gave me a stone, abord his shyp, sayenge, that accordynge to his promesse, he dyd gyve me the fyrst thynge that he found in the newland, wch he gave me openly in presens of 2 yonge gentlemen whome I knowe not; but Rowland York was then in the shyp, and they for the straungenes therof brake of a pece wch they caried awaye wth them.

Within the space of one monthe after, I gave a small pece thereof to Mr. Williams, saymaster of the Towar, not tellynge what nor whens. He made proffe, and answered that it was but a markesyte stone. And another small pece to one Whelar, goldfyner, by Mr. Williams order. He answered also that he made proffe and found it but a markesyte stone, and another small pece to George Nedam; he answered also, that he made prooffe and could fynd no mettal therin.

Herewithall I stayed, makynge small account of the stone.

And at more leysure musynge more theron, in the begynnynge of Januarie, I delyvred a small pece thereof to John Baptista Agnello, not tellynge what nor from whens, but prayed hym to prove what mettall was therin; and within iij dayes I came to hym for answere. He shewed me a very litle powder of gold, sayenge it came therout, and wylled me to gyve hym a better pece to make a better prooffe. I dyd so, and within iij dayes agayne he shewed me more powdar of gold. I told hym I would not beleve yt without better prooffe. He asked another pece to make a better prooffe, sayenge that he would make anatomie therof. I gave it hym, sayenge that I marvayled moche of hys doynges, sythe I had gyven peces to other iij to make prooffe, who could fynd no suche thinge therin. He answered me, 'Bisogna sapere adulare la natura'. And so I departed.

The xviij day of Januarie he sent me by his mayde this lytle scrap of paper hereinclosed, wrytten, No. 1, and therinclosed the grayne of gold wch afterwardes I delyvred to yor Matie.

And herevppon I had large conferens dyvers tymes wth hym parsawdynge, exhortinge, and conjuringe hym by many causes of great importaunce betwene us, to tell me the trewthe hereof. He satisfyed me by all dewtyfull meanes of honesty and of Christianitie that it was trew. Whervppon he entred into many discourses wth me, yf we might have sum quantyte therof, for our owne use, and ernestly exhorted me to secreatnes, and greatly pressed me to knowe where it was had I desyred respyte of a few dayes, to consyder what were best to be done in the matter.

The xxiiij day of Januarie, havynge resolved my sellfe of my dewtye towardes yor Matie I dyd retorne to John Baptista, to avoyde suspicion of doble dealyng wth hym, at wch tyme he entred agayne wth me, to have sum quantyte therof for our owne accountt. Then I delt wth hym sumwhat playne, and told hym, that it would be a hard matter for us to have ytt, for that in trewthe it was had in the new land discovred by Mr. Furboisher, wherof there is priviledge graunted to a companye. Wherto he answered, that sum devyse might be made to lade it as stones, for ballast of the shyp. Whervppon agayne I toke furder tyme to consyder what might be done therin. And at my departynge he exhorted to secreatnes, and specially to concealle his knowledge hereof.

The next day Mr. Furboisher at my table at dynner, was very desyrous to know what was found in the stone he gave me. I answered, that I had gyven prooffes to iij or iiij, and they found nothinge in ytt, savynge one man found tynne and a litle sylver therin, wch was worthy of the fetchynge awaye, wherat he was very glad.

The xxviij day, I delyvred to yor Matie in wrytynge, the very trewe information of all that I had knowen herein. And the same daye Mr. Secretary Walsyngham, in yor Maties name sayd unto me, that in my wrytynge I dyd promes a thinge wch I had not delyvred. I answered the very trewthe of my meanynge, that bycaus the bulke therof was sumwhat great, I dyd reserve it to a second speche wth yor Matie, at which tyme I dyd purpose to have declared more of this matter, and presently I dyd delyver it to hym. And he said yor Matie had told hym theffect of my wrytynge, and therfore he wylled me to tell hym the circumstance of this matter. I told hym presently theffect of all this herebefore wrytten, and that John Baptista was the man, but that he would not be acknowen to be the man. Neverthelesse I sayd he might know the matter of hym by others then by me. Whervppon he answered me, that he dyd thynk it to be but an alchamist matter, such as dyvers others before had byn brought to yor Matie by others without trewethe. And in my presens he brake the stone into iij or iiij peces, wh^ch^ he sayd he would delyver to dyvers men to make prooffes. And so he lycensed me to depart to London that night.

The xxxi day of Januarie, John Baptista sent for me agayne, as shall appere by his second wrytynge hereinclosed, at wch tyme he devysed that a ship might go secreatly out of sum place, and brynge the thynge to another place farre from London. But I answered that was not possible, for that none knowe the place but C. Furbisher and the ship master, who would not be corrupted. Then he thought to revele it to the captayne. I said I thought he would reveale it to yor Matie, but I devysed wth hym, that I would send a ship to the place in company of the captayne under culler of fysshynge, and when the captayne were gone throughe to Kathai, the ship should lade this thinge for ballast, and retorne hether. He allowed well of this devyse, and so I departed for that tyme.

The i day of Februarie, I retorned to Mr. Secretarie, who sayd to me that he had gyven peces of this ure to certayne very excellent men, and that sum found nothinge therein, but one found a litle sylver, and that Mr. Dyar had made prooffe therof, and found the lyke, and that hym sellfe had seene the proffe made, wherby he was parswaded to be so, and that Baptista dyd but play the alchemist wth me. I answered that yesterday I had spoken agayne wth Baptista, and that he dothe styll confyrme to me his former sayenges, and wyll justefie the same, but Mr. Secretarie would not beleve me. Wheruppon I prayed hym to consyder better of the matter, for that I was well assured that it was trew, wheruppon he lycensed me to retorne to London.

The iiij day of Februarie, I went agayne to John Baptista, as well to intertayne hym wth sum matter to avoide suspicion of doble delynge untill I might have answere of Mr. Secretarie of yor Maties plesure herein, as also to urge more matter wherby more tryall of the trewthe might be had. And I moved hym to know how he would deale wth me, yf I should fynde meanes to send a shyp for this ure. After longe discoursynge he resolved, that he had a frynde that would furnishe a ship at his charges, and that yf I would gyve hym a man to shew hym the place where he might have 100 tons hereof, he would gyve me £20 of money for every ton, within iij monthes after the arivall therof here, and would put me in good assurans for the parformans therof, and at the arivall in London he woulde teache me the art, yf he should chaunse to dye. I told hym I would take tyme to consider whether he should send a ship, or I send a shyp.

The vi day of Februarie, I retorned to Mr. Secretarie, and gave hym in wrytynge, this offer made to me by Baptista. He answered, it was but devyses of alchamists, for that Baptista was but poore, and not able to put suretyes, nevertheless he sayd he would consyder of the matter. And so I retorned to London.

The xiij day I went to Baptista, and put of tyme, hopinge for better answere of Mr. Secretarie. I said to Baptista that I was informed by a frynd learned in the lawes, that we have a lawe termed tresor trouvee, wherby it is not lawfull for any subject to dealle in suche a matter as this, without lycens of the prynce, and therefore (meanynge to dryve hym to dyscover the matter to yor Matie, wherby you might be certiffied of the trewthe) I sayd ther must be sum meanes found, to have a lycence of yor Matie for a ship to passe thether, or ells there is daungier bothe of yor Maties dyspleasure, and also of the companye who are privileged therin, wherin I sayd I woulde travaylle, yf he could not. He answered, he had a frynd in the courte by whose meanes he would move yor Matie therof.

The xvj day I went agayne to Mr. Secretarie for answere. He sayd the matter had no good foundacion, excepte good suretyes might be put for parformance, also that he had agayne caused others to make prooffe, and that there was no suche thinge found therin. I said that I did marvaile moche therat; for that Baptista dothe styll justifie the matter, and for prooffe therof. I would become bound to her Matie for the same. He said he would not wishe me to venture so farre uppon the worde of an alchamist. And so the matter rested untyll answere might be had from yor Matie.

The xxvij day I had a letter from Baptista, wch is the third writinge hereinclosed, wherby yor Matie may parceave, what answere he receved uppon his sute to have lycens for a ship to passe thether. Wheruppon he would have proceded wth me, that I should send a ship thether in secreat, accordinge to our first talke. Nevertheless, I parswaded hym that he should wryte a lettar to yor Matie, wherby to gyve you knowledge of his meanynge in sendynge a ship thether, and to dysclose part of the matter to yor Matie. Wherin my meanynge was, that uppon this occasion I thought yor Matie would have appointed sum to hvae had full conferens wth hym, to have serched the trewthe of this great matter to your satisfaction.

The vi day of Marche, I went agayne to Baptista, to know what answere he had from yor Matie to his lettar. He said the answere was dilatorie, so as he had no more courage in that sute. Wheruppon I said to hym, that I had a frynd in the court, by whose meanes I would attempt to have a lycens to send a shyp, for that without that lycens I durst not deale therin. He bad me prove.

The vii day of Marche, I went agayne to Mr. Secretarie, and told hym theffect that I had passed with Baptista, and he said that yf Baptista would put good suretyes for the parformans of payment, he would warrant me to have lycense for a shyp of 100 tons to fetch this ure. Wheruppon I retorned to Baptista as in myne owne name to know what assurans I should have. He said I should have very good assurans to my contentement; but named no man, which I reported agayne unto Mr. Secretarie, and offred my owne bond, and the ure to be delyvred into yor Maties custodye at the arivall.

In this mean tyme entringe more deepely into the matter, and consyderynge that the weightynes therof would be myne utter undoynge yf the matter were not good, I went agayne unto Baptista, and more effectually dyd enter into talk of the maner of the contract to delyver hym c. tons of this ure. Wheruppon he offred me to pay xxx li. a ton, being delyvred here at my charges, and the best assurans that I could then gett was to have the ure in myne owne custodye, and for the rest I must credit his honestie. That the ure was of sufficient valew to make me ryche, I was so well perswaded of his honestie, that I was fully resolved to put the whole matter in hasard, theruppon makynge this account wth my selfe that the charges of the ship and the men to dyg the ure would cost me x li. the ton, and I would gyve to yr Matie for the lycens x li. the ton, and the other x li. the ton should be to reliefe me and my children, yf that yor Matie would not deale wth this matter for yoursellfe.

The xi day, I came agayne to Mr. Secretarie, and gave hym my request in wrytinge. That yf it were so that yr Matie could not be satisfied of the trewthe of this matter, and were not mynded to deale therein for yoursellfe, that for the triall of the matter I would venture on it at my charges. Yf it would plese yr Matie to gyve me lycens to bryng hether iij c tons of this vre at my costes and charges, I would pay yr Matie iij m li. of money within one yere after the arivall, and for assurans would gyve my bond, and the ure into yr Maties custodie. He said this demand was to great. I remytted it to his owne moderation. He promysed he would move yor Matie, and said I should have lycens for a reasonable quantyte, which I dyd beleve verely to obtayne, consyderynge the manyfold refusall had more then x tymes; and the great dyscredite of my playne report made of the trewthe of the matter from tyme to tyme, accordinge to my dewtye, and the reportes he had of others to whome he had put the ure to proffe, who found no gold.

The xvi day of Marche I came agayne to Mr. Secretarie for answere. He said he had no leysure as yet to move yor Matie thereof, but he would doo. He asked of me yf M. Furbisher knew of this matter, I said no, nor none other parson by me, but onely yor Matie, and he and Baptista, which is the very trewthe.

In this meane tyme I was dayly urged by Baptista to fynishe the contract betwene us as yor Matie may parceave by his iij billes, No. 4, 5, 6, hereinclosed. Whereuppon, at the xix day of Marche I fynished and subscribed the same, as yor Matie may parceave by the same contract hereinclosed, hopynge that eyther by yor Maties lycens, I should be able to parforme the same, or ells that by yor Maties favour I should be dyscharged therof againste the said Baptista, sythe I dyd it onely for the better tryall of the trewthe of this great matter, and dyd declare theffect of all my doynges therin dayly vnto Mr. Secretary. And when I came to the housse of Baptista to subscribe the same contract, I found thereat subscribed the name of Sir John Barkley as suretye for Baptista to parforme the covenantts ; a thynge very straunge unto me, for that I never in all my lyffe had spoken wth Sir John Barkley, neyther before nor after.

The xx day, I came agayne to Mr. Secretarie, to know yor Maties plesure. He said he had moved yor Matie in the matter, but had no answere. He asked agayne yf M. Furbysher knew of the matter, I said no. He wylled me to imparte it to hym. I said I would, and so I dyd. He prayed me to get hym another pece of the vre. I said I would.

The xxii day, I came to Mr. Secretarie, and brought hym another pece of the vre. He wylled me in his name to carrye it to one Geffrey, a Frenchman, and to tell hym that it came out of Ireland, and to wyl hym to make a proffe therof, and he to bringe reporte to hym. I dyd accordingly delyuer it wth the message, and synce that howar I never saw the same Jeffrey, nor never beffore, but Mr. Secretary hathe told me that he found nothinge therin, but a little sylver, as I remember.

The xxviij day Marche, I was wth others at howsse of Sir William Wyntar in commyssyon by letters from yor Maties Honourable Privye Councell had by my procurement to consyder vppon all matters requisite for the furnyture and dyspache of M. Furbisher for Kathai, wch busynes beinge done for that daye, Sir William Wyntar wylled me to come to hym the next mornynge to talk wth hym in a matter of importannce. I came. He entred wth me in secret, prayenge me to tell hym what I had found of the state of the vre brought home by M. Furbisher. I refused that conferens, sayenge I knew not the matter, nor dyd vnderstand his meanynge. He said he knew the matter as well as I, and that he desyred this conferens vppon good meanynge towards me. I refused hym agayne, sayenge I knew nothynge, nor would tell nothynge. Then he vrged me agayne, sayenge that yf I would not tell hym, he would tell me. Then I asked hym, yf he had spoken wth Sir John Barkley of late. He sayd yea. Then I said, he might know moche of the matter. He answered that he dyd know the whole matter. I answered I thought he dyd not know all. And then presently he told me the whole effect of all my contract made wth John Baptista in wrytinge, and furder told me that Sir John Barkley had opened the same to Sir William Morgan, and Sir William Morgan vnto hym, and thervppon he and they and others had made proffes therof in a howsse at Lambethe, and also hym sellf in his owne howsse wth his owne man, the prooffes wherof in gold he shewed me presentlye in his chamber wyndowe, sayenge that it was moche rycher then I was informed of, and that it was a matter to great for hym and me to deale withall, and belonged onelye to the prynce. Then I told hym that I was of hys mynde, and that therfore accordynge to my dewtye I had alredy informed yor Matie therof longe before, accordinge to the trewthe that was informed me by Baptista, but that I was not credited therin, and that this was (as styll it is) vnknowen to Baptista and to Sir John Barkley. Wherat he was moche abashed, and sayd God hathe brought us together this daye for suum good, for otherwyse I should have done sumwhat herein that should have hurte bothe you and me. And then he sayd that it was our dewty that yor Matie should knowe hereof, and that hym sellfe would certyffy yor Matie hereof, so as you should be right well assured that it was trew. And said agayne that it was a farre greattar treasour then was knowen ; which thynge in deed synce that tyme I have parceaved, by a prooffe therof wch I have seene made by the same workmen, wch holdethe more than iiij onces of golde in c.lb. weight of vre, wch at iij*li*. of money the once amounted xii*li*. of money the c wch is ccxl pounds of money for every ton of the vre. And it is very likely that where this vre laye on the face of the earthe, there is farre more ryche vre vndar the earthe. But of this matter I thynke yor Matie have byn fully certyffyed by Sir William Wyntar and C. Furboisher, but onely I put in mynd of yor Matie parte of my first wrytynge delyured, that yor Matie gyve order in this matter in secreto quanto si puo et con fortessa, et con expeditione, least forayne prynces sett footte therin. Whervppon that yor Matie may the better consyder, I beseche yor Matie to beholde the situation of the world in this small carta herewithall presented trewly thoughe grossely made accordynge to my skyll.

And thus by this means of the doynges of Sir John Barkley and Sir William Morgan dealynge therin wth others their parteners, and wth the Douchemen their workmen vtterly without my knowledge, or ells by the meanes of others, who have pece of the vre for prooffes of others, and not of me, the secreatnes of this great matter is discoured so as it is abroade.

And bycause that I doo vnderstand, aswell by aletter hereinclosed received from Baptista dated the iiij April, as also by credable report of others, that the blame is layed all on me, as author of the speche that now is abroade of this great treasour. I doo by this wrytynge purge my sellfe of that vntrew surmyse. And I doo call to wytnes heaven and earthe, that herein I have symply and trewly sett downe in wrytynge, the maner of all my procedynges in this matter. And I do yelde into the handes of yor Matie all my goodes and my lyffe at yor pleasure, yf other then this can be proved to be done by me in this matter.

And most humbly cravynge pardon of yor Matie for this my presumtion and besechynge the same to accept my dewtyfull trew meanynge. I beseche the lyvynge God to preserve yor Matie longe to raygne over us, wth all happynes.

I humbly beseche yor Matie to restore me the wrytynges of Baptista when as you are well satisfyed in this matter.

Yo^{r} Maties most humble subjecte,

MICHAEL LOK.

The 22 April, 1577.

[*Colonial* 34. *Domestic Eliz.*, cxii, No. 25, i.]

No. 1.

Questo poco oro e cauato fuori di quell poca minera mi mandacti, ó vero mi donasti, di sorte che si truoua esser in ogni cento lib. oz j. $\frac{1}{4}$ di oro finissimo, et largamente.

Jno Bap^{ta} Agnelo, Italiano, in Santa Helena in Londra. Adi 18 Janaro 1576. De la Mimera di Tramontana Maistro nuovo.

No. 2.

Mag^{co} et Honor^{do} Sr Lok se vi piacera venir fin qui, io ho di gia pensato al caso nostro et sara di sorte che penso ne restarete satisfatto; et mi vi racco^{do}.

H^{o} V^{o} Giouābatt^{a} Agnello.

Adi 31 Janaro.

No. 3.

Mag^{co} et Honor^{do} Sr Loko. Hieri hebi risposta da sua Mag^{ta} quale mi fĕce dire che se io gli hauessi dimandato quello io disiderauo, inanci che lei hauessi concesso el priuilegio che uoi sapete, mi hauerebe concesso quanto io desiderauo, ma che essendo detto priuilegio passato et confirmato, non lo volena romper, per tanto sara buono lasciar l'impresa. Credo che il primo acordo facemo voi et io sarebe stato buono per voi et per me senza cerchar fini oltra. Ogni cosa per il meglio. Et con questo me vi racco^{do} di cuore. Questo di 27 Feb^{o}. del. 77 à natiuitate.

H^{o} V^{o} Giouābatt^{a} Agnello In Casa.

No. 4.

S^{r} mio Mag^{co} mi sara grato intender quello habiamo à fare accio l' amico mio si possi preparare per la giornata ha do fare, per tanto vi prego faciamone vna fine, et mi vi racco^{do}.

H^{o} V^{o} Giouābatt^{a} Agnello. In Casa.

No. 5.

Sr Lok. Hieri vi scrissi l'vltima mia resolutione desiderandoui di darmi vltima risposata dell' animo v^{o} et nō ho poi inteso altro. Hora perche l' amico mio se ne va damatina in paese, son forciato al risoluerlo del tutto per tanto se per tutto hoggi nō mi rissoluete del tutto : pretendo che tutto quello habiamo di tempo in tempo conferito insieme sia del tutto ancillato, riō vi delete poi di me et mi vi racco^{do}.

H^{o} V^{o} Giouābatt^{a} Agnello.

No. 6.

Mag^{co} et Honor^{do} Sr Lok.

Considerando mediante le parole vostre mi dicesti hieri, circa la confidencia hauete in me per il negocio nostro, non posso mancare di darui causa che non siate del tutto satisfatto ancora che voi non lo rechiedete, vidi io adonque che quando farete fare l' obligo d'accordo tra noi, fate vi sia vn spacio di foso con vn sigillo per vn amico mio quale sara sigurta per me di supplire, et m . . . tenire tutto quello io vi ho promesso, accio non parsata che morendo io non possiate ottenese el desiderio vostro. Et questo vorei fussi fatto dimane mero marti alla piu longa, et farmi sapere à che hora protrete esser qui 'accio io anco possi fare che l' amico mio sia qui, et con questo mi vi racco^{do} si cuore.

H^{o} V^{o} Giouābatt^{a} Agnello. In Casa.

[*Colonial* 34. *Domestic Eliz.*, cxii, No. 25, ii.]

Contratto fatto d'accordo fatta tra Michele Lok, Inghilese, mercante, di Londra di vna parte, et Giouan Baptista Agnello, Vinetiano, residente in Londra di altra parte, come seguita.

Detto Michele Lok ha venduto et vende a detto Giouan Baptista Agnello la quantita di cento tonelli (si tanto si puo haueve) di vinti centanari il tonello, cioe di libre cento et duodeci del peso di Londra per ogani centanaro di terra o altra materia minerale di sorte tale quale detto Michele ha datto al detto Giouan Baptista vna pezza per mostro, laquale pezza sta posta in vna scatoletta sigillata con sigilli lori, et detta scatoletta sta posta dentro vna cassetta serrata con due serrature et le chaue di quelle sono in custodia loro, et detta cassetta e dato in mano et custodia di   per guardarla a vso loro per mostro quando sara bisogno. La quale detta terra, o meteria minerale di sorte sopra detta, o altrimenti di sorte tale quale detto Giouan Baptista o altro per lui uuole truouare et eligere in la terra nuouamente discorparta per Martin Forbiser, o in gli altre terre circonvisine da discoprire, detto Michele promette a sua speza fare cauare di terra, et carigare in nave, et portare a Londra (Dio mandando la naue a saluamento) et iui consignare a detto Giouan Baptista, lui pagando si come promette di pagare al detto Michele il precio di trenta lire moneta d'Inghilterra per ogni tonello di quella, fra termino di xij mesi dipoi consignata in Londra, cioe ogni tre mesi la quarta parte dela valuta al precio sopra detto di quanto montara la quantita consignata di tempo in tempo. Et per piu causione et securta di detto Michele, il detto Giouan Baptista, promette et si contenta che la detta terra et altra materia minerale restara in mano et puotere di detto Michele come roba sua propria fin che il sia satisfatto di detto pagamento di danari, eccetto la parte di quella che detto Michele ha da consignare al detto Giouan Baptista auanti mano per lauorare sopra il quale di poi si hauera di fare il pagamento sopra detto di tempo in tempo. Et piu il detto Giouan Baptista promette al detto Michele di insegnare a liu o a un altro che per lui sara assegnato, il uero muodo et Arte che detto Giouan Baptista vsara per cauare gli metalli dal detta terra o materia minerale, fra termino di sei mesi dipoi la consignatione dela prima parte di detta terra o materia minerale in Londra, et ancora dareli in scritto gli regoli et vero muodo di detto arte. Et per complire questo contratto il detto Michele et il detto Giovan Baptista mutualmente l'un a l'altero se obligamo loro persone et heredi et tutti beni di complire quanto di sopra hamo promesso. et vogliano che questo contratto et seritto sia di tanto vigore et forsa quanto saria il meglio et piu valido scritto et obligatione che si puoteria fare secondo gli leggi d'Inghilterra o di altri leggi qual si vuoglia per accomplimento di quello. Et in fede di tutto ambe due hanno sotto scritti et sigillati questo et un altro simile copia con loro mano proprio. Fatto adi xix di Marco, Anno 1576, in Londra, by me,

MICHAEL LOK.

Jo. Giouābatta Agnello prometo confirmare quanti di sopra è detto et per fede de cio ho sotto scritto di mia propria mano et sogelato di mio sugello. [*Seal.*]

I, Jhon Barkeley, knighte, doo bynde my selfe, my heyres, executors, and assings to fullfyll all these covenants, articles, and agremts here above written to, and wth one Mychaell Lock, of the citie of London, merchant, wch one Jhon Baptyste Agnello hath promysed and here above covenanted to and wth the above said Mychell Locke, yn as ample mannr as the sayde Jhon Baptyste ys bownde by the aforesayde covenants to fullfill the same, yn wytnes wherof I have wrytten and syngned thys byll wth my owne hande and name, and sealyed the same also wth my seale of armes even the 19th day of Marche, Ano. Dm. 1576, by me,

JHON BARKELEY.

[*Seal.*]

[*Colonial* 34. *Domestic Eliz.*, cxii, No. 25, iii.]

Magco et Honordo Sr Lok,

Mic venuto all' orecchie vn certo tuono inaspettato quale mi ha fatto inarassiglia re molto, considerando la promessa vostra mi facesti di tener el negocio nostro secreto la qual cosa non hauete fatto, anci hauete cercato per tutta Londra se vi fassi qual cuno sapessi far quello ch' io ho fatto, senza proposito alcuno, et mancando della fede: piu oltra hauete pro ferto alla Regino di dargli piro dieci per tonello di vna ccerta vostra minera, fondandossi sopra le mie parole, il chi fu contra quello mi prometesti. Et se pur si doucua prescrire à sua Ma^ta qualcosa, io l' harei potuto farc con piu fondamento et honesta che non hauete fatto voi, perche io gli harei parlato con sostancia et del mio et non farni bello delle mie forti che come hauete fatto voi. Et piu mi prometeste di venir à sugellare el vostro scritto, et non l' hauete fatto. Et anco mi prometeste di portarmi el restante della vostra minera, et non l' havete maneo fatto, per il che io considerando tutte queste vostre qualita io son deliberato di non sequitar piu oltra, per tanto vi prego rimandatemi el mio scritto, et cercato chi meglio vi parera, ch' io per me non me ne voglio piu impaciare in conteniuno, et vi protesto oli non mandar per detto minera per mio conto ch' io non la voglio hauer in conto niuno. Et conquesto miracco^do di cuore, et prego Dio vi dia meglior fortuna con altri. Questo di x Aprile del 77. In casa.

H^o V^o Giouābatt^a Agnello.

Al Molto Mag^co Sr Michiel Lok.

1577 Jno Bap^ta Agnello. Rec^d the 3 April, 1577.

[*Colonial*, *East Indies*, No. 28. *Domestic Eliz.*, cxix, No. 31.]

MR. MARTIN FURBISHERS PETITION TO HER MAJ^IE TO BE GRANTED TO HER IN RESPECT OF HIS TRAVAIL ALLREADY, AND HERE-AFTER TO BE BESTOWED IN DISCOVERIE OF NEW LANDS.

THAT it maye please yor Matie in respect of the late discoveries I have made to the north west, and my greate charges and travaill performed therin to graunte to me and myne heires, for ever under yor Maties letters pattentes, the high Admirall-shipp by sea, as well of all those seas alreadie by me discovered or hereafter to be discovered as also duringe my life the government and order by land of suche people of what nature soever they shalbe that shall inhabitt in any parte of those discoveries made or to be made by me and the same to be executed by my self or sufficiente deputie w^h suche consideration of fee or allowunce for thexecucion therof as shall beste please yor Matie to bestowe on me for the same.

Ffurther, that it maie please yor Matie to graunte me duringe my life for my travaill and service performed in thies discoveries fyve powndes of the cleare gaine of every c^um. that shalbe brought owte of the landes or islandes discovered or hereafter to be discovered by me to the north-west. And after my deathe to myne heires forever xxs, 5, of every c^um. of cleare gaynes to be brought as aforesaid.

Item, that I maie make free yerelie, duringe my life, of this voyage, six persons, so that for the firste yere they comme not in wth above c[um]. stocke, and after to adventure as all others shall do by order.

And also that every shipp fraighted yerelie into thies new discoveries in consideracion of the greate care I must take of theim bothe in appointinge apte men to take charge of their shipps, and also must instruct theim by sondrie orders and observacions how to holde companie wth me shall, duringe my life, give me one toone fraight of every c. toones to be brought from those places to be paide me in monie by the owner or owners of those shipps accordinge to the value they shall receave or to carrie me the fraight of one toone at my choise.

[*Colonial, East Indies*, No. 32. *Domestic Eliz.*, cxix, No. 33.]

A BRYEF NOTE OF THE COSTE AND CHARGE OF THE IIJ SHIPPS AND FURNYTURE FOR THE SECOND VOYAGE FOR CATHAI, ETC.

	li.		
For the shyp Ayde, to the Quenes Ma[tie] . . .	750	0	0
For the ij barkes Gabriel and Michael, wth almaner furnyture and ordonans	400	0	0
For new buyldinge and translating the same ships and for new tackelyng and implementes	650	0	0
For ordenans and munytion new bought . . .	550	0	0
For vyttelles	950	0	0
For wagys of men	650	0	0
For necessaryes, for the mynes and workmen . .	150	0	0
For marchandyse, for traffyke, and provision . .	300	0	0
Sum of all . .	*li.*4400	0	0

This account is but gessed very nere the trewthe for that thaccounttes are not yett brought in parfectlye.

And the whole stock of the venturars sett downe in certayntye as yet dothe amonte but	3000	0	0
Wherof is yet received but	*li.*2500		
And so thear lachethe in stock of the venturars to supplye this whole charge	*li.*1400	0	0

Ffor the w[ch] summe of *li.*1400, the venturars are to take order presentlye to dyscharge the debt owinge to dyvers men for thinges had for the furnyture of the said shyps and voyage, whiche is most humbly beseched by Michael Lok, who hathe gyven his promesse to them for the payment therof by order of the Commyssyoners.

[*Colonial*, No. 29. *Domestic Eliz.*, cxi, No. 49.]

A NOTE OF THE PROVISION AND FURNYTURE NECESSARYE FOR THE SECOND VOYAGE FOR THE DYSCOURYE OF KATHAI, &c.

A great peece of this charge cut of, for thare went but one shippe and two barckes in this viage.

	li
A shyp of cxl ton burden, wth tackelinge, ordenans, and munition	1000
A ship of cxx ton burden, wth tackelings, ordenans, and munition	800
ij barkes of burden xxv ton eche, wch were in the first voiage, wth their tackelinge, ordenans, and munition, wch now they have, and others to furnyshe and repayer the same, all .	450
v shallop, botes, wth their takle and furniture, wherof ij wth close overlops at xx*li* the pece, and iij open, at x*li* the pece	70
A ship of c ton burden, to be fraighted for fysshinge, in the Straytts where Furbysher was, and from thens to retorne, wth one of the barks in valew	—
Sum of the shyps	2320*li*

The vyttels for the 180 men, for the said ij shyps and ij barkes, for xviij monthes, at xiij*s*. iiij*d*. the monthe, for eche man	*li*2160	0	0
To say in one ship 70 maryners, another 60 maryners, in the ij barkes 30 mariners, and twenty men of offyces and artyfyce.			
The wages before hand of those 180 men, for their provision, at iiij*li* the man, one wth another . . .	*li*600	0	0
Sum of all this—	*li*5080	0	0
The marchandyse for stock, clothes, 50 carseys, 200 cottons, 40 frizes, 10 tyn ijm. leade, ijm. coppar, and kettels ijm. and all other marchandyse	*li*1200	0	0
Sum of all—	*li*6280	0	0

That it would please the Quenes Matie to graunt her letters patentts of priviledge in the Corparation to the fyrst Venturars and their successors, in ample maner.

That it would please Her Matie also to graunt auctoritie to Mr. Frobysher, for the governement of the men in obediens.

That warrant may be graunted to take vyttells at reasonable pryces, and to prest men at reasonable wagys, and to take shyps at reasonable praysement for the sarvyce of this voyage, yf the Quenes Maties shyps doo not sarve.

That order may be taken by agrement of the venturars for offycérs for the good governement of the Company, and the mayntaynans of their pryvyledges, and to take the charge of the whole busynes and accountts.

That the shyps may be redy to departe on their voyage by the x day of Marche.

That men may be named by secret commission, to supplye the charge of Mr. Frobysher and Mr. Hawlle, vppon any myshappe, and to be kept secret vntill tyme of nede.

[*Colonial* 33. *Domestic Eliz.* cxi. No. 48, i.]

26 of Marche, 1577.

SHIPPINGE THOWT MYET TO BE EMPLOYDE IN THE VIAGE ENTENTED BY MR. FURBUSHER, viz. :—

	Tons.	Mariners.	Soldiers.	
The Ayde . .	200	65	25	Men 115
The Gabriell . .	15	10	3	
The Myghell . .	25	10	2	
	240	85	30	

A preportion of vittouls for the said 115 men.

(Bysket 16 tons.) Item bysket for v monthes of 28 daies to the monthe contayneng 140 daies after the rate of 1*li* per man per diem, xiiij[mt] iij c. iij quarters at xiij s. iiij ct. per c.	95	16	8
(Meale 30 tons.) Item meale for xiij monthes contayneng 364 daies, 240 barrells contayneng eche barrell iiij bushels w[ch] maketh 960 bushells at iiijs per bushell, 192*li* mor for the barrells and gryndinge at xxs. per barrell 19 10 0 mownts	211	10	0
(Biere 80½ tons.) Biere for vi monthes conteyaneng 168 daies after the computation of one gallone aman per daie 80½ ton at 2*li* 5s per ton wth caske iron whoopes and chardges	181	2	6
(Wyne 5 tons.) Malmsey and secke v tons at xx*li* per ton	100	0	0
(Biefe 5 tons.) Bieffe for iij monthes having fleshe daies 48, at 1*li* a man per diem, vij[ml] iiij[c] weight grose at xiiijs per c weight 51 16 0. Item for baye sawlte to preserve the same 55 bushels at ijs per bushell 5 10 0.			

Item for iiij tons ij hogsheads of caske to packe the same in at xs per ton, 2 5 0 mounts	59	11	0
(Porke 15½ tons.) Porke for 15 lieke monthes contayneng 240 daies after the rate of 1*li* weight aman per diem, xxiiij^{ml} viij^{c} xxiiij *li* weight at xiiijs per c weight 173 15 0 more for 186 bushels sawlte at ijs. 18 12 0 more for 15 ton ½ of caske at 10s per ton, 7 15 0 mownts	200	2	0
(Peasee 10 tons.) Item pease for 288 fleshe daies in the 18 monthes as afore saied allowinge to iiij men 1 q^{t} of pease per diem, 258 bushels at 3s per bushell 38 14 0 more for 10 tons of caske, 5 0 0	43	14	0
(Stocke fyshe 2½ tons.) Stocke fyshe for 108 daies in ix monthes, as afore saied at 1 quarter of a fyshe aman per diem, iij^{ml}c fyshes at ij*li* per c	62	0	0
(Butter 2 tons.) Butter for the saied 108 daies at 1 quarter of a *li* weight eche man per diem, 3,105 *li* weight salte at iiijd per *li* 51 15 0 more, for waste 250 *li* weight at iiijd per *li* weight 4 3 4	55	18	4
(Chiese 4 tons.) Chiese for other 108 daies at half-a-*li* weight aman per diem 6210 *li* weight subtill more 500 *li* weight for allowans of waste amowntinge in the wholle to 6710 *li* weight at xvjs viijd per c weight . .	55	18	4
(Otmeale 1½ ton.) Item otemeale 40 bushels towardes the suplyenge the want of fyshe	10	0	0
(Riese 1½ ton.) Item riese for the lieke cause 2000 oz. .	26	13	4
Item caske to stowe the saied otmeale and riese in iij tons at xs per ton.	1	10	0
Item honney ij barrells at iij*li* vis viijd per barrell .	6	13	4
Item sallet oyle 1 hogshead	10	0	0
(Provision for store 8 tons.) Item vyneger 1 ton . .	8	0	0
Item aquaviete ij hogsheads	10	0	0
Item musterseed iiij bushels	1	10	0
Item candles xij c weight at iijd per *li* . . .	15	0	0
Item baye sawlte v ton at iiij*li* per ton . . .	20	0	0
(Woode 14 tons.) Item wood xiiij^{ml} at xiijs iiijd per j^{ml} .	9	6	8
(Sea coales 30 tons.) Item sea coales 20 chawders at xiijs iiijd per chawders	13	6	8
(Charcoale 1 ton.) Item charcoales 1 loade . .	1	6	8
Item fyshinge nets, fyshinge lyenes, hooks, harpinge irons and suche lieke nessesaries	8	0	0
(Provision for sick men.) Item to allowe the surgeone towards the furneshinge of his cheaste . . .	6	13	4
Proynes 2 firkens	2	0	0
Item reasons, almonds, liccores, etc. . . .	2	0	0

(Provision for the apparelling of the men.) Item wollinge

clothe for jirkens, breche and hose, canvas and lynnenge clothe for dublets and sherts, hats, caps, and shewes, etc.	100	0	0
(Chardgs). Item for land carrage, wharffage, labras packinge of bief and pork, water carriage, and other extraordennarie, etc.	30	0	0
(Rigging, wages, and vittailes.) Item for the rigginge, wages, and vittails of lx men for ij monthes to end the last of Aprille next at xxijs vid aman per mensem .	127	10	0
(Prest monney vppon wages.) Item for the too monthes wages to be emprested to the company at ij*li* per man .	230	0	0
(Merchandizes). For provision of merchandizes . .	500	0	0
(The Ayde). The Ayde the firste penny	750	0	0
Item more to furnishe her wth ordenance, takle, apparrell and monytions, etc.	450	0	0
(The Gabriell.) The Gabriell throwghly perfected in all respects	180	0	0
(The Myghell.) The Myghell in lieke sorte . .	180	0	0
Item ij shallopes	24	0	0
2582 3 4			
Sum of tons 226 tons ½			
Sum of monney 3778*li* 2 10			
Item for the hier of a ship of cxx tons to waight vpon the ships to the Straight, etc. . . 500*li*	4278	2	10
Item for divers extraordinary chardges . . .	221	17	2
Sum of all *li*4500	£4500	0	0

[*Colonial East Indies*, No. 50. *Domestic* cxix, No. 41.]

THE NAMES OF THE VENTURARS IN THE SECOND VOYAGE FOR CATHAIA, &C, BESYDES THEIR VENTURE IN THE FIRST VOYAGE.

The Quenes Ma[tie]	1000
The Privie Counsell.	
The Lord Highe Treasorer, 50	100
The Lord Highe Admirall	100
The Lord Chamberlayne, 50	100
The Erle of Warwyke, 50	100
The Erle of Bedford, 25	50
The Erle of Leycester	100
Mr. Treasorer	50
Mr. Controller	50
Mr. Secretarie Walsyngham	200

Other Venturars.

The Erle of Pembroke	150
The Countesse of Warwyke	50
The Countesse of Pembroke, 25	50
The Lady Anne Talbot, 25	50
The Lord Hounsdon	50
The Lord Charles Howard	50
Sir Henry Wallop, and others	200
Sir Thomas Gresham	100
Sir Leonell Duckett, 25	50
Sir William Wynter	200
Mr. Phillip Sydney, 25	50
— William Pellam	50
— Thomas Randolphe, 25	50
— George Wyntar	100
— Edward Dyar, 25	50
— Symon Boyer, 25	—
— Anthonye Jenkynson, 25	50
— Mathew Smythe	50
— Geffrey Turvyle	50
— William Payntar	50
— Richard Boylland	50
— Mathew Ffyld, 25	50
— Edward Hogan	50
— Richard Yonge	50
— Thomas Allyn	50
— Christofer Huddesdon	50
— William Ormshamc, 25	50
— Robert Kyndersley	50
— Michael Lok	200

[*Colonial*, 33. *Domestic Eliz.*, cxi, No. 48, II.]

YERE 1577.

THE VENTURERS IN THE SECOND VOYAGE FOR CATHAY, ETC.

In the first voyage as folowethe.		*li.*
	The Quenes Matie	500
50	My Lord Highe Treasuror	100
50	My Lord Highe Admirall	100
50	The Erle of Sussex	100
	The Erle of Bedford	25

50	The Erle of Warwyke		100
50	The Erle of Leycester		100
	The Erle of Pembroke		100
	My Lord Hounsdowne		50
	My Lord Charles Howard		50
	Sir Frauncys Knowells		50
	Sir James Croft		50
25	Mr. Francys Walsyngham		50
25	Mr. Phillip Sydney		50
	My Lady Anne Talbot		25
	Mrs. Mary Sydney		25
100	Sir Thomas Gresham		200
25	Sir Leonell Ducket		50
	Sir Henery Knevet		25
	Sir William Wyntar		50
25	Mr. Thomas Randall		50
	Mr. George Wyntar		50
25	Anthony Jenkynson		50
	William Sakford		25
	William Kyllygrew		25
	Symon Boyer		25
	Geffrey Turvyle		25
	William Payton		25
	Richard Boyland		25
100	Michael Lok		300
50	Edmond Hogan		100
	Mathew Fyld		50
100	William Bond, yonger		200
25	Mathew Kyndersley		50
	Robert Kyndersley		50
25	Christofer Androwes		50
25	Robert Martin		50
All 850	Henry Lok		25
	Thomas Marshe		25
	William Ormshaw		25
	Olyffe Burre		100
	Thomas Chester	Of Bristow	25
	Thomas Kelke		25
	Thomas Aldworthe		25
	Robert Halton		25
			1225
		All the 30 Marche 1577 *li.*	2000

[*Colonial*, No. 33. *Domestic Eliz.*, cxi, No. 48.]

REPORT UPON THE OUTFIT FOR THE SECOND VOYAGE.

It may please yor Lordships to understand, wheras you have by yor letters, beringe date the xviith of this present, requyred that wee shulde take vppon vs the care of the thoroughe and speedy settinge furthe to the seas of Mr. Furbusher, wth the shippinge thought meete to passe wth him for the discoverie pretended. So it is that accordinge to your honnorable comawndement wee have travelled in the same, and do perceave as followeth: First, wee do finde as well by thexamynacion of the said Mr. Furbusher, as also of the master that was wth him in his last viadge, and other of that company whom wee have particulerly examyned a part one from the other, and also vppon dyvers and sundry other matters whiche wee have pervsed and weyed, that the supposed Straight whiche Mr. Furbusher doth sett out is so farr fourth as we can gather and judge a trueth, and therfore a thinge wurthie in our opynyons to be followed. The nomber of shippes and other vessells wth the men to go in them, the provicion of victuells to be made, wth all other necessaries fitt for to serve for the said viadge, wee have throughly considered therof, and haue sett downe the same in particulers whiche dothe amount to iiijm vc *li*, as by the said particulers subscribed wth our handes may plainely appere, towardes the whiche wee do likewise finde that there is in provicion as well in shippinge as other necesseries to the valewe of one thowsand powndes little more or lesse. So there is to be levied vppon suche as are, or will be the Adventurers, the residewe of the fornamed some of iiijm vc *li*, mowntinge to three thowsand five hundred powndes, whiche matters before rehersed, wee have thought it meete and our dueties to reveale the same, not only to yor Honnors, but also to other partners of that Company of Adventurers in the forsaid viage to thend that the same beinge knowen, speedy order may be taken that every person who hath entred into the Company, and sett downe suche somes of money as they will adventure, that they do forthwith take order to bringe in their said somes to Mr. Huggins, beinge appointed Treasorer of that Company, withoutt the whiche nothinge can be donne, and if it should be any longer deferred, tyme wolde not serve this yere to take the viage in hand. And thus havinge shewed to yor Lordshipps, and the rest, our travaills and opynyons in that behalf the whiche we leave to yor honnorable consideracions, wee rest prayinge God to preserve you. London, the xxxth of Marche, 1577.

The charge to set 3 ships fourth 4500*li*.

There is already in shipping about 1000*li*.

To be yet levyed 3500*li*.

Order to be taken that such as ar entred into the Company, and set down their somes, may bring up the same wt speede.

Nothing can be don vnlesse it be brought yn.

The tyme passeth, almost past for yis yere.

Yor Honnorable LL. to command,

W. Wynter, Tho. Randolph, G. Wynter,
A. Jenkinson, Edmond Hogan, Michael Lok.

[*Colonial*, 30. *Domestic*, *Eliz.*, cx, No. 21.]

ARTICLES OF GRAUNT FROM THE QUEENE'S MAJESTIE TO THE COMPANYE OF KATHAI.

A corporation and all things to be omitted to be compressed in form of the charter of Moscovia.

That A, B, C, D, etc., all the names of the fyrst venturers, with M. Lok and M. Frobisher, in the fyrst vyage and attempt made for discovery of Kathai and other newlandes by the north west wards, shalbe one Companye and corporation for ever to them and their successors.

That the sayd Companye shalbe named the Companye of Kathai.

That they shall haue pour and auctoritie to admytt others into their corporation at their pleasurs.

That they shall haue poure and lybertie to assemble them selves and to kepe courtes when and wher they will.

That thosse which shalbe assembled being xv in nomber at the lest, shall haue poure and auctoritie to chosse a governore, ii consulls, and xii assistantts, to continew in their offyce for terme of iij yeres now nex comyng, and afterwards to chuse ij governors, iiij consulls, and xxiiij assystantts, to contynew in offiyce for iij yeres, and so to be renewed or changed from iij yeres to iij yeres.

That thosse which shalbe so assembled in court in nomber and order afforsayd, which the governor, ij consulls, and xij assystantts, shall haue full poure and auctorytie to make lawes and ordynances and actes from tyme to tyme as they seme good for the good order and govermentt of the Companye, which shall bynd all the Companye to the dew obsarvation therof, and maye sett fynes and penalties uppon the transgressors therof, and comytt them to pryson and attache their goodes untill they performe them.

That they may haue a sargant or offycer or twayne to execute their lawes and actes.

That they maye revok their former lawes, and mak other new lawes, as maye seme good from tyme to tyme for the good goverment of the Companye.

That they may haue a common seal.

That they may sewe and be sewed in all the courtes of the reame by name of the Company of Kathai.

That they may purchase landes and tennements to the vallewe of one hundreth poundes rent, and may sell and mak leases.

That they may possesse and enjoye all their goodes.

That they may do all thinges in as ample and beneficiall manner as any other corporation may doo.

That they and their successors, and their factors, servantes, deputies, and assignes, shall haue free lybertie, poure, auctoritie, and pryveledge for ever at their will and pleasure, to sayll, goo, and otherwysse by any

As much herof as shall not be contrary to the former charters of the merchants of Moscovia to be accorded.

meanes to passe to and from all seas, waters, iles, landes, countryes, etc., saylinge or otherwysse passinge from England, northward, westward, or sowthewarde, or by any other poynt of the sea compasse, betwene the northe, the west, and the southe, aswel under our banners and ensignes as otherwysse without them, with whatsoever shipps and other vessells, and with all manner vythuall, munition and furnyture and necessaryes for the same, and with all manner of marchaundis and goodes for to seke, dyscover, and fynd whatsoever seas, waters, iles, landes, regions, countryes, provences, and other places whatsoever, of whatsoever gentells, heathen, infidells, or other nations, sett and beinge in whatsoever part of the world, which before this tyme, and before the late vyage of discovery made by Martyne Frobysher to the northewestwards hath been unknowne, or not commonly frequented, by the subjects of our reame of England for trade of marchaundise. And also in the same seas, waters, iles, landes, countryes, regions, provences, and other places, and to and from the same, shall and may frelye at their wills and pleasurs from hence fourth for ever, use traffic and trade of merchaundise, and otherwise doo what soever bussines and thinges to them shall seme good and convenyent for their owne proper vantage, comoditie, and proffyt, without theirby incuringe any manner of penaltie, forfayture, or other molestation or trowble whatsoever, notwithstandinge any pryveledge or other actes, lawes, or thinges whatsoever to the contrarye herof in any wysse.

That none other parson, subject, nor denyson of our realmes and domynions, nor any other of whatsover nation, not beinge free of this Corporation or Companye, shall passe by any meanes to nor from any of the sayd seas, landes, etc., nor vyssett the same, nor therin do any featt of marchaundise, nor other business, without the speciall consent and lycence of the sayd Companye, under their common seall, graunted in courtt or otherwysse then for the affayres and bussines of the sayd Companye by their order, uppon payne of losse and forfeyteure, *ipso facto*, of all shipes, vessells, and goodes whatsoever, transported to or from any of the sayd countryes or places, or the vallew therof, the halfe to the Quene, the other halfe to the Company, to be taken or sewed for by seasure or accyon of debt or otherwysse in any of the Quene's courtes, etc.

That to their best poure and abilitie they may forbyd, withstand, and repullse all other parsons of whatsoever nation that shall dysturbe or interrupt them, or intermedle in their trade of marchaundis or otherwysse in any their attemptes in any of the sayd seas, lands, countrys, or other places before sayd, without therby incuringe any penalties or daunger of our lawes, etc.

That of all the marchaundis which they shall carrye out of our reames and domynions they shall pay no more nor greater custome, subsedy, nor other dewties unto us nor to our heyres, then is now dewe or shall be dew to us by our subjectes by the lawes and customes of our reame.

That of all the goodes, wares, and marchaundis which they shall bringe into England from the countryes afforsayd they shall pay but halfe custome, subsedyes, and other dewties now dew to us, for the terme of xx yeres now next ensewinge, and afterwardes shall paye to us and our heyres for ever no more but 5£ of 100£ of the vallew of the same marchaundise that they shall bringe in, accordinge to the vallewation therof, now vsed in our customs in London.

That they shall freeley and at their pleasure transport out of our reames and domynions into any other reames and countryes all such wares and marchaundis as they shall bringe in and not sell in our domynions, free, without payment of any custome or dewtie to us outwardes for the same, notwithstandinge any lawes to the countrarye, etc.

And furdermore, in consideration of the industry, good direction, and great travayll of Michaill Lok of London, mercer, in the fyrst voyage latelye attempted by Martyne Frobysher, gent., for dyscovery of Kathai and other new landes by the northewestwards, we doo grauntt and will that the sayd Michaill Lok shalbe the fyrst governore of the sayd Companye, to contynew in that offyce for terme of his lyffe, except he will resygne the same. And also in consideration of his great cost, charges, and venture for the provision and furnyture of the forsayd fyrst voyage of dyscoverye, we do graunt and will that he shall haue, receive, and tak of the sayd Companye to his owne vse and behoffe for ever, the rate of one of every hundreth of all the wares, goodes, and marchaundis that shalbe browght into England or other countryes for accountt of the sayd Companye, accordinge to the rate and vallewe therof in the payment of custome to the Quene's majestie.

And lykewyse, in consyderation of the industry, good order, and great travayll of Martyne Frobysher, gent., in the execution of the fyrst voyage latly made in his own parson for the dyscovery of Cathai and other new landes by the northewestwardes, we do grauntt and will that the sayd Martyne Frobisher, dewringe terme of his naturall lyffe, shalbe High Admyrall of all seas and waters, countryes, landes, and iles, as well of Kathai as of all other countryes and places of new dyscovery. And also in consyderation of his good sarvyce theirin, we do grauntt and will that he shall haue, receave, and tak of the sayd Companye to his owne proper vse and behoffe for ever the rate of one of every hundreth of all the warres, goodes, and marchaundise that shalbe brought into England or other countryes for accountt of the Companye, accordinge to the rate and vallew therof in the payment of custome to the Quene's majestie.

That all the malle chyldren of all the fornamed parsons, which weare fyrst venturers of the Companye, and also the heyres malle of every of the sayd malle chyldren for ever, shalbe admytted into the lyberties and pryveledges of the sayd Companye gratis from tyme to tyme.

[*Colonial*, No. 31. *Domestic Eliz.*, cx, No. 22.]

ARTICLES CONSENTED AND FULLY AGREEDE BY THE COMPANY OF KATHAYE.

Thatt the Company shalbe named the Company of Kathay.

Thatt Michaell Lok shalbe Governour for vj yeres next ensuinge.

Thatt A. B. shalbe Consullor for iij yeres.

Thatt A. B. C. shalbe Assistante for iij yeres.

Thatt A. shalbe Agente for iij yeres, to doe all the buisness of the Company, according to order of the Company, and shall have stipende yerely.

Thatt Edmond Hogan shalbe Tresourror for iij yeres.

Thatt in consideracion of the industry, good direction, & payns takinge of Michaell Lok in the first viage latly attempted for discouery of Kathay & other new landes, by the Northwestwardes, and also of his great cost, charges, and venture for the provision and furnyture of the same, he shall haue, receiue & take of the Company to his owne vse for ever, the rate of one of every 100, of all the wares, goodes and marchandise, thatt shalbe brought into England or other countries for account of the Company, accordinge to the rate & valew therof in the payment of custome to the Quene's Ma[tie].

Thatt in consideracion of the like industry, good order & great travile of Martyn Frobisheir, gent., in the execution of the firste viage latly made in his owne parson, for the discouery of Kathay and other new landes by the Northwestwardes, he shalbe general Captayne by sea and Admyrall of the shipps & navie of the Company duringe his life, and shall have stipend yerely duringe his life, and also shall have, receve & tak of the Company to his owne vse for ever the rate of one of 100 of all the wares, goodes and marchandise thatt shalbe brought into England or other countries for accounpt of the Company, according to the rate & valew therof in the payment of custome to the Quene's Ma[tie].

Thatt from hence forth for ever the some of one hundreth poundes of Englishe money shalbe accompted one single parte or share in stok of the Company.

Thatt every parson of this Company as well those w[ch] now are the first ventures as all others w[ch] hereafter shalbe free of this Company & wilbe venturers, shall put into accompt for their stock one hundredth poundes of English money, w[ch] shalbe accompted for one single parte and as many more lik single partes as they please, nott beinge above five single partes, and as the traffick from tyme to tyme will suffer to occupy great stock.

Thatt every one of the first veuturers shall haue liberty to put in stock doble nomber of single partes of any other of the venturers from tyme to tyme.

Thatt every new stock of new account from tyme to tyme shall contenew for iij yeres, and att thatt tyme thaccompt therof shalbe clearly made up and fynyshed, & therof divydent then shalbe made, & shalbe paid to every one of the Company venturars in thatt accompt, or to the heires or executors of those wch shalbe deade in thatt meane tyme all thatt shalbe founde dew to them vpon thaccompt accordinge to the rate of their stok therin put.

Thatt all such parsons as shalbe admitted into the fredom of this Company att this next cominge viage & venture to be made, shall paye for a fyne xxx poundes towardes the charges and losses sustayned by the venturers of firste viage made for discouery.

Thatt Mychaell Lok and Martyne Frobusher shalt haue libertye to assigne x parsons to be admytted into the fredome of this Company gratis, att their pleasure, thatt is to saye, each of them to assigne five parsons.

Thatt no parson shalbe admitted into the fredome of this Company after this next coming viadge to be made vntil the ende of iij yeres & fynyshinge of that accompt.

Thatt all other parsons wch shalbe admitted into the fredome of this Company by redemption after the saide iij yeres tyme & ende of thatt accompt, shall paye for afyne tow hundreth poundes of money, to the vse and benefitt of the whole Company.

Thatt a competent howse and warehowse shalbe highred for the buisnes of the Company, and officers & servantts nedfull for the same.

Thatt all the goodes & marchaundise of the Company shalbe marked wth the mark in the margent.

Thatt all the male children of all the forenamed parsons which weare first venturers of this Company, & also the heires male of every of the said male children for ever, shalbe admitted into the liberties & priveledge of the saide Company (gratis) from tyme to tyme.

Thatt such of the Company as shall dye w^{th}out male children may give and assigne over his fredome of this Company to one other parson, by his last will and testament.

[*Colonial, East Indies*, No. 26. *Domestic*, cxix, No. 32.]

A BRIEFF NOTE OF ALL THE COST AND CHARGE OF THE IJ SHIPPS GABRIELL AND MICHAELL AND THEIR PYNNASSE WITH ALL THEIR FURNYTURE FOR THE FYRST VOYAGE TO CATHAY, ETC., SENT WITH MARTYN FFURBISSHER IN JUNE, ANNO 1567.

	li.		
Ffor the hull of the new shipp Gabriell	83	0	0
For the new pynnasse of vij ton	20	0	0
For the shippe Michaell, with old takle and furnyture .	120	0	0

For new takling and rigging them all . . .	229	16	10
For ordenance and municion	100	8	4
For instrumentes of navegacion	50	14	0
For vittuall for the whole voyage	387	14	10
For men's wages paid before hand	213	17	8
For marchandiz for traffick	213	5	8
Sume of all the charge appering particularly bie account	*li*1418	17	4
And nowe wages and charges paid syns the retorne of the shipps home untill the end of December anno. 1576	*li*195	1	11
Sume totall	*li*1613	19	3
And all the whole stok of the adventurers sett down in certaintie were but	875	0	0
So there lakid in stok of the adventurers to supply the whole charge	*li*738	19	3

Ffor the which sume of *li*738 19*s* 3*d* the venturers are to consider towardes Mighell Lok, who did pay and disburse the same for them so long tyme as thaccountes shall declare to his great hyndrans and great danger if it had been lost.

After the retorne home of the said ij shipps were sold divers parcells of the merchandiz and vittuall which wold not kepe good as particularly by account apperith	*li*117	18	7

And all the rest of the shipps and goodes remayning were valued and sold to the account of the second voyage as follows:—

For the marchandiz	*li*148	5	5
For the ij shipps	400	0	0
For wyne and other victualls an divers implementes	147	15	3
Sume of the remayner sold and charged in account of the second voyage	696	0	8
Sume of all the discharge of this first voyage amountes	813	19	3
So ther was left clare by this first voiage with the stok of the adventurers must bere untill God send better successe	800	0	0

And so restith still good in stok of this first voyage *li*75 0*s* 0*d* which is putt to the account of the second voiage with god increasse *li*75 0 0

ACCOUNT OF THE SECOND VOYAGE, WITH THE INSTRUCTIONS GIVEN TO MARTYNE FURBISHER.

Colonial, 39. *Domestic Eliz.*, cxiii. No. 12.

INSTRUCTIONS GIVEN TO MARTYNE FFURBISHER, GENT., FOR ORDERS TO BE OBSERVED IN THE VIAGE NOWE RECOMMENDED TO HIM FOR THE NORTH WEST PARTS AND CATHAY.

1. FYRST, you shall entre as Captayne Generall into the chardge and government of theise three vessels, viz., the Ayde, the Gabriell, and the Michaell, wth all that appertaynethe to them whatsoever.

2. Item, you shall appoint for the furnishing of the sayd vessels, the nomber of 120 persons, whereof xc shalbee maryners, gonneers, carpenters, and other necessarie men to serve for the vse of the shippes: the other xxx to bee moyners, fyners, merchants, and other necessarie persons bothe to accompanie and attend vppon you wch nomber you shall not in any wise exceade.

3. Item, the victualls for vij monthes wch is delivered into the sayd shippes for the provision of the foresayd persons, you shall carefully see the same expended and preserved without spoyle or hurt taking by negligence.

4. Item, you shall not receave into yor companie anie disordred person as neere as you may, and vppon knowledge had to remove them except such as you have received by our order yat were prisoners and condemned persons.

5. Item, you shall vse all diligence possible to depart wth yor sayd vessells from hence before the xxth of this

present, and to takê yor course by north or the west as the wynds will best serve you.

6. Item, in yor waye outwards you shall, if it bee no hinderance to yor viage, sett on land vppon the coast of Friesland vi of the condemned persons wch you carie wth you, wth weapons and vittualls suche as you maye conveniently spare: and yf hit can not be don outwards, you shall doe yor endeavour to accomplishe the same in yor returne; to which persons you shall give instructions howe they may by their good behaviour wyn the goodwill of the people of that country, and also learne the state of the same: and yf you sett them a land in yor goyng outwards, then doe your best to speake wth them in yor returne.

7. Item, when you shall bee past the lands of England, Scotland, or Ireland, you shall direct yor course withe all yor vessels to the island called Hawls Island, beyng in the entrance of the supposed Straight wch wee name Furbisher's Straight, discovered by yor selfe this last yeare. And in yor jorney thitherwards you shall have a speciall regard so to order the matter as yor vessels doe not loose the companie one of the other: and yf anie wilfulnesse or negligence shall appeare in anie person or persons that shall have chardge (or otherwise) in doyng of the contrarie then you shall sharply punishe the same to the example of the rest.

8. Item, at your arrivall at Hawls Island you shall seek a gôod harborrowe for the shippes as nere the same island as may bee, and there to place yor shippes in saftie; and from thence you shall repare wth suche vessels and furniture as is apt to the place where the mynerall oore was had wch you brought hither the last yeare, and there to place the moyeners and other men to worcke and gather the oores foreseeyng that they maye bee placed as well from danger and malice of the people as from anye other extreamitie that maye happen.

9. Item, when you have placed yor moyeners and other persons as before is sayd, you shall then embarck yor selfe in

one of the small barcks and take the other barck also wth you, leaving the Ayde behind you in the chardge of some discreet person as well to receave and lade the oores wch shalbe gotten as also to succour the worckmen; wth the wch twoe barcks you shall repare towards the place where yor men and boate was taken from you, and in the waye goynge you shall make searche bothe for good harborrowes, and also for other moyennes: and yf vppon prooffe you shall fynd moyenes to bee richer than those from whence you came, then you shall returne to the fyrst worckes and remove the moyeners and shippe to those other moyenes as you shall see cawse: and the worckmen beyng once well settled, then you wth the barcks shall proceade for the searching owt of yor men lost, and also to discover L or a c (so farr) leagues westward more from that place as ye maye be certayne that you are entred into the South Sea; and in yor passage to learne all that you can, and not to tarye so longe from the Ayde and worckmen but that you bee able to retorne homewards wth the shippes in due tyme.

10. Item, to consider what places maye bee the most aptest to make fortification, yf neede requyre to the defence of the moyeners and possessynge of the countrie, and to bring perfect plottes and notes thereof.

11. Item, yf it bee possible you shall leave some persons to wynter in the Straight, givyng them instructions how they maye observe the nature of the ayre and state of the countrie, and what tyme of the yeare the Straight is most free from yse: wth whome you shall leave a sufficient proportion of vittals and weapons, and also a pynnesse wth a carpenter and thinges necessarie so well as maye bee.

12. Item, yf it shall happen that the moyenes do not yeald the substance that is hoped for, then you shall furnishe the twoe barckes wth such as you maye take owt of the Ayde, and therewithall you shall proceade towards the discovering of Cathaya wth the two barcks and returne the Ayde for England agayne.

13. Item, as you shall mistrust rather to muche than any thinge to litle towching the matter of yor salftie, when you happen to come to have conference wth the people of those parts where you shall arrive: so agayne wee requyre you, that in all yor doynges you doe so behave yor selfe, and to cawse yor companie to doe the like towards the sayd people as maye gyve lest cawse of offence, and to procure as muche as in you shall lye to wynne bothe frendshippe and likynge.

14. Item, yf you fynd that the oore bee of that qualitie and quantitie that is looked for, that then you doe procure to lade so muche therof in all yor shipping as maye bee, allthoughe you doe leave owte other superfluouse thinges.

15. Item, you shall make yor returne homewards by the west parts of Ireland, and so by the narrowe seas of England to London, for that wee doe take the same to bee yor salfest course. And because wee doe not knowe what other matters maye happen to you in the tyme of yor jorney, and therfore can not prescribe what is to bee done for yor reliefe in suche a case: wee doe therfore referre the consideration of the same to yor good discretion, not doubtyng but that the order wch you will take therin shalbe agreeable wth the good expectation that is conceaved of you.

16. Item, wee doe not thinke it good you should bringe hither above the nomber of iij or iiij or 8 or tenne at the most of the people of that countrie: whereof some to be ould and the other yonge whome wee mynd shall not returne agayne thither; and therfore you shall have great care howe you doe take them for avoidyng of offence towards them and the contrie.

Lastlie we thincke it verie meete that you geve expresse commaundement vnto the fyners and tryers of the oore that they doe not dyscover the secreats of the riches of suche moynes as by you shall be founde out vnto anie besids your self and such others as to you shall be thought fit should be made acquaynted therwith for her M^{tie} better service in that behalf.

A TRUE REPORTE

OF SUCH THINGS AS HAPNED IN THE SECOND VOYAGE OF CAPTAYNE FROBYSHER, PRETENDED FOR THE DISCOVERIE OF A NEW PASSAGE TO CATAYA, CHINA, AND THE EAST INDIA, BY THE NORTH WEST. ANNO DO. 1577.

BEYNG furnished with one tall-shippe of hir Majesties, named yᵉ Ayde, of two hundreth tunne, and two other small barkes, the one named the Gabriell, the other the Michael, about thirtie tunne a peece, being fitlie appointed with men, munition, victuals, and all things necessary for the voyage, the sayde Captayne Frobysher, with the rest of his companie, came aboorde his shippes riding at Blackwall, intending (with God's help) to take yᵉ first winde and tyde serving him, the five and twentith day of May, in the yeare of oure Lorde God, a thousande five hundred seventie and seaven.

The names of such gentlemen as attempted this discoverie, and the number of souldyoures and mariners in each shippe as followeth.

Boord the Ayde being Admirall, were yᵉ number of one c men of all sorts, whereof xxx or more were gentlemen and souldyers, the rest sufficiente and tall saylors.

Aboorde the Gabriell being Vice-admirall, were in all 18 persons, whereof sixe were souldyers, the rest mariners.

Aboorde the Michaell were sixteene persons, whereof five were souldyers, the rest mariners.

Aboorde the Ayde was: General of the whole company

for hir Majesty, Martin Frobisher; his Lieutenant, George Best; his Aunciente, Richar. Philpot; Corporal of ye shot, Frauncis Forder; the rest of ye gentlemen, Henry Carew, Edmund Stafford, John Lee, — Harvie, Mathew Kynersley, Abraham Lyns, Robert Kynersley, Frauncis Brackenburye, William Armshow; the Mayster, Chrisofer Hall; the Mate, Charles Jackman; the Pylotte, Andrew Dyer; the M. Gunner, Richard Coxe.

Aboorde the Gabriell was: Captayne, Edward Fenton; one gentleman, William Tamfield; the Mayster, William Smyth.

Aboorde the Michaell was: Captaine, Gilbert Yorke; one gentleman, Tho. Chamberlaine; the Mayster, James Beare.

On Whitsonday, being the 26 of May, anno 1577, earely in the morning we wayed ancker at Blackwall, and fell that tyde downe to Gravesende, where we remayned untill Monday at night.

On Monday morning the 27 of May, aboorde the Ayde we receyved all the Communion, by the Minister of Gravesende, and prepared us, as good Christians towardes God, and resolute men for all fortunes: and towards nighte we departed to Tilburie Hope.

Tewsday the 28 of Maye, aboute nine of the clocke at nighte, we arrived at Harwitch in Essex, and there stayed, for the taking in of certaine victualles, until Friday, being the thirtith of May, during whyche tyme came letters from the Lordes of the Counsell, streightelye commaunding oure Generall, not to exceede hys complemente and number appoynted hym, whyche was, one hundred and twentye persons: whereuppon he discharged many proper men, whiche wyth unwilling myndes departed.

The number of men in this voyage.

The condemned men discharged.

He also dismissed all hys condemned men, whyche he thoughte for some purposes verie needefull for the voyage, and towardes nyghte upon Friday, the one and thirtith of May, we sette sayle, and putte to seas agayne. And sayling northwarde alongst the east coastes of Englande and Scotlande, the seaventh day of June, wee arrived in Sainte Magnus Sounde in Orckney Ilandes, called in Latine *Orcades*, and came to ancker on the south syde of the Bay.

The first arrivall after our departure from Englande.

Heere oure companye goyng on lande, the inhabitants of these ilandes beganne to flee, as from the enemie, whereuppon, the Lieutenante wylled everye man to staye togyther, and wente hymselfe unto theyr houses, to declare what wee were, and the cause of oure comming thyther, whyche beeyng understoode, after their poore manner they friendly entreated us, and brought us for oure money, such things as they had. And heere our goldfynders found a mine of silver.

A myne of silver found in Orkney.

Orkney is the principall of the Iles of Orcades, and standeth in the latitude of 59 degrees and a halfe. The countrey is much subject to colde, aunswerable for suche a climate, and yet yeeldeth some frutes, and sufficient mayntenance for the people contented so poorely to live.

There is plentie ynough of poultrey, store of egges, fishe, and fowle.

For theyr bread, they have oaten cakes, and theyr drinke is ewes milke, and in some partes ale.

Their houses are but poore without, and sluttish ynough within, and the people in nature thereunto agreeable.

For theyr fire, they burne heath and turffe, the countrey in most parts being voyde of woode.

They have greate wante of leather, and desire our olde shoes, apparell, and old ropes (before money) for their victuals, and yet are they not ignorant of the value of our coine. The chiefe towne is called Kyrway.

Kyrway the chief towne of Orkney.

In this iland hath bin sometime an abbey, or a religious house, called Saint Magnus, being on the west side of the ile, whereof this sound beareth name, through whyche we passed. Their Governeure, or chiefe lorde, is called the Lord Robert Steward, who at oure being there, as wee understoode, was

St. Magnus sound, why so called.

in durance at Edenburgh, by the Regent's commaundement of Scotlande.

After we had provided us heere of matter sufficiente for our voyage, the eyght of June we sette sayle agayne, and passing through Saint Magnus Sounde, having a merrie winde by night, came cleere, and lost sight of all the lande, and keeping our course west-north-west by the space of two dayes, the wind shifted upon us, so that we lay in traverse on ye seas, with contrarie, making good (as neere as we could) our course to the westward, and sometime to the northward, as the winde shifted. And heereabout we met with three sayle of English fishermen from Iseland, bound homewarde, by whome we wrote our letters unto our friends in England. We traversed these seas by the space of 26 dayes, without sight of any land, and met with much drift woode and whole bodyes of trees. We saw many monsterous fishe, and strange fowle, whyche seemed to live only by the sea, being there so farre distant from anye lande. At length, God favoured us with more prosperous windes, and after we hadde sayled foure dayes with good wind in the poupe, the fourth of July the Michaell (being formost a head) shotte off a peece of ordinance, and stroke all hir sayles, supposing that they descryed land, whyche by reason of the thicke mistes, they could not make perfit: howbeit, as wel our accompt, as also the greate alteration of the water, whiche became more blacke and smooth, dyd playnely declare we were not farre off the coast. Our Generall sent his Maister aboorde the Michaell (who had bin within the yeare before) to beare in with the place, to make proofe thereof, who descried not the lande perfect, but sawe sundrie huge ilands of ise, which we deemed to be not past twelve leagues from the shore, for about tenne of the clocke at night, being the fourth of July, the weather being more cleere, we made the land perfect, and knew it to be Freeseland. And the heigth being taken heere, we founde oureselves to be in the latitude of 60

Great bodies of trees dryving in the seas. Monstrous fish and strange fowle lyving only by the sea.

Water being blacke and smooth signifieth land to be neare.

Islands of ice.

The firste syght of Freeselande.

degrees and a halfe, and were fallen with the southermost parte of this land. Betweene Orkney and Freesland are reckoned leagues.

This Freeseland sheweth a ragged and high lande, having the mountaynes almost covered with snow alongst the coast full of drift ise, and seemeth almost inaccessible, and is thought to be an iland in bignesse not inferior to England, and is called of some authours Weast Freeseland, I thinke, bycause it lyeth more weast than anye part of Europe. It extendeth in latitude to the northward, verie farre as seemed to us, and appeareth by a description set out by two breethren, Nicholaus and Antonius Genoa, who being driven off from Ireland with a violent tempest, made shipwracke heere, and were the first knowen Christians that discovered this lande, aboute three hundred yeares sithence; and they have in their sea cardes set out everie part thereof, and described the condition of the inhabitants, declaring them to be as civill and religious people as we.[1] And for so much of this land as we have sayled alongst, comparing their carde with ye coast, we find it very agreeable. This coast seemeth to have good fishing, for we lying becalmed, let fall a hooke without anye bayte, and presently caught a great fish called a hollibut, which served the whole companie for a days meate, and is dangerous meate for surfetting. And sounding about five leagues off from the shore, our lead brought up in the tallow a kind of corall almost white, and small stones as bright as christall: and it is not to be doubted but that this land may be found very rich

Freeseland described.

An easie kind of fishing.

White corall got by sounding.

[1] The brothers referred to are Niccolo and Antonio Zeno, of Venice, the former of whom, in 1380, made a voyage to the north, and was driven by a storm to the Faroe Islands, whence he dispatched a letter to his brother Antonio, urging him to find means to join him, which he did. The account was published at Venice, in 1558, by Francisco Marcolini, a descendant of the Zeno, and was compiled from the fragments of letters written by Antonio Zeno to Carlo, his brother. The Faroe Islands are the "Friseland," here referred to.

and beneficiall if it were throughly discovered, although we saw no creature there but little birds. It is a marvellous thing to behold, of what great bignesse and depth some ilandes of ise be heere, some seventy some eighty fadome under water, besides that which is above, seemyng ilands more than halfe a mile in circuite. All these ise are in tast freshe, and seeme to be bredde in the sounds thereabouts, or in some land neere the pole, and with the wind and tides are driven alongst the coastes. We found none of these islands of ise salt in taste, whereby appeareth they were not congealed of the ocean sea water, which is always salt, but of some standing or little moving lakes or great fresh waters neere the shore, caused eyther by melted snow from the tops of mountains, or by continuall accesse of fresh rivers from the land, and intermingling with ye sea water, bearing yet the dominion (by the force of extreame frost) may cause some part of salt water to freese so with it, and so seeme a little brackish, but otherwise ye maine sea freeseth not, and therefore there is no *mare glaciale* or frozen sea as the opinion hitherto hath bin. Our general proved landing here twice, but by ye suddaine fall of mistes (whereunto this coast is much subject) he was like to lose sight of his ships, and being greatly endangered with the driving ise alongst the coast, was forced aboord, and faine to surceasse his pretence till a better oportunitie might serve: and having spent four dayes and nights sailing alongst this land, finding the coast subject to such bitter cold and continuall mistes, he determined to spend no more time therein, but to beare out his course towards ye streights called Frobishers straights, after ye generals name, who being the firste that ever passed beyonde fifty-eight degrees to the northwards, for any thing hath bin yet knowen of certainty of New found land, otherwise called ye continent or firme land of America; discovered ye said streights this last yeare 1576, and hopeth that there wil be found a thorough passage into

Monstrous iles of ise in taste freshe, where hence they are supposed to come.

The opinion of the frosen seas is destroyed by experience.

Frobishers straightes.

the sea, which lieth on the back side of ye said New found land called *Mare Pacificum* or *Mare de Sur*, by the which we maye go unto Cataya, China, the East India, and all the dominions of the Great Cane of Tartaria. Betweene Freeseland and the straights we had one great storme, wherin ye Michael was somewhat in danger, having hir steerage broken and hir top mastes blowen over bord, and being not past fifty leagues short of ye straightes by our accompt, we strooke sayle and lay a hull, fearing the continuance of the storme, the wind being at the northeast, and having lost company of the barkes in that flaw of wind, we happily mette againe the 17 day of July, having the evening before seene divers ilandes of fleeting ise, which gave an argument that we were not farre from land. Our Generall in ye morning from the maine top (ye weather being reasonable cleere) descried land, but to be better assured, he sent the two barkes two contrarie courses, wherby they might descrie either the south or north forlande, the Ayde lying off and on at sea, with a small saile by an iland of ise, whiche was the marke for us to meete togither agayne. And aboute noone, the weather being more cleere, we made the North Forlande perfite, which otherwise is called Halles Iland, and also the small ilande bearing the name of the saide Hall, whence the ore was taken uppe, whiche was broughte into Englande this last yeare 1576, the said Hall being present at the finding and taking up thereof, who was then maister in the Gabriell withe Captayne Frobisher. At oure arrivall heere, all the seas about this coast were so covered over with huge quantitie of great ise that we thought these places might only deserve the name of *Mare Glaciale*, and be called the Isie Sea.

The steerage of the Michaell broken by tempest.

The first entrance of the straights.

Halles iland.

This North Forlande is thought to be devided from the continente of the norther lande by a little sounde called Halle's Sounde, whiche maketh it an iland, and is thoughte little lesse than the ile of Wight, and is the firste entrance of

The description of the streyghtes.

the streightes upon the norther side, and standeth in the latitude of 62 degrees, 50 minutes, and is reckned from Freeseland leagues. God having blessed us wyth so happie a lande fall, we bare into the streightes whyche runne in next hande weast, and somewhat to the northwarde, and came as neere the shore as we mighte for the ise, and uppon the eyghteenth day of July our Generall taking the gold fynders with him, attempted to go on shore with a small rowing pinnesse, upon the small iland where the ore was taken up, to prove whether there were anye store thereof to be found, but he could not gette in all that iland a peece so bigge as a walnutt, where the firste was founde, so that it may seeme a greate miracle of God, that being only one rich stone in all the iland, the same should be found by one of our countrymen, whereby it shoulde appeare, God's divine will and pleasure is, to have oure common wealth encreased with no lesse abundance of His hyden treasures and golde mynes than any other nation, and would that the fayth of His Gospell and holy name should be published and enlarged throughe all those corners of the earth, amongst these idolatrous infidels. But oure men whiche sought the other ilandes thereaboutes, found them all to have good store of the ore, whereuppon our Generall with these good tidings retourned aboorde aboute ten of the clocke at night, and was joyfully welcomed of the companie with a volie of shotte. He brought egges, fowle, and a yong seale aboord, which the companie hadde killed ashore, and having founde upon those ilandes ginnes set to catch fowle, and stickes new cut, with other things, he well perceived that not long before some of the countrey people had resorted thither. Having therefore found these tokens of the peoples accesse in those partes, and being in his firste voyage well acquainted with their subtile and cruell disposition, he provided well for his better safetie, and on Friday the ninteenth of July in the morning early, with his best company of gentlemen and

No more gold ore found in the fyrst ilande.

Egges and fowles of Meta Incognita. Snares set to catch birds withall.

souldioures to the number of fortie persons, went on shoare aswell to discover the inlande and habitation of the people, as also to fynd out some fitte harborowe for our shippes. And passing towardes the shoare with no small difficultie, by reason of the abundance of ise whiche lay alongest the coaste so thicke togither, that hardely any passage throughe them might be discerned, we arrived at length upon the maine of Halles greater iland, and founde there also, aswel as in the other small ilands, good store of the ore. And leaving his boats here with sufficient guarde passed up into the countrey about two English miles, and recovered the top of a high hill, on the top whereof our men made a columne or crosse of stones heaped uppe of a good heighth togither in good sorte, and solemnly sounded a trumpet, and said certaine prayers, kneeling aboute the ancient, and honoured the place by the name of Mount Warwicke, in remembrance of the Right Honorable the Lord Ambrose Dudley, Erle of Warwick, whose noble minde and good countenaunce in this, as in all other good actions, gave great encouragement and good furtherance. This done, we retired our companies, not seeing any thing here worth further discoverie, the countrie seeming barren and full of ragged mountaines, in most parts covered with snow. And thus marching towards our boats, we espied certaine of the countrey people on the top of Mount Warwicke with a flag, wafting us backe againe, and making great noise, with cries like the mowing of bulls, seeming greatly desirous of conference with us: wheruppon the General, being therewith better acquainted, answered them again with the like cries, whereat, and with the noise of our trumpets, they seemed greatly to rejoyce, skipping, laughing, and dancing for joy. And hereuppon we made signes unto them, holding up two fingers, commanding two of our men to go aparte from our companies, wherby they might doe the like. So that forthwith two of oure menne and two of theirs mette togither a good space from companie,

The building of a column, called Mount Warwick.

The first sight of the countrie people wafting with a flag.

The meeting apart of two Englishmen with two of that countrie.

neither partie having their weapons about them. Our men gave them pinnes and pointes, and such trifles as they had. And they likewise bestowed on our men two bowe cases, and such things as they had. They earnestly desired oure men to go uppe into their countrie, and our men offered them like kindnesse aboorde oure shippes, but neyther parte (as it seemed) admitted or trusted the others curtesie. Their manner of traffic is thus: they doe use to laye down of their marchandise uppon the ground, so much as they meane to parte withall, and so looking that the other partie, with whome they make trade, shoulde doe the like, they themselves doe departe, and then, if they doe like of their marte, they come againe, and take in exchange the others marchandise, otherwise, if they like not, they take their owne and departe. The day being thus well neare spent, in haste we retired our companies into our boates againe, minding forthwith to searche alongst the coast for some harborowe, fitte for our shippes; for the present necessitie thereof was much, considering that all this while they lay off and on between the two lands, being continually subject, as well to great danger of fleeting yse, which environed them, as to the sodain flawes which the coast seemeth much subject unto. But when the people perceived our departure, with great tokens of affection they earnestly called us backe againe, following us almost to our boates: whereuppon our generall taking his maister with him, who was beste acquainted with their maner, went apart unto two of them, meaning, if they could lay sure holde upon them, forcibly to bryng them abord, with intent to bestowe certain toyes and apparell upon the one, and so to dismisse him with all arguments of curtesie, and retaine the other for an interpreter. The generall and his maister being met with their two companions togither, after they hadde exchanged certaine thinges the one with the other, one of the salvages for lacke of better marchandise, cutte off the tayle of his coate (which is a chiefe ornament

The order of their traffic.

Another meeting of two of our men with two of theirs.

among them) and gave it unto oure general for a present. But he presently upon a watchword given, with his maister sodainely laid holde upon the two salvages. But the grounde underfeete being slipperie, with the snow on the side of the hill, thire handfast fayled, and their pray escaping, ranne awaye, and lightlye recovered their bowe and arrowes, which they had hid not farre from them behinde the rockes. And being only two salvages in sight, they so fiercely, desperately, and with such furie assaulted and pursued our generall and his maister, being altogither unarmed, and not mistrusting their subtilities, that they chased them to their boats, and hurte the generall in the buttocke with an arrow, who the rather speedily fled backe, bycause they suspected a greater number behind ye rocks. Our soldiers (which wer commanded before to keepe their boates) perceiving the danger, and hearing our men calling for shot, came speedily to rescue, thinking there had bin a more number. But when ye salvages heard ye shot of one of our calivers, and yet (having first bestowed their arrows) they ran away, our men speedily folowing them. But a servante of my Lorde of Warwickes, called Nicholas Conyer, a good footeman, and uncumbred with anye furniture, besides a dagger at his backe, overtooke one of them, and being a Cornishman, and a good wrastler, shewed his companion such a Cornishe tricke, that he made his sides ake against the grounde for a moneth after. And so being stayed, he was taken alive, and brought away, but the other escaped. Thus with their straunge and newe praye, our men repaired to their boates, and passed from the maine to a small iland of a myle compasse, where they resolved to tarrie all night, for even now a sodaine storme was grown so great at sea, that by no means they coulde recover their ships. And here every man refreshed himselfe with a small portion of victualles, whiche was laide into the boates for their dinners, having neither eate nor drunke all the daye before. But bycause

The Englishmen chased to their boats.

One of that countrimen taken.

they knewe not howe long the storme might laste, nor how far off the ships might be put to sea, nor whether they should ever recover them againe or not, they made great spare of their victualles, as it greatly behoved them. For they knew ful wel, that the beste cheare the countrey coulde yeelde them, was golden rockes and stones, a harde foode to live withall, and the people more readie to eate them, than to give them wherewithall to eate. And thus keeping verie good watche and warde, they lay there al night upon harde cliffes of snowe and ise, both wette, cold, and comfortlesse.

These things thus hapning with the company on lande, the danger of the shippes at sea was no lesse perilous. For within one houre after the generalls departing in ye morning, by negligence of the cooke in over heating, and the workman in making the chimney, the Aide was set on fire, and had bin the confusion of the whole, if by chaunce a boye espying it, it hadde not bin speedily with great labour and Gods helpe well extinguished.

The Aide set on fire.

This day also, were diverse stormes and flawes, and by nine of the clocke at night the storme was growen so great, and continued suche untill the morning, that it putte our shippes at sea in no small peril, for having mountaines of fleeting ise on every side, we went romer for one, and loofed for another, some scraped us, and some happily escaped us, that the least of all of them were as dangerous to strike as any rocke, and able to have split asunder the strongest shippe of the worlde. Wee had a scope of cleare withoute ise (as God would), wherein we turned, beyng otherwise compassed on everye side about, but so much was the winde, as so little was our sea room, that being able to beare only our fore-coast, we cast so oft about, that we made fourteene bordes in eight glasses running, being but foure houres: but God being our best steresman, and by ye industry of Charles Jackman and Andrew Dyer, then maisters mates, both very expert mariners, and Richard Cox, the maister gunner, with

The great danger of those rocks of ise.

other very carefull saylors, then within borde, and also by the helpe of ye cleare nights which are without darkenesse, we did happily avoyde those present daungers, whereat since we have more marvelled than in the present daunger feared, for that every man within borde, both better and worse, had ynough to doe, with his handes to hale ropes, and with his eyes to looke out for daunger. But the nexte morning, being the 20 of July, as God would, the storme ceased, and the general espying the shippes with his newe captive and whole companie, came happilye aborde and reported what hadde passed a shoare, whereupon, altogither upon our knees, gave God humble and hartie thankes, for that it hadde pleased him, from so speedy peril to send us such speedie deliverance, and so from this norther shoare we stroke over towards the southerland.

Night without darknes in that countrey.

The one and twentieth of July, we discovered a bay, which ranne into the lande, that seemed a likely harborow for our shippes, wherefore our general rowed thither with hys boates, to make proofe thereof, and with his goldfinders to searche for ore, having never assayed anye thing on the south shoare as yet, and the first small iland whiche we landed on, here all the sands and cliftes did so glister and had so bright a marquesite, that it seemed all to be golde, but upon tryal made, it proved no better than blacke leade and verified the proverb—All is not golde that shineth.

Our firste commyng on the southerland of the sayde Straightes.

A myne of blacklead.

Upon the two and twentieth of July, we bare into the sayd sounde, and came to anker a reasonable bredth off the shoare, where, thynking our selves in good securitie, we were greatly endangered with a peece of drift ise, which the ebbe brought forth of the soundes, and came thwart us ere we were aware. But the gentlemen and souldiers within borde, taking great paynes at this pinche at the capstone, overcame the most daunger thereof, and yet for all that might be done, it stroke on our sterne such a blowe, that we feared least it had stryken away our rudder, and being forced to cut our cable in the hawse, were fayne to set our foresaile to runne further up within, and if our sterage had not bin stronger, than in the present time, we feared we had runne the shippe upon the rocks, having a very narrowe channell to turne in; but, as God woulde, al came well to passe. And this was named Jackmans Sounde, after the name of the maisters mate, who had first liking unto the place.

Jackmans Sound.

Smiths Iland.

Upon a small ilande, within this sound, called Smiths Iland (bycause he first set up his forge there), was founde a myne of silver, but was not wonne out of the rockes without great labour. Here our goldfynders made saye of suche ore as they founde uppon the Northerlande, and founde foure sortes thereof to holde golde in good quantitie. Upon another small iland here was also founde a great deade fishe, whiche, as it should seeme, had bin embayde with ise, and was in proportion rounde like to a porpose, being about twelve foote long, and in bignesse answerable, havyng a horne of two yardes long growing out of the snoute or nostrels. This horne is wreathed and strayte, like in fashion to a taper made of waxe, and maye truely be thoughte to be the sea Unicorne. This horne is to be seene and reserved as a jewel, by the Queens majesties commandement in hir wardrop of robes. The form whereof is here set down.

The fynding of an unicornes horne.

Tewsday, the three and twentyth of July, our general with his best company of gentlemen, souldiers and saylers, to the number of seaventie persons in all, marched with aunciente displayde uppon the continent of the Southerlande (the supposed continent of America), where, commandyng a trumpet to sounde a call for every man to repayre to the auncient, he declared to the whole company, how much the cause imported, for the service of hir majestie, our countrey, our credites, and the safetie of our own lives, and therefore required every man to be conformable to order and to be directed by those he shoulde assigne. And appointed for leaders, Captaine Fenton, Captain Yorke, and his lieutenant,

George Beste; whiche done, we cast our selves into a ring, and altogither uppon oure knees, gave God humble thankes, for that it had pleased him of his greate goodnesse in preserving us from such imminent dangers, to bestow so great and hidden treasures upon us his poore and unworthye servants, beseeching likewise the assistance of his holy spirite so to deliver us in saftie, into our countrey, whereby the light and truth of these secretes being known, it might rebound to the more honor of his holy name, and consequently to the advancement of our common wealth. And so, in as good sorte as the place suffered, we marched towardes the tops of the mountains, which were no lesse painful in clyming, than dangerous in descending, by reason of their steepenesse and ise. And having passed about five miles by such unweldie wayes, we returned unto oure ships, without sighte of any people, or likelyhoode of habitation. Here diverse of the gentlemen desired oure general to suffer them to the number of twentie or thirtie persons, to marche up thirtie or forty leagues in the countrie, to the ende they mighte discover the inlande, and do some acceptable service for their countrey. But he not contented with the matter he sought for, and well considering the short time he had in hande, and the greedie desire our countrey hath to a present savour and retourne of gayne, bent his whole indevour onely to finde a mine, to fraight his ships, and to leave the reste (by God's help) hereafter to be well accomplished. And therefore the twentie sixte of Julye he departed over to the Northlande with the two barkes, leaving the *Ayde* riding in Jackmans Sound, and ment (after he had founde convenient harborowe, and fraight there for his shippes) to discover further for the passage. The barkes came the same night to anker in a sound, upon the Northerland, where the tydes did runne so swifte, and the place so subject to indraftes of ise, that by reason thereof they were greatly endangered, and having founde a very rich myne, and got almoste twentie tunne of ore togither, upon the 28 of July, the ise come driving into the Sounde where the barkes road, in such sort, that they were therewith greatly distressed. And the *Gabriell* riding asterne the *Michaell*, had hir cable gaulde asunder in the hawse, with a peece of driving ise, and lost another anker, and having but one cable and ancker lefte; for she has loste two before, and the ise still driving uppon hir, she was (by Gods helpe) wel fenced from the daunger of the rest, by one great iland of ise which came a grounde harde a heade of hir, which, if it had not so chaunced, I think, surely she had bin cast upon the rockes with the ise. The *Michael* mored ancker upon this great ise, and roade under the lee therof: but about midnight, by the weyght of it selfe, and the setting of the tydes, the ise brake within halfe the barke's length, and made unto the companie within bord, a sodaine and fearefull noyse. The next flounde towarde the morning we weyed ancker and went further uppe the straightes, and leaving our ore behinde us which we had digged, for hast, left the place by the name of Beares Sound, after the masters name of the *Michael*, and named the iland Lecesters Iland. In one of the small ilands here, we founde a tombe, wherin the bones of a dead man lay togither, and our savage being with us and demanded (by signes) whether his countreymen had not slain this man and eat his flesh so from the bones, he made signes to the contrarie, and that he was slain with wolves and wilde beastes. Here also was founde hid under stones good stoare of fish, and sundrie other things of the inhabitants: as sleddes, bridles, kettels of fishe skinnes, knives of bone, and such other like. And our savage declared unto us the use of all those things. And taking in his hand one of those countrey brydels, he caughte one of our dogges, and hampred him handsomely therein, as we do our horses, and with a whip in his hande, he taught the dogge to drawe in a sledde, as we doe horses in a coatche, setting himselfe thereuppon like a guide: so that we might

Beares Sound.

Lecester Iland.

A tombe with a dead mans bones in it, found at degrees in latitude.

Bridles, knives, and other instruments found hid among the rocks.

see, they use dogges for that purpose, as we doe our horses. And we founde since by experience, that the lesser sorte of dogges they feede fat, and keepe them as domesticall cattel in their tentes, for their eating, and the greater sort serve for the use of drawing theyr sleds.

They use great dogges to draw sleds and little dogs for their meate.

The twentie ninth of July, about five leagues from Beare's Sound, we discovered a bay, which being fenced on eche side with small ilandes, lying of the mayne, whych break the force of the tydes, and make the place free from any indraftes of yse, did prove a very fitte harborow for our ships, where we came to anker under a small iland, whiche now togither with the sound, is called by the name of that right honorable and vertuous lady, Anne, Countesse of Warwicke. And this is the furthest place that this yeare we have entred uppe within the streyghtes, and is reconed from the cape of the Queen's forelande, which is the entrance of the streights, not above 30 leagues. Upon this ilande was found good store of the ore, which in the washing helde golde plainly to be seen: whereupon it was thoughte beste rather to loade here, where there was store and indifferent good, than to seek further for better, and spend time with jeopardie. And therefore oure generall setting the myners to worke, and shewing fyrste a good president of a painefull labourer and a good captaine in himselfe, gave good examples for others to follow him: whereuppon every man, both better and worse, with their best endevors, willingly laide to their helping handes. And the nexte daye, being the 30 of July, the *Michaell* was sente over to Jackmans Sounde for the *Ayde* and whole companie to come thither. Upon the maine lande over against the Countesse's Iland, we discovered and behelde to our great marvell, the poor caves and houses of those countrie people which serve them (as it shoulde seem) for their winter dwellings, and are made two fadome under grounde, in compasse rounde, like to an oven, being joyned fast one by another, having holes like to a fox or conny

The Countesses Iland.

Thirtie leagues discovered within the Straytes.

A good president of a good captayne shewed by Captain Frobisher.

The maner of their houses in this countrey.

berrie, to keepe and come togither. They under-trench these places with gutters, so that the water falling from the hills above them, may slide away without their anoiance, and are seated commonly in the foote of a hil, to shielde them better from the colde winds, having their dore and entrance ever open towardes the south.

Whales bones used instead of timber.

From the ground upward they builde with whales bones, for lacke of timber, whiche, bending one over another, are handsomely compacted in the toppe togither, and are covered over with seales' skinnes, whiche instead of tiles, fenceth them from the rayne. In eache house they have only one roome, having the one halfe of the floure raysed with broad stones a foote higher than ye other, whereon strawing mosse, they make their nests to sleepe in. They defile these dennes most filthylie with their beastly feeding, and dwell so long

The sluttishnesse of these people.

in a place (as we thinke), untill their owne sluttishnesse lothyng them, they are forced to seeke a sweeter ayre and a new seate, and are (no doubt) a dispersed and wandring nation, as the Tartarians, and live in hords and troupes, withoute anye certayn abode, as may appeare by sundry circumstances of our experience. Here our captive being

A signe set up by the savage captive, and the meaning thereof.

ashore with us, to declare ye use of such things as we saw, stayd himselfe alone behind the company, and did set up five small stickes round in a circle, one by another, with one smal bone placed just in ye middst of all: which thing when one of our men perceived, he called us backe to behold ye matter, thinking that he had meant some charme or witchcraft therin. But ye best conjecture we could make thereof, was, that he would thereby his countreymen should understand yt for our five men which they betrayed the last yeare (whom he signified by ye five sticks) he was taken and kept prisoner, which he signified by ye bone in ye midst. For afterwardes, when we shewed him the picture of his countreyman, which ye last yeare was brought into England (whose counterfet we had drawne, with boate, and other furniture, both as he was in his own, and also in English aparell) he was upon the suddayne muche amazed therat, and beholding advisedly the same with silence a good while, as though he would streyne courtesie whether shoulde begin y° speech (for he thoughte him no doubte a lively creature) at length, began to question with him, as with his companion, and finding him dumme and mute, seemed to suspect him, as one disdaynful, and would with a little help have growen into choller at the matter, until at last by feeling and handling, he founde him but a deceiving picture. And then with great noyse and cryes, ceased not wondering, thinking that we coulde make menne live or die at our pleasure.

The savage captive amazed at his countreymans picture.

And thereuppon calling the matter to hys remembrance, he gave us plainely to understande by signes, that he had knowledge of the taking of our five men the last yeare, and confessing the manner of eche thing, numbred the five men upon his five fingers, and poynted unto a boate in our ship, which was like unto that wherein our men were betrayed. And when we made him signes that they were slaine and eaten, he earnestly denied, and made signes to the contrarie.

The last of July, the *Michael* retourned with the *Ayde* to us from the Southerlande, and came to anker by us in the Countesse of Warwicks Sounde, and reported that since we departed from Jackmans Sound, there happened nothing among them there greatly worth the remembraunce, until the thirtieth day of July, when certaine of our companie being ashore upon a small iland within the said Jackmans Sound, neare the place where the *Ayde* roade, didde espie a long boate with divers of the countrie people therein, to the number of eighteene or twentie persons, whom so soone as oure men perceived, they retourned speedily aboorde, to give notice therof unto our companie. They might perceive these people climbing up the toppe of a hill, where, with a flagge, they wafted unto our shippe, and made great out-

Another shew of twentie persons of that countrie in one boate.

cries and noises, like so many bulls. Here uppon our men did presently man forth a small skiff, having not above sixe or seaven persons therein whiche rowed neare the place where those people were, to prove if they could have any conference with them. But after this small boate was sente a greater, beeyng well appoynted for their rescue, if neede required.

As soone as they espied our companye comming neare them, they tooke their boates and hasted awaye, either for feare, either else for pollicie, to drawe our men from rescue further within their danger: wherefore our men construing their comming thither was but to seeke advauntage, followed speedily after them, but they rowed so swiftly away that our men could come nothing neare them. Howbeit they failed not of their beste endeavour in rowing, and having chased them above two myles into the sea, returned into their shippes againe.

The morning following, being the first of August, Captaine Yorke, with the *Michaell,* came into Jackman's Sound, and declared unto the company there, that the laste night's past, he came to anker in a certaine baye (which sithens was named Yorkes Sounde), aboute foure leagues distant from Jackmans Sound, being putte to lewarde of that place for lacke of winde, where he discovered certaine tents of the countrie people, where going with his companye ashoare, he entred into them, but founde the people departed, as it shoulde seem, for feare of their comming. But amongest sundrie straunge things whiche in these tentes they founde there was rawe and newe killed fleshe of unknown sortes, with dead carcasses and bones of dogs, and I know not what. They also beheld (to their greatest marvaile) a dublet of canvas, made after the Englishe fashion, a shirt, a girdle, three shoes for contrarie feete and of unequal bignesse, which they well conjectured to be the apparell of our five poore countriemen whiche were intercepted the laste

Yorkes Soundes.

The apparell founde againe of oure Englishmen whiche the yeare before were taken captive.

yeare by these countrie people, aboute fiftye leagues from this place further within the straightes. Wherupon, our men being in good hope that some of them might be here, and yet living, the captaine devising for the best, lefte his mind behind him in writing with pen, yncke, and paper also, whereby our poore captive countriemen, if it mighte come to their handes, mighte knowe their friendes mindes, and of their arrivall, and likewise retourne their answer. And so without taking any thing away in their tentes, leaving there also looking glasses, pointes, and other of our toyes (the better to allure them by such friendly means) departed aboorde hys barke, wyth intent to make haste to the *Ayde*, to give notice unto the companye of all such things as he had there discovered: and so ment to returne to these tents againe, hoping, that he might by force or policie entrappe or entice the people to some friendly conference. Which things, when he had delivered to the company there, they determined forthwith to goe in hande with the matter. Hereuppon Caytaine Yorke, with the maister of the *Ayde* and his mate (who, the night before, had bin at the tents, and came over from the other side in the *Michaell* with him) being accompanied with divers of the gentlemen and souldiers, to the number of 30 or 40 persons, in two small rowing pinnesses, made towardes the place where the night before they discovered the tents of those people, and, setting Charls Jackman, being the master mate, ashor with a convenient number over lande, mening to compasse them on the one side, whilst the captaine with his boats might entrap them on the other side. But landing at last at ye place where the night before they left them, they found them with their tents removed. Notwithstanding, our men which marched up into ye countrie, passing over two or three tedious mountains, by chance espied certaine tentes in a valley underneath them neare unto a creeke by the sea side, whiche, bycause it was not the place where the guide had

A good device of Captaine Yorke.

bin the night before, they judged them to be another company, and besetting them about, determined to take them if they could. But they having quickly descried our companie, launched one great and another small boate, being about 16 or 18 persons, and very narrowly escaping, put themselves to sea. Whereupon our souldiers discharged their calivers and followed them, thinking the noise therof being hearde to our boats at sea, our men there woulde make what speede they might to that place. And therupon, indeede, our men whiche were in the boats (crossing uppon them in the mouth of the sounde, whereby their passage was let from getting sea roome, wherein it had bin impossible for us to overtake them by rowing) forced them to put themselves ashore upon a point of lande within the said sound (which upon the occasion of the slaughter there was since named the Bloudie Point) whereunto our men so speedily followed, that they hadde little leysure lefte them to make any escape. But so soone as they landed, eche of them brake his oare, thinking by that meanes to prevent us in carying awaye their boates for want of oares. And desperately retorning upon our men, resisted them manfullye in their landing, so long as theyr arrows and dartes lasted; and, after gathering up those arrows which our men shot at them, yea, and plucking our arrowes out of their bodies, encountred afresh againe, and maintained their cause, until both weapons and life utterly failed them. And when they founde they were mortally wounded, being ignorant what mercy meaneth, with deadly furie they cast themselves headlong from off the rocks into the sea, least perhaps their enemies shoulde receive glory or praye of their dead carcasses; for they supposed us be like to be canibales, or eaters of mans flesh. In this conflict one of our men was dangerouslie hurt in the bellie with one of their arrowes, and of them were slayne five or sixe. The rest by flight escaped among the rockes, saving two women, whereof the one being old and ougly, our men

The savages have boats of sundry bignesse.

The Englishmen pursue those people of that countrey.

The swift rowing of those people.

The Bloudy Point.

Yorkes Sound.

A hot skirmish between the English and them of that countrey.

The desperate nature of those people.

thought she had bin a divell or some witch, and therefore let her go: the other being yong, and combred with a sucking childe at hir backe, hiding herselfe behinde the rocks, was espied by one of oure men, who, supposing she had bin a man, shot through the heare of hir head, and pierced through the child's arme, whereupon she cried out, and was taken, and our surgeon, meaning to heale hir child's arme, applyed salves therunto. But, she not acquainted with such kinde of surgerie, plucked those salves away, and, by continuall licking with hir own tongue, not much unlike our dogges, healed uppe the child's arme. And bycause the daye was well neare spent, oure menne made haste unto the reste of oure companie, which on the other side of the water remained at the tents, where theye founde by the apparell, letter, and other English furniture, that they were the same companye whiche Captaine Yorke discovered the night before, having removed themselves from the place where he left them. And now considering their sodaine flying from our men, and their desperate manner of fighting, we beganne to suspect that we hadde already heard the laste newes of our men, whiche the laste yeare were betrayed of these people. And considering, also, their ravenesse and bloudy disposition in eating anye kinde of rawe flesh or carrion, howsoever stinking, it is to be thoughte that they had slaine and devoured oure men. For the doublet whiche was founde in their tentes had many therein, being made with their arrowes and darts.

The taking of the woman and hir child.

A prettie kind of surgerie which Nature teacheth.

But nowe the night beinge at hande, our men with their captives and suche poore stuffe as they found in their tentes, returned towardes their shippes; when being at sea there arose a sodaine flawe of winde, which was not a little dangerous for their small boates. But, as God would, they came all safely aboorde. And with these good newes they retourned (as before mentioned) into the Countesse of Warwicks Sound, unto us, and betweene Jackmans Sound, from whence they came, and the Countesse of Warwicks Sound, betweene land and land, beinge thoughte the narrowest place of the straightes were judged nine leagues over at leaste: and Jackmans Sounde being uppon the Southerlande, lyeth directlye almoste over againste the Countesses Sound, as is reckned, scarce thirty leagues within the straightes from the Queenes Cape, whiche is the entrance of the Straightes of Southerland, being the supposed continent of America. This Cape being named Queene Elizabeths Cape, standeth in the latitude of degrees and a halfe to the northwardes of Newefound lande, and uppon the same continent, for any thing that is yet knowen to the contrarie.

The narrowest place of the straightes is 9 leagues over.

The Queenes Cape.

Having now got a woman captive for the comfort of our man, we brought them both togither, and every man with silence desired to beholde the manner of their meeting and entertainment, the whiche was more worth the beholding than can be well expressed by writing. At theyr first encountering, they behelde eache the other very wistly a good space, withoute speeche or worde uttered, with greate change of coloure and countenance, as though it seemed the greefe and disdeyne of their captivitie had taken away the use of their tongues and utterance: the woman at the first verie suddaynely, as though she disdeyned or regarded not the man, turned away and beganne to sing, as though she minded another matter: but being agayne broughte togyther, the man brake up the silence first, and with sterne and stayed countenance beganne to tell a long solemne tale to the woman, whereunto she gave good hearing, and interrupted him nothing till he had finished, and, afterwards being growen into more familiar acquaintance by speech, were turned togither, so that (I think) the one would hardly have lived without the comfort of the other. And, for so muche as we could perceive, albeit they lived continually togither, yet did they never use as man and wife, though

The manner of the meeting of ye two captives and their entertainment.

the woman spared not to do all necessarie things that apperteyned to a good huswife indifferently for them both, as in making cleane their cabin, and every other thing that apperteyned to his ease: for when hee was seasicke, shee would make him cleane, she would kill and flea ye dogges for their eating and dresse his meate. Only I thinke it worth the noting the continencie of them both; for the man would never shifte himselfe, except he had firste caused the woman to depart out of his cabin, and they both were most shamefast least anye of their privie parts should be discovered, eyther of themselves or any other body.

The shamefastnesse and chastitie of those savage captives.

On Monday, the sixth of August, the lieutenante, wyth all the souldyers, for the better garde of the myners, and the other things a shoare, pitched their tents in the Countesses Ilande, and fortified the place for their better defence as well as they could, and were to the number of forty persons: when being all at labour they might perceyve uppon the toppe of a hill over against them a number of the countrey people wafting with a flagge and making great outcries unto them, and were of the same companie whiche had encountred lately our men upon the same shoare, being come to complayne their late losses and to entreate (as it seemed) for restitution of the woman and chylde, whyche our men in the late conflict had taken and brought away: whereuppon the generall taking the savage captive with him, and setting the woman where they mighte best perceyve hir, in the highest place of the ilande, wente over so talk with them. Thys captive at the fyrste encountrie of hys friends, fell so out into teares, that he could not speake a worde in a greate space; but, after a while, overcomming his kyndnesse, hee talked at full wyth hys companyons, and bestowed friendly uppon them suche toyes and trifles as we had gyven him, whereby we noted that they are verie kynde one to the other, and greatly sorrowfull for the loss of their friendes. Oure generall by signes requyred his five men whyche they

Another appearance of that countrie people.

took captive the last yeare, and promised them not only to releasse those whyche hee hadde taken, but also to reward them wyth greate giftes and friendship. Our savage made signes in answere from them, that oure men shoulde be delyvred us, and were yet living, and made signes lykewise unto us, that wee shoulde write oure letters unto them; for they knewe very well the use we have of writing, and receyved knowledge thereof, eyther of oure poore captive countreymen whyche they betrayed, or else by thys oure newe captive who hathe seene us dayly write and repeate agayne such wordes of hys language as we desired to learne: but they, for thys nyght, bycause it was late, departed without any letter, although they called earnestlie in haste for the same. And the nexte mornyng earelie, beeyng the seaventh of August, they called agayne for the letter, whyche beeyng delyvred unto them, they speedily departed, making signes wyth three fingers, and poyntyng to the sunne, that they meante to returne wythin three dayes, untyll whyche tyme we hearde no more of them: and, aboute the tyme appoynted, they returned in suche sorte as you shall afterwardes heare.

These people know the use of writing.

A letter sent to the five English captives.

Thys nyghte, bycause the people were very neere unto us, the lieutenant caused the trumpet to sounde a call, and everie man in the ilande repayring to the auntiente, he putte them in mynde of the place so farre from theyr countrey wherein they lived, and the danger of a multitude whyche they were subject unto, if good watche and warde were not kepte; for at everie lowe water the enimie myghte come almost dryfoote from the mayne unto us, wherefore hee wylled everye man to prepare hym in good readynesse uppon all soddayne occasions, and so giving the watch their charge, the company departed to rest.

I thought the captaynes letter well worth the remembring, not for the circumstance of curious enditing, but for the substance and good meaning therein contayned, and therefore have repeated heere the same, as by himselfe it was hastilie written.

THE FORME OF MARTIN FROBISHERS LETTER TO THE ENGLISHE CAPTIVES.

In the name of God, in whom we al beleve, who, I trust, hath preserved your bodyes and souls amongst these infidels, I commend me unto you. I will be glad to seeke by all meanes you can devise, for your deliverance, eyther with force or with any commodities within my shippes, which I will not spare for your sakes, or anything else I can do for you. I have aboord of theyrs a man, a woman, and a childe, which I am contented to deliver for you; but the man I carried away from hence the laste yeare, is dead in England. Moreover, you may declare unto them, that if they deliver you not, I wyll not leave a manne alive in their countrey. And thus unto God, whome I trust you do serve, in haste I leave you, and to him we will dayly pray for you. This Tuesdaye morning, the seaventh of August, anno 1577.

Yours to the uttermost of my power,

MARTIN FROBISHER.

I have sent you by these bearers, penne, incke, and paper, to write backe unto me agayne, if personally you can not come to certifye me of your estate. Postscript.

Now, had the generall altered his determination for going any further into the straights at this time, for any further discoverie of the passage, having taken a man and a woman of that countrey, whiche he thought sufficiente for the use of language; and having also mette wyth these people heere, which intercepted his men the last yeare (as the apparell and Englishe furniture whiche was found in their tentes very well declared), he knewe it was but labour lost to seeke them further off, when he had found them there at hand. And considering, also, the shorte time he had in hande, he

The cause why Frobisher entred no further within ye straightes this yeare.

L 2

thoughte it best to bend his whole endevour for the getting of myne, and to leave the passage further to be discovered hereafter. For his commission directed hym in this voyage only for the searching of the gold ore, and to deferre the further discouverie of the passage untill another tyme.

On Thurseday, the ninth of August, we beganne to make a small fort for our defence in the Countesse Iland, and entrenched a corner of a cliffe, which on thre parts like a wall of good heygth was compassed and well fenced with the sea, and we finished the rest with caskes of earth to good purpose, and this was called Bestes Bulwarke, after the lieutenants name, who first devised the same. This was done for that we suspected more least the desperate men might oppresse us with multitude than any feare we had of their force, weapons, or policie of battell, but as wisdome would us in such place (so far from home), not to be of our selves altogither carelesse: so the signes whiche oure captive made unto us of the comming downe of his governoure or prince, which he called Catchoe, gave us occasion to foresee what might ensue thereof; for he shewed by signes, that this Catchoe was a man of higher stature far than any of our nation is, and he is accustomed to be carried upon mens shoulders.

Bestes Bulwarke.

Their king called Catchoe.

How he is honoured.

Aboute midnighte the lieutenant caused a false alarme to be given in the iland, to prove as well the readynesse of the companie there a shoare, as also what help might be hoped for upon the suddayne from the shyppes if neede so required, and every part was found in good readynesse upon such a suddayne.

Saturday, the eleventh of August, the people shewed themselves agayne, and called unto us from the side of a hil over against us. The generall (with good hope to heare of hys men, and to have aunswer of his letter), wente over unto them, where they presented themselves, not above three in sight, but were hidden in deede in greater numbers behynde the rockes, and makyng signes of delay with us to entrappe some of us to redeeme theyr owne, did onely seek advantage to trayne our boate aboute a poynte of lande from sight of our companie: whereupon, our menne justly suspecting them, kepte aloofe without their danger, and yette sette one of our companie a shore, whyche tooke up a greate blather whiche one of them offered us, and leavinge a looking glass in the place, came into the boate agayne. In the meane while, our men whyche stoode in the Countesses Iland to behold, who might better discerne them, than those in the boate, for that they saw divers of the savages creeping behynde the rocks towards our men: whereuppon the generall presently returned without tidings of his men.

A blather changed for a looking glass.

No news of the English captives.

Concerning this blather which we receyved, our captive made signes that it was given him to keepe water and drinke in; but we suspected rather it was gyven hym to swimme and shifte away withall; for he and the woman sought divers times to escape, having loosed our boates from a sterne our shippes, and wee never a boate lefte to pursue them withall, and had prevayled verie farre, had they not bin verie timelie espyed, and prevented therein.

To what end the blather was delivered.

After our generalls comming away from them, they mustered themselves in our sight uppon the toppe of a hill, to the number of twentie in a rancke, all holdyng handes over theyr heads, and dauncing, with greate noyse and songs togither, wee supposed they made thys daunce and shew for us to understand, that we might take vew of theyr whole companyes and force, meaninge belike, that we should doe the same. And thus they continued uppon the hyll toppes untyll nighte, when hearinge a peece of oure greate ordinance, whiche thundered in the hollownesse of the hygh hylles, made unto them so fearefull a noyse, that they hadde no greate wyll to tarrie long after. And this was done, more to make them knowe oure force, than to do them anye hurte at all.

Those people daunsing upon the hill toppes.

On Sunday, the twelfth of August, Captayne Fenton trayned the companye, and made the souldyoures mainteyne skyrmishe among themselves, as well for theyr exercise, as for the countrey people to beholde in what readynesse oure menne were alwayes to bee founde; for it was to bee thoughte that they lay hydde in the hylles thereaboute, and observed all the manner of our proceedings.

A skirmish showed to those people.

On Wensday, the foureteenth of August, our generall, wyth two small boates, well appoynted; for that he suspected the countrey people to lye lurking thereaboute, wente up a certayne bay wythin the Countesses Sound, to search for ore, and mette agayne wyth the countrey people, who so soone as they saw our men, made greate outcryes, and with a white flagge made of blathers, sowed togyther wyth the guttes and sinewes of beastes, wafted us amayne unto them, but shewed not above three of theyr companye. But when wee came neere them, wee myght perceyve a greate multitude creeping behynde the rocks, whyche gave us good cause to suspecte theyr trayterous meaning: whereuppon we made them signes, that if they would lay their weapons aside, and come forth, we woulde deale friendly with them, although theyr intente was manifested unto us: but, for all the signes of friendship we coulde make them, they came still creeping towards us behinde the rockes to get more advantage of us, as though we had no eyes to see them, thinking belike, that our single wittes could not discover so bare devise and simple drifts of theyrs. Theyr spokesman earnestly persuaded us, with many enticing notices, to come, eat, and sleepe ashore, with great arguments of courtesie, and clapping his bare handes over his head in token of peace and innocencie, willed us to do the like. But, the better to allure our hungry stomachs, he brought us a trimme bayte of raw flesh, which, for fashion sake, with a boathooke, we caught into our boate: but when the cunning cater perceived his first cold morsell could nothing sharpen our stomacks,

Their flags made of blathers.

he cast about for a new trayne of warme fleshe to procure our appetites, wherefore he caused one of hys fellowes in halting manner, to come forth as a lame man from behind the rockes, and the better to declare his kindnesse in carving, he hoysed him uppon his shoulders, and bringing him hard to the water side where we were, lefte him there lymping, as an easie pray to be taken of us. His hope was, that we would bite at this bayte, and spedily leape ashore within their danger, whereby they might have apprehended some of us, to ransome theyr friendes home againe, which before we had taken: but, I doubt, our flesh is so sweete meate for them, that they will hardly part from so good morsels, if we come once nere their handling. The gentlemen and souldiers had great will to encounter them ashore, but the generall more carefull by processe of time to winne them, than wilfullie at the first to spoyle them, would in no wise admitte that any man shoulde put himselfe in hazarde ashore, considering the matter he now entended was for the ore and not for the conquest: notwithstanding, to prove this criples footemanshippe, he gave liberty for one to shoote: whereuppon, the criple having a parting blowe, lightly recovered a rocke, and went awaye a true and no fained criple, and hath learned his lesson for ever halting afore suche criples againe. But his fellows whiche lay hid before, full quickely there appeared in their likenesse, and maintained the skirmishe with theyr slings, bowes, and arrowes very fiercely, and came as neere as the water suffered them: and with as desperate minde as hath bin seene in any men, withoute feare of shotte or any thing, followed us all along ye coast, but al their shot fel short of us, and are of little danger. They had belayd al ye coaste along for us, and being dispersed so, were not wel to be numbred, but we might discerne of them above one hundreth persons, and had cause to suspect a greater number. And thus, withoute losse or hurte, we returned to our shippes againe.

Great offers.

Nowe, our worke growing towardes an end, and having onely with five poore miners, and the helpe of a fewe gentlemen and souldiers, brought aboorde almost twoo hundreth tunne of golde ore, in the space of twentie dayes, every man therewithal wel comforted, determined lustily to worke afreshe for a bone voyage, to bring our laboure to a speedie and happy ende.

And upon Wednesday, at night, being the one and twentieth of August, we fully finished the whole worke. And it was now good time to leave; for, as the men were wel wearied, so their shoes and clothes were well worne, their baskets bottoms torne out, their tooles broken, and the shippes reasonably well filled. Some with over-straining themselves received hurtes not a little daungerous, some having their bellies broken, and others their legges made lame. And about this time ye ise began to congeale and freese about our ships sides a night, whiche gave us a good argument of the sunnes declyning southwarde, and put us in minde to make more haste homeward.

It is not a little worth the memorie, to the commendation of the gentlemen and souldiers herein, who leaving all reputation aparte with so great willingnesse, and with couragious stomacks have themselves almost overcome in so short a time, the difficultie of this so great a laboure. And this to be true, the matter, if it be well wayed, without further proofe, nowe brought home, dothe well witnesse, God graunt for their forwarde mindes, they may be as well considered, as theyr honest merites have well deserved.

Thurseday, the 22 of August, we plucked downe oure tentes, and every man hasted homewarde, and making bonefires uppon the toppe of the highest mount of the iland, we gave a vollie of shotte for a farewell, in honour of the right Honourable Lady Anne, Countesse of Warwicke, whose name it beareth, and so departed aboorde.

The 23 of August, having the wind large at west, wee

sette sayle from out of the Countesses Sound homeward, but the winde calming, we came to anker within the poynt of the same sound agayne.

The 24 of August, about three of the clocke in the morning, having the winde large at west, we sette sayle agayne, and by nine of the clocke at nighte, we left the Queenes Forland asterne us, and, being cleere of the Straytes, we bare further into the mayne ocean, keeping our course more southerly, to bring our selves the sooner under the latitude of oure owne climate.

The wynde was very greate at sea, so that we laye a hull all night, and had snowe halfe a foote deepe on the hatches.

Snow half a foot deepe in August.

From the 24 untill the 28, we had very muche winde, but large keeping our course south south-east, and were lyke to have lost the barkes, but by good happe we mette agayne. The heygth being taken, we were in degrees and a halfe.

The nine and twentieth of August, the winde blewe muche at northeast, so that we coulde beare but onely a bunt of our foresayle, and the barkes not being able to carrie any sayle at all.

The *Michaell* lost company of us, and shaped hir course towardes Orkney as we supposed, bycause that way was better knowne unto them.

The thirtieth of August, with the force of the wynd, and a surge of the sea, the mayster of the *Gabriell* and the boateson were stricken both overboorde, and hardly was the boateson recovered, having holde on a roape hanging overboorde in the sea, and yet the barke was laced fore and afte with ropes a breast high within boorde.

The mayster of the *Gabriell* striken overboorde.

Thys mayster was called William Smyth, beeying but a yong man, and a very sufficient maryner, who beeing all the morning before exceeding pleasaunte, tolde hys captayne hee dreamed that he was cast overboorde, and that the boateson hadde hym by the hande, and could not save hym, and so immediately uppon the ende of hys tale, hys dreame came right evelly to passe, and indeede the boteson in like sorte helde hym by one hande, having hold on a rope with the other, untill hys force fayled, and the mayster drowned. The heygth being taken we found oure selves to bee in the latitude of degrees and a halfe, and reckoned our selves from the Queene's Cape homeward, about two hundreth leagues.

The last of August, aboute midnighte we had two or three great and suddayne flawes or stormes.

The firste of September, the storme was growne very greate, and continued almoste the whole day and night, and lying a hull to tarry for the barkes, our ship was much beaten with the seas, every sea almoste overtaking oure poope, that we were constrained with a bunte of oure saile, to try it oute, and ease the rolling of oure shippe. And so the *Gabriell* not able to beare any saile to keepe company with us, and oure shippe being higher in the poope, and a tall shippe, wheron the winde had more force to drive, went so fast awaye, that we loste sighte of them, and lefte them to God and their good fortune of sea. The seconde daye of September in the morning, it pleased God of hys goodnesse to sende us a calme, whereby we perceived the rudder of oure shippe torne in twaine, and almost ready to fall away. Wherefore taking the benefite of the time, we flung halfe a dozen couple of our best men overboard, who taking great paines under water, driving plancks, and binding with ropes, did wel strengthen and mend the matter, who returned the most parte more than halfe deade out of the water, and as God's pleasure was, the sea was calme untill the worke was finished. The fifth of September, the height of the sun being taken, we founde ourselves to be in the latitude of degrees and a half. In this voyage commonly we took the latitude of the place by ye height of the sun, bycause the long day taketh away the light not only of the polar, but also of all

The rudder of the *Ayde* torne in twaine.

other fixed starres. And here the north starre is so muche elevated above the horizon, that with the staffe it is hardly to be wel observed, and the degrees in the Astrolobe are too small to observe minutes. Therefore we alwaies used the staffe and the sunne, as fittest instruments for this use.

How the latitudes were alwayes taken in this voyage rather wyth the staffe than Astrolobe.

Having spent foure or five dayes in traverse of the seas with contrarye winde, making oure souther way good as neare as we could, to raise our degrees to bring ourselves with the latitude of Sylley, we tooke the height the tenth of September, and founde ourselves in the latitude of degrees and ten minutes. The eleaventh of September about sixe a clocke at night the wind came good southwest, we verde short and sette oure course southest.

And upon Thursday, the twelfth day of September, taking the height, we were in the latitude of and a halfe, and reckened oure selves not paste one hundred and fiftie leagues short of Sylley, the weather faire, the winde large at west-south-west, we kepte our course southest.

The thirteenth daye the height being taken, we founde ourselves to be in the latitude of degrees, the wind west-south-west, then being in the height of Sylley, and we kept our course east, to run in with the sleeve or channel so called, being our narrow seas, and reckoned as shorte of Sylley twelve leagues. Sonday, the fifteenth of September, aboute foure of the clocke, wee began to sounde with oure lead, and hadde grounde at sixty-one fadome depth, white small sandie grounde, and reckned us upon the backe of Sylley, and set our course easte and by north, easte north-easte, and north-east among.

The sixteenth of September, about eight of the clocke in the morning sounding, we had sixty-five fadome osey sande, and thought ourselves thwart of Saint Georges Channell a a little within the bankes. And bearing a small saile all nighte, we made many soundings, whiche were aboute fortie fadome, and so shallowe that we coulde not well tell where we were.

The seaventeenth of September we sounded, and had fortie fadome, and were not farre off the landes end, branded sande with small worms and cockle-shells, and were shotte betweene Sylley and the landes ende, and being within the baye, we were not able to double the pointe wyth a south and by east way, but were fayne to make another boorde, the wynde beeyng at southweast, and by weast, and yet could not double the poynte, to come cleere of the landes ende, to beare along the Channell: and the weather cleered up when we were hard aboorde the shore, and we made the landes ende perfite, and so put up alongst Sainte Georges Channell: and the weather beeyng very foule at sea, we coveted some harborough, bycause our steerage was broken, and so came to anker in Padstowe roade in Cornewall. But riding there, a very dangerous roade, we were advised by the countrey to put to sea agayne, and of the two evils, to choose the lesse, for there was nothing but present perill where we roade: whereuppon we plyed along the Channell to gette to Londy, from whence we were agayne driven, being but an open roade, where our ancker came home, and with force of weather put to seas agayne, and aboute the three and twentith of September, arrived at Milforde Haven in Wales, whyche beeyng a very good harborough, made us happy men, that we hadde receyved suche longe desired safetie. And more happie we helde our selves, not for the safetie of ourselves so muche, as the comforte that we had that our countrey shoulde thereby have perfecte knowledge of oure discoverie, to the greate benefyte of oure common wealth.

The arrivall of the *Ayde* at Padstow in Cornwall.

Our coming to Milford Haven.

Aboute one moneth after oure arrivall heere, by order from the Lordes of the Counsell, the shippe came up to Bristowe, where the ore was committed to keeping in the castell there. Heere we founde the *Gabriell*, one of the barkes, arrived in good safetie, who having never a man within boorde very sufficient to bring home the shyppe, after the mayster was lost, by good fortune, when she came upon

The arrivall of the *Gabriell* at Bristow.

the coast, mette with a shyppe of Bristowe at sea, who conducted hir in safetie thither.

Here we heard good tidings also of the arrivall of the other barke called the *Michaell*, in the north partes, whyche was not a little joyfull unto us, that it pleased God so to bring us to a safe meeting agayne, and lost in all the voyage only one man, besydes one that dyed at sea, whiche was sicke before he came aboord, and was so desirous to followe this enterprise, that he rather chose to dye therein, than not to be one to attempte so notable a voyage.

The *Michaell* arrived in the north partes.

Only one man dyed in this voyage.

FINIS

[Another account of this Voyage was written by Master Dionisi Settle, and will be found in the xii volume of *Pinkerton's Voyages and Travels.* London, 1812.]

STATE PAPERS SUBSEQUENT TO THE SECOND VOYAGE.

STATE PAPERS RELATIVE TO THE TRIAL OF THE ORE SUBSEQUENT TO THE SECOND VOYAGE.

STATE PAPERS
SUBSEQUENT TO THE SECOND VOYAGE.

[*Colonial* 51. *Domestic Eliz.*, cxv. No. 35.]

MR. LOCKES MEMORIAL.

To the Quenes Maties most honorable Privye Councell.

Maye it please yor honors in most humble maner to be advertised.

THE ij ships *Ayde* and *Gabriell* are arived at Bristowe in saffetye accordinge to yor honors orders.

The disposing of the ore at Bristowe.

And yf yor honor's doo think it good to dyscharge the ure on land there yt may please yor honors to gyve order (under correction) that Mr. Furbisher may delyver the same by weight, and that the same may be kept in the castell or other saffe place there under iiij severall lokes and keys wherof one wth the Mayor of Bristowe, one wth Sir Richard Barkley, one wth Mr. Furbisher, and one wth Michael Lok, or any other of the venturars as shall seeme good to yor honors. Also there is to be payd presently uppon the dyschargenge of the ships ladynge viij^{c} poundes of money or more for the wagys of the souldiars and maryners wch have sarved wch doo remayne styll at charges of the companye for meate and wagys untill they be payd the wch sayd money cannot be found in London uppon interest nor exchange notwthstandinge the dyllygens used by the Commyssyoners to take up the same wherefore for the provision therof restethe but ij meanes. Fyrst yf it would please the Quenes Matie to prest the same for iij monthes untill the ore may be melted downe wch may be receved of her Maties customars or other offycers at Brystowe or ells the same must be ceassed and collected of all the venturars accordinge to the rate of their stok ventured wch would be but xxv^{li} for every c^{li} of their venture, but it would be very longe tyme and moche dyfficultye in collection.

The wages to furnishe money for ye discharge of the manners and sowldiers.

Also the ship *Michael* is now in the Ryver of Tamys arived in saffety by Gods grace and the same ship must be presently dyscharged at London. And the Comyssioners doo think good to put the ure in saffe custodye in the howsse wher Sr Willm Wynter dothe now dwell at St Katheryns hyll, where they have alredy made a furnace to melt downe the same. May it please yor honors to gyve order that the same may also be delyvred by weight and kept under iiij loks and keys to be

The disposing of the ore at London.

[in] the custodye of Sr William Wyntar, Mr. Thomas Randall, Mr. Furbisher, and Michael Lok or others as to yor honors [semethe] good. Also yt may please yor honors to gyve ordre and auctoritie to the sayd Commyssioners wch are Sr William Wyntar, Mr. Thomas Randall, George Wyntar, Anthonye Jenkynson, Edmond Hogan, Michael Lok. And to joyne to them Mr. Furbisher or any others that yor honors shall lyke that they maye consulte and determyne for the spedye meltynge downe of the ure bothe at London and at Brystowe.

To appoynte commissioners to looke to the melting.

Michael Lockes request.

And wheras Michael Lok for the advancement of this voyages for the space of iij yeres of his own goodwyll hathe taken the paynes and charge of kepyng all accountts to reconynces frely wthout any recompence, and hathe byn named tresourer, thoughe he hathe had but lytle tresour in his kepynge. Now yf yor honors doo lyke well of his doynges acordinge to his small power yt may plese yor honors to ratifye hym in that offyce to take charge of the money and treasour of this companye to account and hereafter to consyder of hym for his paynes and doynges as to yor honors shall seeme good or his defectes shall appere.

[*Colonial*, 55. *Domestic Eliz.*, cxvi, No. 14.]

XIIJ DIE OCTOBRIS, 1577. A NOTE OF MONEY PRESENTLYE TO BE DISBURSED FOR MARYNERS WAGES OF THE THREE SHIPPES RETORNED WTH MR. FURBUSSHER.

In primis for the wages of an hundreth maryners in all the three shipps, at severall rates, from xs to vjli xiijs iiijd for a man the moneth, amounteth to the some of cxlli the moneth, wch from the ffirst of June unto the last of October beinge ffyve monethes, amounteth in the hole to ye somme of DCCli.

At xs ye men. Item for wages for xxvjli souldiors, for the sayde ffyve monethes, as it shall please yor honors.

Item for recompence for xiiij gentilmen duringe the tyme aforesayde, as it shall please yor honors to consyder of some reasonable porcion of the adventure.

This money cannot yet be found to be taken uppon interest nor exchange.

And yf it be levied by collection uppon the venturars, it will come to xxvli uppon everie cli of ther venture.

Item, whether their honors doo continew in mynde that the ij ships at Bristow shall discharge there or come to London.

Item, whether Michael Lok shall contynew still Treasorer or not.

Indorsed. For the ships of Cathai to be unladen.

[*Colonial*, 55. *Domestic Eliz.*, cxvi, No. 24.]

AN ORDER OF MY LORDS, THE 16 OF OCTOBER, 1577, FOR PAYMENT TO BE MADE TO THE MARINERS AND SOULDIERS OF THE CATHAY VIAGE.

Where as there is presently to bee disbursed for the discharge of suche mariners and souldiers as have been employed in the viage towards the northwest under Captayne Ffurbisher, the somme of eight hundrethe powndes, yt is ordred that suche as have been adventurers in the sayd viage should contribute toward the discharge of the sayd somme of viijcli, after twentie in the hundrethe, *pro rata*, wch sayd contribution is thought also meete to bee delivered into the handes of Michaell Locke, Treasorer for the Companie, beynge appointed to give bylles signed under his hand for suche sommes as he shall receave.

[*Colonial, East Indies*, 54. *Domestic Eliz.*, cxix, No. 44.]

These desyre now to be venturars in the goodes now come home, wch may be graunted uppon the whole stok now come home, or ells in the next adventure, as yor honors shall thynk good.

My Lord Kepar	*li*25
Therle of Bedford	*li*25
Mr. Controllar	*li*25
The Erle of Oxford	*li*25
My Lord Hunsdon	*li*25
My Lord Charles Howard	*li*25
My Lord of Comerland	*li*25
My Lord Cobham[1]	*li*25
My Lord Wharton	*li*25
Mr. Hatton	*li*25
Mr. Hennage	*li*25
Mr. Horsey	*li*25
Sr Humfrey Gilbart	*li*25
Mr. Woolley	*li*25
William Kyllygrew	*li*25
Thomas Dudley	*li*25
Raffe Lane	*li*25
Hew Smythe	*li*25
John Dee	*li*25
Jeffrey Ducket	*li*25
Thomas Nyccolls	*li*25
	*li*500

[1] Added in another ink.

Francis Mylles[2]	*li*25
Laurens Tomson	*li*25
Arture Dawbney	*li*25
John Capelin	*li*25
Thomas Cesar	*li*25

[*Colonial East Indies*, 70. *Domestic Eliz.*, cxxx, No. 35.]

THE NAMES OF THE VENTURARS OF BOTHE THE VOYAGES MADE BY MARTIN FURBUSHER, GENT. TO THE NORTH-WEST, ANNO 1576 AND ANNO 1577.

		Stok.	Cessement.
	The Quenes Matie	*li*1000	*li*200
In the first voiage			
*li*50	The Lord Highe Treasorer	*li*100	*li*20
	The Lord highe Admyrall	*li*100	*li*20
*li*50	The Lord highe Chamberlayn	*li*100	*li*20
50	The Erle of Warwyke	*li*100	*li*20
50	The Erle of Leycester	*li*150	*li*30
	Mr. Treasorer of the Q. Matie household	*li*50	*li*10
25	Mr. Secretarie Walsingham, for hym sellffe and others	*li*400	*li*80
	Mr. Secretarie Wyllson	*li*50	*li*10
		£2050.	
	The Erle of Pembroke	*li*150	*li*30
	The Countesse of Warwyke	*li*50	*li*10
	The Countesse of Pembroke	*li*25	*li*5
	The Lady Anne Talbot	*li*25	*li*5
25	Mr. Phillip Sydney	*li*50	*li*10
100	Sr Thomas Gresham	*li*200	*li*40
25	Sr Leonell Duckett	*li*50	*li*10
	Sr William Wyntar	*li*200	*li*40
	William Pellham	*li*50	*li*10
	Edward Dyar	*li*25	*li*5
25	Thomas Randolphe	*li*50	*li*10
	George Wyntar	*li*50	*li*10
	Mathew Smythe	*li*50	*li*10
	Symon Boyer	*li*25	*li*5
25	Anthony Jenkynson	*li*50	*li*10
	Jeffrey Turvile	*li*50	*li*10
	William Payntar	*li*50	*li*10
	Richard Bowlland	*li*50	*li*10

[1] This and the following names are added in another ink.

	Robert Kyndersley . . .	*li*50	*li*10
50	Edmond Hogan	*li*100	*li*20
25	Mathew Fylld	*li*50	*li*10
	Richard Yonge	*li*50	*li*10
	Thomas Allyn	*li*50	*li*10
	William Ormshawe . . .	*li*25	*li*5
*li*500		*li*3575	*li*715
	Christofer Hudson . . .	*li*50	*li*10
	Thomas Owen	*li*25	*li*5
	John Dee	*li*25	*li*5
	Julio Cesar	*li*50	*li*10
	Eleazar Lok	*li*25	*li*5
	Gerson Lok	*li*25	*li*5
	Martin Furbusher . . .	*li*100	*li*20
100	Michael Lok	*li*1000	*li*200
		*li*1300	*li*260

£275	100	William Burde	*li*	*li*55	*li*20
	100	William Bonde	*li*		*li*20
	25	Mathew Kyndersley	*li*		*li* 5
	25	Christofer Androwes	*li*		*li* 5
	25	Robert Martyn	*li*		*li* 5
	£375				*li*315
	£500				*li*715
Stok	875	Received the first voyage			*li*1030 Cessement.

*li*1300	*li*800 first voiage spent
*li*3575	*li*4350 second voiage paid
Venturars *li*4875 of first and second voiages	*li*5150 paid all the stok outward.
li 275 of first voiage onelie	
Received *li*5150 all the stok.	

Received by the said cessement of 20^{li} for 100^{li} stok *li*1030	paid for the mynes	*li*198
	paid for mens wagys of the iij ships come home	*li*1044
	Somme paid the 24 Decembar 1577	*li*1242

[*Colonial, East Indies*, 75. *Domestic Eliz.*, cxix, No. 30.]

MICHAEL LOK HATHE DONE FOR THE VOYAGE OF CATHAI, AS FOLOWETHE.

XXV yeres studye and travaylle to satisfye his knowledge thereof.

M^{l} poundes spent for thinges necessarie for his satisfaction of knowledge therof in bookes, maps, cartes, instrumentes, and gyftes to men for conference therof; wherof is not one peny put to the account besydes all his howshold charges.

M^{l} m^{l} $cccc^{li}$[1] powndes paid and disbursed by hym sellfe for furniture of the first and second voyages over and above all that he received of all other venturars w^{ch} *li*2400 he dyd beare venture of in the same first and second voyages untill the shyps retorned home.

W^{ch} said *li*2400 was for the sums of venture wrytten uppon the names folowinge, wch was not paid hym untill the ships retorned, and is not yett all paid hym, but he had promes of sum of them.

My Lord Highe Treasuror	*li*50
My Lord of Leycester	*li*100
Sr Thomas Gresham.	*li*100
Mr. Doctor Wyllson	*li*50
George Wyntar	*li*50
Symon Boyer, he gave pledge	*li*25
Richard Owen	*li*25
Julio Cesar } M. L.	*li*25
Thomas Cesar } M. L.	*li*25
Eleazar Lok } M. L.	*li*25
Summe	*li*475
For Michael Lok hym sellfe	*li*1000
For the second voiage. Sum	*li*1475
M. Lok in first voyage	*li*825
Ventured by M. Lok	*li*2300

Indorsed. Mr. Lock's privat memorandum. 1576 and 1577.

[*Colonial*, 94. *Domestic Eliz.*, cxxvi, No. 32.]

THE BRYEF ACCOUNT OF THE SECOND VOYAGE, 1577.

Recd of all the venturers for their stokes	*li*5150	Paid for the *Ayde*	*li*850		
		For the *Michaell* and *Gabriell*	*li*400		
		For taklinge and rigginge	*li*577	7	
		For ordnance and munition	*li*467	5	1

[1] The last c appears to have been erased.

For vyttells . .	*li*963	18	3
For wages outwards .	*li*600	12	9
For necessaries for the mynes . .	*li*123	8	4
For marchandyse .	*li*346	5	0
	*li*4328	17	6
For charges outwards .	21	2	6
	*li*4350	0	0
For losse spent first voiage	*li*800	0	0
Sum paid .	*li*5150	0	0

Rec^d of supplie of venturers to paye wages at retorne of the shippes	*li*1030		
lackinge .			
My Lady Anne Talbot .	*li*6		
The Erle Sussexe .	*li*10		
Sr Wm Winter .	*li*1	12	4
Sr Tho. Greshame	*li*6		
	*li*22	12	4

Paid after the ships retorne; paid outwards above the stoke received . .	*li*1	15	2
Paid for wages of mariners and gentlemen at retourne of the shipps .	*li*1582	15	5
Paid for wages of myners and charges of tryall of the ewer at London	*li*324	1	8
	*li*1908	12	3
For not rec^d in the supplie of *li*1030	*li*22	12	4
	*li*1931	4	7
	*li*1030	0	0
Rest paid .	*li*901	4	7

THE ACCOUNT OF MONY FOR THIRD VOYAGE 1578.

Received as follow^th				Paid as fallow^th			
Of the Q. Matie .	£1350	0	0	To W^m. Kerin, frestone	£4	0	0
Of Mathew Feld .	£35	0	0	To Christ. Hawlle, bristole . .	£13	0	0
Of M. Kindersley .	£15	0	0	To J. Roberts, bellows	£3	10	0
Of Sr Frances Knolles	£67	10	0	To Ff. Grene, bucher	£100	0	0
Of Ed. Hoggan .	£40	0	0	To Mr. Frobiser .	£400	0	0
Of L. Pembroke .	£202	10	0	To Pointell, baker .	£100	0	0
Of La. Pembroke .	£33	15	0				

Of Mr. P. Sydney .	£67	10	0
Of Mr. S. Walsingham	£182	7	0
Of Jo. Somers .	£67	10	0
Of Mr. S. Willson .	£67	10	0
Of Sr Henry Wallop	£67	10	0
Of Roberte Kyndersley	£67	10	0
Of M. Kindersley .	£18	15	0
Of Erle Warwyke .	£135	0	0
Of Countesse Warrike	£35	0	0
Sum received	£2452	7	0

To Mr. Fenton .	£50	0	0
To Willson, caper, 5. 20. 20.	£45	0	0
To Jonas, a quarter .	£25	0	0
To Olyver Skiner, iron	£3	5	0
To John Gonne, iron	£3	8	0
To Jo. Roberts, bellowes	£2	0	0
To Jo. Fysher, smithe	£4	0	0
To N. Chanselar, 20.25.	£35	0	0
To F. Shawe, buttar	£50	0	0
To C. Hawle, bristowe	£20	0	0
To Ed. Selman, bristowē	£62	0	0
To F. Lee, shippe .	£50	0	0
To Ro. Denam, a quarter lent .	£12	10	0
To P. Barnston, aquavite .	£6	0	0
To Mr. Fenton, targats	£10	0	0
To Ed. Selman, bristowe	£16	0	0
To Thomson, carpenter	£30	0	0
To Hitchecoke, carpenter . .	£20	0	0
To Poyntell, baker .	£50	0	0
To Mathew, baker pynnasses .	£12	0	0
To Ro. Denam, additamentes .	£10	0	0
To Jeronias stoves .	£10	0	0
To S. Burow, pynnasse	£6	0	0
To Thomson, carpentar	£30	0	0
To other od charges, 17 Aprill . .	£16	0	0
To Mr. Frobiser acc. 40. 10. 10. .	£60	0	0
To Vyllers colles .	£10	0	0
To Croker, smithe .	£5	0	0
To L. Admiralls man	£2	0	0
To Stanley, currier, for lether .	£5	0	0
To Mr. Fenton . .	£10	0	0
To Mr. Furbusher, bristowe . .	£596	5	0
To the shippe *Hopewell*	£50	0	0
To Jonas . .	£10	0	0
To John Hayles, Dartford . .	£12	0	0

Somme paid	£1958	18	0
Paid for second voyage	£901	4	7
Sum paid	£2860	2	7

Receyt as ffollow^th

Of Mr. Dowgle .	£33	15	0
Of Sr Tho. Gresham.	£70	0	0
Of Lord Tresorer .	£100	0	0
The 2 May 1578			
Of Sr Thomas Gresham	£100	0	0
Of Rich. Young .	£50	0	0
Of Christofer Hudson	£67	10	0
Of Ed. Hogan .	£95	0	0
	£416	5	0

Payments as follow^th

Sum paid reste .	£407	15	7
The 23 Aprill 1578.			
To Thomas Willson, coopar . .	£10	0	0
To Thomson, carpentar	£20	0	0
To Augar, chaundler	£20	0	0
The 24 said [month].			
To Mr. Hawle, Bristowe	£12	0	0
To Chanselor purser.	£8	0	0
To Shawe for butter	£20	0	0
To Mr. More shipe *Foy*	£48	0	0
To Mr. Rasheley shipe *Foy* .	£48	0	0
The 26 said			
To Whitnall, coopar	£10	0	0
To Morris, tente maker	11	9	0
To Baker, shipwright	£16	0	0
To Poyntell, baker, of Lymehouse .	£50	0	0
To Thomson, carpentar	£30	0	0
To Vyllars, secoles .	£21	5	0
To Mr. Fenton .	£30	0	0
To Chanselar purser	£13	18	0
The 30 April, 1578.			
To men of Judeth wagys . .	£110	0	0
To Shaw, for fyst .	£40	0	0
To Thomson, seacoles	£18	10	0
The 3 May.			
To Whitnall, coopar	£5	0	0
To Willson, cans .	£4	4	0
To Ffrances Lee, ship	£30	0	0
To Thomson, carpenter	£20	0	0
To Eliot, brise .	£5	0	0
To Maryners, Judethe wages . .	£55	0	0
To Dowd, coopar .	£2	0	0
To Newson, clokemaker	£6	12	0
To J. Roberts, bellowes	£4	16	8
To Hopkins, smyth .	£2	15	6
To Morris, tent maker	£15	0	0
	1095.		

STATE PAPERS RELATIVE TO THE TRIAL OF THE OGRE SUBSEQUENT TO THE SECOND VOYAGE.

[*Colonial* 77. *Domestic Eliz.* cxxii. No. 3.]

JANUARY 6TH, 1577. FROM DR. BURCOTT OF HIS PROCEEDYNGE IN THE TRIALL OF THE OWRE.

My humble commendacions Sir Walsingame I gyve you most hartie thankes for your laboure and delygence for the performinge of your promysse and the sendinge of my pattane and, I hope by Godes grace to performe my promysse towardes you I had or this sertefied your Honore the trewe matter off this ower and my proves butt God hathe towched me so hard wth the gowte that I have kept my bed this thre wekes and do yett, and I sertefie your Honore off a trothe that I have mayd a hundred dyvers sayes of sondry owers out of that lande, and I fynde not such goodnes in yt as I thought to have founde not prosedinge half an ounce, as the blacke ower ys and I have prepared and rosted accordinge as I promyssed you, and I cane bringe yt to no clay but I have moltine doune a pounde, and hathe founde in yt fouer graynes as by this prove I have sent you dothe appere that will come in a houndred weight almoste thre quarters of an ounce that is nere xiiii ounces in a tonne and I dout not when yt is right prepared yt will faule out in the great fyere very well and I am mynded as sone as I am able wth my bodye I will melte doune a hundred weight and send you that prove accordinglye. So I byd you fare well wth my humble disier to advertyse the Quene's Heignes and my Lord of Lasyter. Datum the vi of January.

Your humble to comand,

Burchard Kranrych.

To his honorable and singuler good frend Sir Fraunces Walsing e,

Seecret e to the heigh.

[*Colonial*, 79. *Domestic Eliz.*, cxxii, No. 9.]

JANUARIE 19TH, 1577. FROM MR. MICHAELL LOOKE, WHAT CHARGES SR WILL'M WINTAR AND THE REST VPPON THEIR MEETYNGE HAVE THOUGHT PRESENTLY TO BE REQUISITE FOR THE FYNYNG OF THE EWRE AND OTHER THINGES TO BE FURTHER DONE IN THAT BEHALF.

Right Honorable, accordinge to the commission of Her Maties Privie Councell directed to Sr Wm. Wyntar, Mr. Randolph, Mr. Dyar, Mr. Mr. Yonge, Mr. Furbusher, and my sellf we all (except Mr. Randolph) have syttogethers theise ij days past, and have consydered thervppon as the tyme would suffer, and have had before us all the chyef workmasters for the erectinge of the howsse and furnaces at the mylls at Dartford, for the meltyng of the ewr brought by Mr. Furbusher, and as nere as we can esteme the charges therof wylbe thus :—

For bryk, stone, tyles, lome, lyme sand, lathes, naylles, and workmanship therof, for all the myllehows and iij furnaces, wth makinge the groundwork	*li*160
For tymber, and all other stuffe and the carpentars workmanship therof; for the howsse of 84 foote long and 36 foote wyde, and for the myll wheles, and the stampinge mylls, and the iij paier bellowes for the furnaces, wherof ij for melting, and i for drivinge or fynynge, and for all other engynes belonging therto	*li*240
Sum hereof	*li*400
For charges of a man to go to Germania, for ij chief workmasters of meltinge and fyndynge mynes, and another man into Yorkshier, for stuffe to melt the ewr	*li*40
For to provyde wood and coles, and other thinnes extraordinarie by estymation	*li*100
For wagys paid to the maryners and men come home wth the shyps, above the estimation made of 20 on 100 collected sum *li*360 includynge therin *li*200 alredy paid for charges of buyldinges of furnaces and small proffes and sayes of the ewr made by the handes of dyvers men. Sum	*li*360
Sum of all	*li*900

The said sum of *li*900 of money muste be provyded presentlye by collection of the venturars or otherwysse.

And for the better and more spedye provision of the stuffe and men requysyte for the buyldynges forsaid, it is thought requysyte to have comyssion in Her Maties name to be directed to Sr Will'm Wyntar, who hathe alredy a sufficient warrant for the marine affaires.

This beinge done all the woorkes wilbe fynyshed and the ewr molten and fyned wth in vi or viij wekes tyme by Godes help.

We have also thought it good that Her Matie be moved for her favourable letter to the Duke of Saxonia declaring the staying here of Jonas for Her Maties' sarvyce wch shalbe sent by the parson that goethe for workmen.

Of other matters this bearer Mr. Furbusher can informe yor Honor at large, wherfore I end, and comytt yor Honor to the tuition of Almighty God. From London the xix Januarie, 1577.

Yo^r Michael Lok.

To the Right Honorable Sr Francys Walsingham, Knyght,
Her Maties Chyef Secretarie, etc.,
Delyured at the Court.

Commission from Her Matie
Letter to the Duke of Saxonie.

[*Colonial*, No. 80. *Domestic Eliz.*, cxxii, No. 10.]

JANUARIE 19TH, 1577. FROM MY L.L.'S TO MR. MICHAEL LOCKE FOR THE COLLECTION OF 900LI. AMONG THE ADVENTURERS OF THE NORTH-WEST, ETC.

After o^r hartie comendations where as for the buildyng of certayne furnaces, and sendyng for skilfull men owt of Germanie for the tryall and refinyng of the oore brought owt of the north-west: as also for the payment of certayn wages due to the mariners and others employed vnder Mr. Furbusher, gent., in the last viage to the sayd north-west parts, we are enformed the somes of 900*li* to be verie necessarie and presently required. Theise are to require you to repayre to all such as be Adventurers in the sayd viage, as from vs so appointed, and to collect and gather of everie of them for the vses, purposes aforesayd, the sayd some to be levied amongst them after xx*li* in the hundredth, accordyng to the rate of and proportion of everie of their Adventurers, and this to be done wth as convenient speede as you can, to the end that so good an enterprise and proffitable as this viage is hoped will prove after so great charges allreadie bestowed thereon; nowe be not hindered either for want of so small a som or not prosecutyng the triall of the sayd oore. And so we, etc.

[*Colonial* 81. *Domestic Eliz.*, cxxii, No. 17.]

JAN. 24TH, 1577. FROM MASTER GEOFFREY LA BRUM, TRYAL METAL FALLING NOT OWT.

Monseigneur jay considere toutes ces espesses de mineres quil vous ha pleu me bailler et ay trouve que la pluspart ne sont que marchasites et non mineres lesquelles marchasites ont ordinairement tant de souphre que quant il y auroit quelque peu de bon metal on ne le pourroit tirer sans grant peine et perte quant a en faire preuve et essay et essay certain il en faudroit bien davantage et ne se fault fier a deux ny troys preuves seullement car tel morceau quon pense estre bon et habonder dor ou dargent je ne sy trouve rien ou peu de chose le plussouvent et au contraire tel quon mesprise se trouve le meilleur quelquefloys avec ce quil y ha divers noyens por extraire le pur metal et le fault tenter par diverses voyes par ce que'n aulcunes minieres le metal est encore tant crud quil sesauvit et perd en le fusion sil nest retenu par cementations fixatives en oultres il est accompaigne de tant dordures et impurtes comme de souphre, terre, pierre loppes et semblables qu'on ne le depart qu'a grant peine de facon quon le doit esprouver tantost avec selz tantost avec vifargent tantost la bruslant tantost la cementant ou fondant avec savon plomb selpitre borax ou aultres qui sont en grant nombre et trop longues a escrire. Quant a celle du Cappe Forbisher elle merite bien tant de diverses preparations et sy jen eusse eu nombre jen eusse tente plusieurs facons mais pour sy peu on nose asseurer de rien seullement je vous peux tesmoigner que jen ay brusle lave et purge puis joint avec plomb et mis en cendres ou couppelle qui n'ha rien laisse de fin qui ne faict juger quil ny ha grant profit par ce que les aultres voyes et moyens dextraire le fin sont de grant coust et labeur sur quantite, vray est que sy j'en auvis ugne livre ou deux jessairoys quelque preparation par sel commun prepare avec aultres ingrediens par le quel jay aultrefoys reduit des metaux fort calcines que le borax, saint de verre nitre et aultres semblables ne pouvoient reduire. Je nay baille a Monsr Marchant les fuses mineres a vous reporter par ce que je ne scay sy les desires sy je peux recouvrer davantage de celle du Cappe Forbischer jen feray amples preuves jattendois a vous en cscrire jusque a ce que jeusse ven le fin de quelques euvres que jay commences y a ja longe temps pour ta medecine affin de vous en faire part en tesmoignage de tant de courtoysies et bienfaictz que jay recevez de vous pour lesquels je suis et seray a tousjours oblige a vostre grandeur, mais lesditz euvres restantz parfaites et mons^r^ Marchant maiant adverty a midi de vostre commandement jay fait te presenter po^r^ mexcuser et vous suplier me tenir au rang de voz treshumbles et tresfideles serviteurs je prie Dieu Monseigneur quil augmente vostre grandeur et tentretienne en longue et tresheureuse sante pour de plus en plus servir a sa glore. Amen. De Londres ce Samedi 24e jour de Janvier 1578.

Vostre treshumble et obeissant serviteur,

Geoffroy Le Brumen.

A Monseigneur

Monseigneur de Vualsinghant.

[*Colonial*, 83. *Domestic*, *Eliz.*, cxii, No. 53.]

21 FEBRA, 1577, FROM D. BURCCOTTE, WITH A PROOFFE HOW MUCH GOULD AND SILUER A POWND, AND ONE HUNDRETHE POWND WEIGHT YELDETH, W^CH HE WILL WARRANT TO HOULD THROUGHOUT THE WHOLE EWRE.

Please your Honore, I perceyve by your letter that you cane not come so quycklye as I thought, therfore, I send you here by the captayne the sylver and gold of a pound, and a hundred weight, wher by I will abyde by yt off my credyt and honestlye that I will bringe twenttie tymes so mvche out of every toune in the lyke ower as that was and never vnder butt reyther above that rayte, and I pray you showe to the Quene that Her Matie will be good lady vnto me, for I am lyke loss mvche this springe to go about that busynes; and that you will be ameane that the captayne may be spedeley sett fourthe agayne wth suche teachine and instrucktine, as I have gyvene him, for yf he shall not go spedelye and speciallye now this yeare yt wilbe the worste that ever came to Ingland, and that the ower may be brought hastelye frome Brestoo to Detforthe, and that the detymente that the captayne cane showe you be prepared, and your Honore to come so hastelye as you cane to vysett the place where the meltinge house shall stand. Now I send you the trothe by the captayne how the house shall be; and I trust to se you shortleye. So I comytte your Honore to God in haste, this xxi^th day of Fabruary, 1578.

Your Honore to command,
Burchard Kraurych.

To his honorable and singuler good frend, Sir Fraunce Walsingame, Secretorye to the Quenes Heighnes, delyver this.

[*Colonial*, 86. *Domestic Eliz.*, cxxii, No. 62.]

A LITTLE BUNDLE OF THE TRYEING OF Y^E NORTHWEST EWRE. BY D. BURCOT, JONAS SCHUTZ, BAPTISTA AGNILLO, ETC.

The doings of Jonas Shote in the newe mynes of golde.

In Januarye 1576, Jonas Shuts was brought acquainted wth John Baptista Agnello, by the meanes of Sr John Barteley, and Sr William Morgaine, knights, soo as in the same monthe of Januarye by the meanes of the learninge of the sayd Baptista in alchimia and the knowledge of the said Jonas in myneralls and mettalls handelinge, the verye firste golde was founde and discoueryed by them too bee in the first peace of ewer whiche Mr. Furbusher brought home in his first voyadge, the valwe of oz. 1¼ in c. weight of the ewer, whiche ewer Mr. Locke had delyvered too the sayde Baptista in the same moneth of Januarye too make prooffe thereof, wch prooffe being made, hee, Mr. Locke, delyvered the same too Her Matie ymediately.

And afterwardes, in the same monethe, and in February, and Marche, Baptista and Jonas made diuers other smale proofes thereof whereby still they founde golde, whiche afterwardes was discouered too Sr William Winter by Sr William Morgaine.

In the beginninge of Aprill 1577, when Sr William Winter was assured from Sr William Morgaine and Sr John Bartley, by the handy woorke of Jonas, by prooffes wch he made in their owne presens, too theire owne satisffaction that this was trewly the ewer of a mynd of golde; the sayde Sr William Winter justified the same too bee trewe too my Lordes of the Queenes Maties Honorable Pryvie Councell.

And vppon commission directyd from Her Maties Pryve Councell too Sr William Winter, Mr. Thomas Randolphe, George Winter, and others, by the procurement of Mr. Locke, they weare certyfied by wrytinge geven by Jonas, of the riches of this ewer and order and chardges of meltinge the same, and the buyldings and workemen thereof, and of all things necessarie for the furniture of the mynes, wch all other couninge men coolde fynde owte, wheareby Her Matie & Her Pryve Councell weare content too sett owtt a secounde voyadge for discouerye vnder chardge of Mr. Furbusher.

And therevpon allsoo agreament was made wth Jonas Shots by indenture vnder the hand wrytinge of Sr William Winter and Mr Locke, that he shoulde goe one the voyadge as cheife master of the mynes, and too bee cheife workemaster therein above all other as well abroade as att home, and att his retorne home too bee made denyson and too have a good pencion for his enterteynement duringe his lyfe.

Allsoo in the newe lande all the voyadge Jonas made all the smale prooffes, & sayes of all the ewer that was founde in the mynes theare wch was laden in the shipps and brought home and certyfied the valewe of the ryches thereof, whiche nowe is founde trewe.

Allsoo the ships beinge retorned home intoo Englande in the monethe of September 1577, wth their full ladinge of that riche ewer too the quantitie of clx tonnes, Jonas hathe wrought and donne therein as followethe.

The fyste of October, 1577, Jonas begane too builde the furnaces at Sr William Winter's house, and fynished the same and all things necessary of his own devise.

The theirtie of October he had molten one hundrethe weight of the ewer prepared and handelyd after his owne devise and order. And thereby was founde that a tonne thereof doeth holde of fyne golde more then the valewe of xl*li.* of mooney by wittnes of Sr William Winter, Mr. Furbusher, Mr. Locke, and Robert Denhame & others whoe sawe the woorkinge and prooffes theareof made, besydes the remnants of gold remayninge in the slags, wch Jonas sayd coulde not be well brought outt, butt in the greate woorkes, furnasses, wch he desyred might bee builded accordinge too the plate that he woolde giue, that theirby hee woould thorowghely knowe the nature of this ewer.

The theirde of November reporte was made too the Queenes Maties Privie Councell by Sr William Winter and others of the Commissioners what was founde by Jonas workes, whiche did not satisfie them, and althoughe Jonas him selfe and Denhame, declared too them this furnace too bee not sufficient for the great works, yet woolde have them to melte downe an other hundrethe weight for better tryall of the trewthe, wch Jonas sayde he was willinge, becavse this furnance and bellowse was too smale and place not fytte yet for their commanndement he woolde doe yt.

The twentethe of Novemb', Jonas had newe repaired his furnace at Sr William Winters howse as well as that place woolde suffer. And in that meane tyme Baptista had taken vpon him too prepare the ewer too greate effecte wch did not succeade well in the great works. And also George Woolfe had taken vpon him to prepare the ewer too melte easely wth out any adetaments in the greate works wch did not succede well.

The fourthe of Decemb' Jonas, wth danger of his lyffe, throughe the smoke, had molten doune the secounde hundrethe weight of ewer wch provyd as the fyrst hundreth did better then xl*li.* the tonne in pure golde as was certyfied to Her Matis Councell by lre of Sr Winter and other Commissioners, whoe sawe the prooffes made, but still remayned more gold in the slags whiche Jonas sayd onely the great woorkes must bringe owt.

And hearevpon Jonas hade his patente graunted too him, he promised too delyver at the leaste halfe oz. golde in everye hundrethe weight, wch ys tenn oz. tonn and allsoo hoaped too cleare all chardgs of the workinge and yf he did better yt should bee their proffitt and his creditt. And thearevppon hee should have gonne to Brystowe too have builded the furnaces theire for the greate woorkes.

Butt the syxte of Decemb' the Lords weare not yet well satisffied wth this seconnde prooffe made becavse of the remayners in the slags wch was enformed coolde not bee gotten owtt, butt chiefly they beleaved nothinge that was donne becavse the goldesmithes and goldefyners of London and manye other namyd counynge menn had made many prooffes of the ewer and could fynde noe whitt of goolde therein, and therefore they vouched too the Councell that theare was noe whitt of gold in this ewer vppen gage of their lyffe and goodes.

And vppon the arryvall of Mr. Furbusher at the Coorte retorned from Brystowe, the seventhe of Decemb' he stayed the woorks of Brystowe and turned yt too farther prooffe too bee made at London, wherefore yt was thought good too have conference therein wth Mr. Burcott, whose doings shall appere in articles a parte.

The tenthe Decemb', Jonas was browght too conference wth Mr. Burcott, and w^th^in iij or iiij dayes he dislyked the dealinge of Mr. Burcott boethe for his evell manners and allso his ignorannce in divers points of the works and handelynge of this ewer soo as as woold not anye more deale wth him.

The syxetenthe Decemb' Jonas wth Mr. Furbusher and Mr. Locke ryde too diuers placs too see water mills, for the workeinge of the ewer. And at Deartforde Jonas lyked the mills best of all others for the comodious water and place.

The twelfte of Januarye vpon newe comission too procead in the great works, wee went again too Dertforde, wth Jonas and Henricke, the mason, and Sebastian, the carpenter, whoe was then newe come too London, and measured owte the platt of grownd for errection of the buildings and furnacs.

The fowertenth Januarye Jonas delyuered too all the Comissioners a platt drawen by him of all the buildings of the howsse and mills and furnasces wch was well lyked. But uppon newe conference wth the workemen yt was founde that the tyme woold bee soo longe in the buildings, that yf that should be taryed for the money it woold be too late too sett owt a newe voyadg this yeare.

The 21 dict. Mr. Furbusher devised that x tonn of ewer shold be molten at Keswicke for the expedicon of the tryall.

The 22 dict. Jonas alleadged manye lacks theare and the Comissioners sawe great coaste and longe tyme hanginge theare one allsoo soo that Jonas offeryd too make a theird great prooffe at London of a toonn by cc weight at a tyme wch shoold suffice for a tryall of all wch was well lyked.

And the 23 daye Jonas gave informacon of certaine merkesytes too melt the ewer wth all wch Denham or Coole shoold have sought abroade.

The 23 sayd John Baptista made offer of a newe great prooffe too be wrought wth a winde furnace, whearevpon he had a c weight of ewer but yt succeeded not well.

The 30 of Januarye John Broed at Crepelgate had cc ewer too make prooffe wch did succead well.

The 12 of February Jonas did melt doune cc of the ewer for his theird prooffe at the howsse of S^r^ Wm. Winter wth great danger of his lyffe thoroughe the smoke, and at the second tyme of the mealtinge thereof The 18 February yt succeaded verye well in presens of the Comissioners, found oz. 6⅓ of silver and goold myxt, wheareof oz. 3½ was perfytted and browght 1 qr. of oz. of fyne golde and oz. 3¼ of fyne sylver, the rest of the oz. 6⅓ of silver mixt was sonke intoo the test becavse the test was not drye made thoroughe the hast of tyme and much gold remayned in the slegs.

All the doings of Jonas from the tyme too tyme was donn openly, and Mr. Furbusher caryed all the secrets thereof too Mr. Burcott, too healpe him, and all Mr. Burcott doings was in secrett, soo as none knoweth yt but him selfe.

THE DOOINGS OF MR. BURCOTT IN THE NEWE MYNES OF GOLDE.

Mr. Burcott had delyvered too him certaine smale peces of ewer in Novemb' 1577, whereof he made sayes and proffes and founde golde therein as Jonas had donne before.

The 9 Decemb' 1577, Mr. Locke and Mr. Furbusher brought lres from Mr. Secrytarye Walsingham vntoo Mr. Burcott and delyvered them too him, and had suche conference wth him therevpon that Mr. Burcott grew too full promes too delyver halfe a oz. of fyne gold out of everye hundreth of the ewer vppon certaine condicons sett doune in wrytinge, wch condicons was sertyfied too Mr. Secrytarye who dislyked of them.

The 10 Decemb' Jonas was brought too conference wth Burcott and wthin iij or iiij dayes he dislyked the dealinges of Mr. Burcott boethe for his evell manners and ignorance in diuers points of the woorks and handelynge of the ewer soo as he woold nott anye more deale wth him.

In fewe dayes afterwards when Mr. Burcott sawe that wee did cleave still too Jonas dooings and made little accompt of his doings he made great styre of his owne cunninge and soo wrought wth Mr. Furbusher that he cam too bee harde againe and putt him selfe too the consideracon of Her Matie and Councell, and vouched too warrant the ewer too hold soo muche gold as ys sayd and too gage his lande, goods, and lyffe for the delyverye thereof whereby he was credityed and the matter lyked.

The 20 Decemb' Mr. Furbusher woold that Mr. Burcott shoulde melt a c of ewer in Jonas furnace as Sr Wm Winters howsse, wch Jonas would not suffer, therefore he willed Jonas too pull doune the secrets of his furnace that Burcott might build a new after his manner, wch Jonas did soo.

The 20 Januarye, or their about, Robert Denham told Mr. Locke that he had a peace of strange ewer, wch he proved too doo muche good in the meltinge of our ewer and that he had yt from Mr. Burcotts housse by his man wch Mr. Burcott knewe not of, and shortly after, when Mr. Lock sawe yt in his howse, of Denham, he sayd yt was yt and that he had made prooffe thereof too hold syluer, copper, and leade, &c., that Mr. Burcott called yt ewer of antymonie, &c., and had sayd too him that yt held noe manner of mettell at all.

And Mr. Furbusher told Mr. Locke that Mr. Burcott sayed that Jonas bellowes laye too hie, & that yt should lye lowe too blowe right into the hole on the fore parte, and in fewe days afterwards he sayd againe that Mr. Burcott woold have the bellowes lye more then a fote aboue the hole as Jonas bellowes did lye.

Alsoo afterwards, Mr. Furbusher did shewe too Mr. Locke a paper platt, made by Mr. Burcott, of the furnace that he woold have wth a pott by yt. And nowe, at the 20 February, he shewed an other platt thereof made cleane contrarye too the first, but even iust the coppie of the plate of the furnaces, wheorwth Jonas doethe woorke.

The 21 Februarye, Mr. Burcot shewed too the Commissioners and too my Lorde Treasorer, his proofes made of ½ once of gold and 2 onces & ½ siluer, in one c of ewer, but the same was melted in potts wth additaments by halfe pounds in a pott wch is not the order of the great workes, nor noe man but him selfe knoweth wate he puts in his additaments.

And alsoo he shewed a peace of antimonye ewer vouching that there in was noe manner of mettal, but a prooffe thereof was made by the Commissioners one the 22 daye, and they founde therein boethe siluer, 30 ouncs in a tonn, and some copper, and verye muche leade.

And Mr. Burcott sayed that wth out that antimonnye ewer he woold not stand to his former promys of the golde and siluer too be delyveryed. And Denham saythe that Burcott did not knowe of the mettalls that are in that antymonie ewer vntill Thorsday, beinge the 20 Februarye, when as he tolde yt too Burcott, and was the mann that did first fynde yt too bee soo by his owne tryalls.

And Mr. Burcott was ignoranut of the weight of gold and sylver accoumpted after xxiiii too a pennye weight, and xxiiii pennye weight too the ounce vntill that Denham did shewe him his errore therein.

And yt is manyfest that Burcott was more ignorant then Jonas, in the knowledge of the nature and workinge of this ewer; for him selfe Burcott doeth confesse that he hathe made more then fortye sondrye wayes of tryall thereof, and yet is not well satisffied therein; but Jonas hathe made onely syxe proffs thereof, and those after the order of the greate woorks.

[*Colonial* 82. *Domestic Eliz.*, cxxii, No. 44.]

FEBRUARY 19, 1577. FROM DOCTOR BURCHART TO MR. SECRETARY WALSYNGHAM TOUCHYNGE THE NORTH-WEST OWRE.

Your Honore remember what appountement you mayd wth me in my house and what promysse I mayde you. I have done so and moltine doune a hundred weight and a pound, and I have the proves to show you to save your Honores credyte and my honestye, therfore appounte the day and ower to ride to Detforthe to bring me ther to gyve me your Honores countenance in the first enterpryce, for yff you do not go I will not go that all worke mene and so many off the comyssioners as yt shall pleasse you to se and here oure determynacon and show suche things as I have and cane do before them all for I esteame your Honore and credytt as moche as my owne lyffe, and I doute yf I cane ryde ytt wthout a horse lytter, and yet I know of no eassye horse. And yff your Honore will humble your selfe so muche and yett come once in my house an ower or twane before we departe out off London, I will show your Honore bothe the pattrone off the meltinge house and the sayes prevelye betwene you and me that my doinges be not onely wordes butt deeds. So God increasse your Honore in health. Datum the xix off Fabruary 1577.

Your humble to comand,

Burchard Kranrych.

To his Honorable and very good frend Sr Frauncis Walsingame,
Secretary to the Quenes Heighnes.
Delyver this.

[*Colonial* 84. *Domestic Eliz.*, cxxxi, No. 52.]

FEBRUARY 21, 1577. A DECLARATION OF THE VALUE OF THE NORTH-WEST EWRE BOTH FOR GOLD AND SILVER BY PROOFS THEREOF MADE BY DR. BURCOT.

My dutie done in most humble maner. It maye please yor Honor to be advertized that Burchart hathe fynished and certefied two proffes of the gold ewer, wherof one was of 1*li* w^{eit}, and thother of one c w^{eit}. Dennam hathe bene thonly man that he hathe admytted to be present and privie wth him in the seid proffes, and the same Dennam hathe bene the fyar workman therof, and by vouchem^{t} of them two ye same falle out as folowth :

That proffe of the *li* w^{eit} holdithe in silver . .	xxi gr.
And in golde	iii gr.
W^{ch} is vii parts silver and an viiith parte gold	
After that rate i c w^{eit} holdithe in silver . .	iiii oz. dr. viii^{d} w^{eit}
And in gold	xiiii^{d} w^{eit}
And after the same rate the tonne holdithe in silver	xxxviii oz.
And in gold	xiiii oz.
So the silver of a tonne at v^{s} ii^{d} the oz. amth to .	xxv^{li} vi^{s} iiii^{d}
And the gold at lix^{s} viii^{d} the oz. amountithe to .	xli^{li} xv^{s} iiii^{d}
Sum lxvii^{li} xx^{d}	
That proffe of the c w^{eit} holdithe in silver,	ii oz. dr. i^{d} w^{eit} vi gr
And in gold	xiii^{d} w^{eit} drs.
W^{ch} is nere about iiii parts silver and a vth parte gold	
After that rate the tonne holdithe in silver . .	li oz. v^{d} w^{eit}
And in gold	xiii oz. dr.
So the silver of a tonne amountithe to . .	xiii^{li} iiii^{s} ix^{d}
And the gold of the same to	xl^{li} v^{s} vi^{d}
Sum liiii^{li} x^{s} iii^{d}	

So a tonne answerable to the proffe made of the *li* w^{eit} ys richer then that of the c w^{eit} by xiii^{li} xi^{s} v^{d}.

[*Colonial* 85. *Domestic Eliz.*, cxxii, No. 61.]

FEBRUARY 27, 1577. FROM D. BURCOTT TOUCHYNGE HIS CUNYNGE AND OFFER ABOUT TRIYNGE THE EWRE.

My humble comendacons as I am informed that your Honore and the rest be displeased wth me as thoughe I had mayd a false prove, wch I will stand to the contrary to deathe and lyffe that yt is as I have sent onto you, therfore I have done yt wth my owne hand, and cane do yt agayne save the last quarter wch Dennan hathe moltine doune, and brought yt in agood way after the rayte, and in the dryvinge utterly mared yt for ingnorance and unskylfulnes; yt was in the night when I cold not be at yt, and I wold not trust him further and I mayd that quarter good after the rayte I sent, as for the detymente yt hathe layd this ten yeare in my house and were my invension to melt yt the eassier and the soner, and ther is not so moche sylver in yt as you are informed you for the ower is sufficient in him selfe to augement your Honore and save my credytt because yt is comed to suche an exclemacion agaynest my honestie and deface me wthout acause I will make my selfe cleare, let my have delyvered out of hand in hast to hundred weight be brought in my house of that ower, and ij honest men and not false men to see the rostinge, be cause I have fornace mete for yt, and in the meane tyme when yt is in rostinge I will ryde to the Tower-hill by your comandement to mend Jonas fornace, or to make an new one, that will serve me, and lett thes ij sworne men carrye the ower to the Tower when yt is rosted and see yt moltine doune and puryfyed as I have sent you; and lett me have althings nessessarye; and yf yt please you and Sir Will'm Wynter may be the men appounted to do yt, and Dennan be the workeman because he is the causer of this exclemacion as I cane tell you previlie, when your Honore please betwne you and me you shall here the very trothe. And yf I do not prove yt trewly comed out then take my body and goods to your owne pleasure, and Jonas nor the captaynes, nor any of ther confederats shall not come nye me, but them whome you please, for yf Jonas had any couninge yt had longe since appered; therfore he shall learne nothinge of me untill yt be knowen that I am aperfytt master above them all. Datum the xxvii of Fabruary. God encrease your Honore.

This prove wilbe done out of hand after yt is rosted, havinge all things nesessary prepared. Appount atyme when you will and I will come to you.

Yo^{r} humble to comand,
Burchard Kranrych.

To the Honorable Sir Fraunces Walsingame, Knight,
Secretary to the Quenes Matie, my singuler and very frend.

[*Colonial*, 92. *Domestic Eliz.*, cxxiii, No. 7.]

11 MARCH, 1577. TO THE LORD TREASURER AND LORD CHAMBERLAYNE ABOWT THE NORTH-WEST VIAGE.

My verie good Lordes, Hir Matie havyng been made acquaynted wth the certificats sent lately from the Commissioners appoynted to surveighe the sondrie proffs and trials made of the north-west ewre, and understandyng therby to hir good contentation that the richness of that earth is like to fall owt to a good reaconyng is well pleased that a third viage be taken in hand the plotte wherof consisteth chiefly in theise two poynts, to witte, the charges of the shipping outwardes and some provision to be made for a 100men to inhabite in those north-west partes: what the whole charges of theise two matters will amount to yor Lordships may perceave by that their bearers Mr. Ffurbesher and Mr. Locke will shew you in writyng and otherwise by word. I have allreadie acquaynted my Lord of Leceister how m^ch^: and yf it might please yor Lordships to send yor opinions of the two sayd points I would after impart it further to others of my Lords and the Cowncell to thend that all their advises beyng had hir Matie may grow to some resolution for this newe and third viage and so order accordingly be given for warrant of hir owne part and other men's likewise that be adventurers in this matter.

[*Colonial*, 90. *Domestic*, *Eliz.*, cxxiii, No. 5.]

THE ACCOMPT TAKEN AT MOSKOVIE HOUSE THE VIIJ^TH^ OF MARCH 1577, OF CC WEIGHT OF Y^E^ EWRE BROUGHT BY MR. FFORBISHER MOLTON AND TRIED BY JONAS SHUTTZ AN ALMAINE ADSISTED BY IIJ ENGLISHMEN, VIZ., HUMFREY COLE JOHN BRODE AND ROBERT DENHAM.

Of the said cc weight of ewre so molton and tryed as aforesayd there proceaded in silver vj oz. vij^d^ weight xiij graines di, wch valued at v^s^ the oz. maketh in money	xxxj^s^ x^d^ ob.
And of the same ewre proceaded in gold v^d^ weight v graines wch at iiij^s^ the penny weight maketh in money	xv^s^ vij^d^ ob.
Sum, xlvij^s^ vj^d^.	
So at that rate j c weight of the said ewre will make in money	xxiij^s^ ix^d^
And a toone of the said ewre by like accompt will make in money	xxiij^li^ xv^s^

The charges of getting and fetting the said ewre into the realme as by particulers delivered by Mr. Fforbiser doth appere will not excede the tonne .	viij*li*
So uppon view of this accompt for every viij*li* defrayed the venturers shall gaine v*li* wch ariseth uppon every hundred poundes above . .	lx*li*

W. Wyntar
Edward Dyer
Martin Frobiser
Rich. Young
Mathew Field
Edmond Hogan
Michael Lok
Andrew Palmer

The charge for furnishing shippes for this next voyage as followeth :—

The charge to furnish iiij or v shippes wth 120 men, viz., solidiors, mynars, smithes, carpenters, and other men of necessarie occupacions and to bring home viijc tonne of ewre as appeareth by particulers therof delivered by Mr. Ffrobisher will amount to . . .	vjml iiijc
Wherof must be defrayed presently for the furniture nowe owtward thone moytie or els this yeres voyage wilbe lost, viz.	iijm ijc *li*.
And at the retorne of the shippes must be payed other .	iijm ijc *li*

Ffor the provision of wch money a levie must be made amonge thadventurers after such rate that every one of them wch did before adventure jc *li*. must now put in jc xxx*li* the moytie therof to be payed forthwith. And the other moytie to be readie against the retorne of the shippes to clere the freight and men's wages.

STATE PAPERS CONCERNING THE TRIALL OF THE EWR PREVIOUS TO THE THIRD VOYAGE.

STATE PAPERS CONCERNING THE TRIALL OF THE EWR PREVIOUS TO THE THIRD VOYAGE.

[*Colonial*, 56. *Domestic Eliz.*, cxvi. No. 25.]

OCTOBER 17, 1577. M^M FROM MY LORDS TO THE WARDEN AND WOORKMASTER OF THE MYNT TOVCHYNGE THE ORE BROUGHT OWT OF THE NORTH-WEST.

After our hartie commendacions. The Queenes Ma[ties] pleasure is that certain oore brought into this realme by our loving frend Martin Ffurbusher, gent. out of the north-west partes, shalbe caryed into the Tower and layd in some convenyent place by you to be appointed for that purpos, the said oore to be by him delyvered unto you by weight and so by you receaved. And further that to the doare of the place where the same shall lye there be fower severall lockes and keyes made, whereof the said Ffurbusher to have one, you her heighnes officers two, and Michaell Locke, tresorer of the Company of Adventurers into the said north-west partes, the fourth. Whereof her Ma[tie] hath appointed us to geve you knowledge to thend you maie take present order therein accordinglie.

And so we bid you ffarewell. Ffrom the Castell of Windesor the the xvj[th] of October 1577.

Your loving friudes,

[no signatures.]

And further yt is meant y[t] you shall from tyme to tyme make delyverye of the seyd ore to be melted downe accordyngley as you shall be dyrected by the commyssyoners by us deputed to have the oversyght of the seid meltyng.

Ric. Martin, Warden of the Mint. (Blank) Samson, worckmaster of the same.

[*Colonial*, 59. *Domestic Eliz.*, cxviij. No. 40.]

25 NOVEMB. 1577. FROM MR. EDWARD FENTON. OF THE UNLADYN OF THE OORE IN THE AYD AND GABRIELL, AND HOW MANIE TOONES OF THE SAYD OORE IS IN ETHER OF THE SAYD VESSELS. TO HAVE ORDER FOR THE DISCHARDGE OF THE MARINERS AND UNRIGGINGE THE SAYD VESSELS.

My dutie to yor hono^{r} most humblie used. Receaving letters from my Generall the viijth of this instante to make deliverie of such oure as remayned in the *Ayde* and the *Gabriell*, whereof I had charge wth letters in like manner from yor ho^{r} and others of her Maties counsoull directed to Sr Richarde Barkely, Knight, Hugh Smith, Esquier, and the Maior of Bristoll wch I presentlie sent unto them notwthstanding they mett not abowt the receipte thereof till the xiiijth daye of this instante (by reason Sr Richard Barkely was ymployed elswhere in her Maties service) Sithens wch time they have wth care and diligence attended that service and have now in effect ended the same saving that a little porcion of oare founde in removing of the caske[s] in the *Ayde* is not yet weighed, but wilbe furthwth dispatched. The quantitie of the oare in the *Ayde* (I suppose) wilbe nigh 124 toones. And in the barke *Gabriell* 16 toones rd'i. All wch is saffelie bestowed in the castell of Bristoll according to order appointed therin. Thus much having so conveniente a messenger I thought good to ymparte wth yor honor beseching the same to further the dispatch of all such persons as have served in this action, whose service now rather is chargeful to her Matie and other thadventurers then liking or beneficeall to theim selves. Wherof the gentlemen ymployed in that action (and attending heare thies ij moneths to small purpose) is greatlie charged by the same whose good government in this service both before and sithence their cominge hither on lande deserves (in myne opinion) at her Maties handes bothe favo^{r} and recompence for so greate vertues showed in so honorable an action wherin yor honors good favo^{r} and furtherance is chieflie to be craved for their comforts in the same. And am (amongeste the rest speciallie) to recommende this gentleman and bearer Mr. Carew to yor honors favo^{r} and countenaunce whose readinesse wisdome and good government in this service deserves greate commendacion and rather to be ymployed wth charge then to serve as now he hath donne wch I humblie besech yo^{u} to consider of hereafter for his better advauncemente and creditt ; no lesse (I suppose) my Generall will deliver of him for his desertes (wthout flatterin) doth thus much challenges me to reaporte unto yor honor having well marked his behavio^{r} in this service. It were to good purpose order were taken for the unriggingc of the shippes (if heare they shall remayne) for wante wherof their tackle and other thinges grow to dekaie wch would be saved if order were taken what should be done in the same: Besides the superfliius mariners and others now attending would be discharged wch if it had been performed vij weekes agoo had saved greate charges and been to verie good purpose. All wch I committ to be delt in as shall seeme beste to yor honor craving pardon for my boldnesse do praye daielie to thalmightie to bless you in all yor honorable actions and sende you much encrease of honor. From Brystoll the xxvth of November 1577.

Yor honors most humblie to commaunde,

Edward Fenton.

Postscript.—I am bolde to sende unto yor honor hereinclosed the opinion of Mr. Doctor Dodding towching the death of the man we brought over wth us who often visited him in the time of his sicknes.

To the honorable my verie good frinde Fraunois Walsingham, Esquier, principall secreatarie to her Matie and one of her highnes privie counsaill at the Corte.

[*Colonial* 59. *Domestic Eliz.* cxviii. No. 40, i.]

DOCTOR DODDYNGS REPORT OF THE SICKNESSE AND DEATH OF THE MAN AT BRISTOLL WC CAPT. FURBISHER BROUGHT FROM THE NORTH-WEST: AND OF THE NATURE OF THE WOMAN OF THAT CONTRIE YET LIVYNGE.

In cadaveris dissecti latore sinistro, primum sese michi inspiciendæ obtulerunt costæ duæ (casus vi et impetu quando capiebatur) præfractæ dehiscentes adhuc, nec invicem agglutinatæ: quarum cura (uti in tam turbulentissimis temporibus rerumque navalium augustiis usu venire solet) vel neglecta, vel (quod potius suspicor) morbus a nemine perceptus inflamationem concitaverat: et pulmonis contusio (temporis progressu) putredinem itidem contraxerat: quibus libere indies serpentibus, cum per frigoris externi injuriam commotis tum per malam victus rationem auctis, nec tamen per artem chirurgicam exterius interim emendatis neque per medicamenta interius retusis in pulmonis ulcus insanabile precepe. Ruebat et ingruebat malum puris putrefacti afluentia maxima ejusdem materia viscida et tenaci, ita pulmonis pars sinistra undique scatebat ut nihil omnino per omne morbi tempus expuerit et anhelitus fere retentus est constrictus, huc accedebat natura morbi infestantis furore admodum debilitata: liberior apud nos victus ratio quam aut hujus morbi fert malitia aut hominis fortassis consuetudo quotidiana patiebatur, quod summa ducis summi cura liberalitasque immensa eorum apud quos divertebat effecerat: deceptis omnibus opinione potius morbo latilate et indulgentia stulta quam male volentia, sed cum asthma paulo ante mortem, morbus iste haud obscure similitudine expressit, tum ab hydrope non omnino immunis erat : in thoracis enim spacio sinistro ingens aquæ opia et abundantia (qualis a perspicacibus et industriis anatomicis raro visitur) conspiciebatur fluitare : hæc moto corpore (uti rei eventus fidem fecit) agitabatur, et pulmonis expirationem impediebat, pulmo denique ipse costis firmius quam quisquam putaret hærebat. Cerebri vulnerati apostematique (ut ita loquar) præter surditatem doloremque capitis intensissimum (quibus nunquam non affligebatur) signa infinita extant et emergunt apertissima quæ nunc silentio (ne sim prolixior) sunt involvenda, quod lieni minutissimo detraxerat natura justæ molis id stomacho capacissimo adjecisse visa est cum fœnore qui aqua refertus et distentus multo major quam nostrorum hominum propter insanabilem (credo) ingluviem apparebat. In cæteris partibus Anglium diceres metum in quo a primo ejus adventu erat quamvis vultus hilarior et simulatus non mediocri arte occultabat et ementiebatur, tamen gestus ipsius (sigillatim singula penitius introspiciente me et suspicante omnia) vel eundem aperte prodiderunt det exeruntque vel morbum letale (quod sepius sed surdis cecini) præsagiverunt impendentem, quæ ex pulsibus dilucidiora quam ex ipso innotuerunt et confirmabatur qui semper minores tardiores et debiliores, quam rariores extiterunt, et rariores tamen quam aut etas juvenilis, aut temperatura ejus biliosa eflagitabat. priori morbi insultu, cum vires adhuc integre essent accersitus sanguinis missionem magna contentione suadebam, quo et inflammationis aculei extincti, et materia diminuta jacerent : sed vetuit viri barbari, barbara nimis, insulsaque timiditas, et eorum consilium quibuscum una navigabat, apud me prævaluit. Denique ea hora vocatus quæ proxime horam præcessit in qua de vita discescit reperi cuncta mortem brevi minitantia, nimirum, interruptam fereque omnino abscissam loquelam dejectum appetitum, pulsum nullum, Quid multa! Vires omnes facultatesque prostratas prorsus aliquantulum recreatus ad se quasi e somno profundo redit nosque ejus familiares agnovit. animadverti sed medicamento, et verba nostra quæ ediscerat pauca ut potuit, eloquebatur vicissimque ad interrogata respondit satis apposite, cantelenamque eandem (uti referunt qui utramque audiverant) clare cecinit, qua astantes in littore ejus loci atque ordinis socii ultimum ipsius discessum vel deplorabant vel celebrabant. Haud aliter atque cigni qui providentes quid in morte sit boni cum cantu et voluptate moriuntur, sed vix discesseram quando vitam cum morte commutaverat ; in hæc ultima et nostra lingua edita virumpens verba 'Deus sit vobiscum.'

Angebar et vehementer dolebam non tam ipsius morte quam quod reginæ nostræ serenissimæ spes tanta ejus videndi quasi e manibus nunc secundo exciderat. At mœrore multo majore afficiuntur hujus novi honoris solidique heroes vere Herculei tanquam exantlati laboris præmiis et spoliis spoliati quos jure optimo (dicam enim quod senceo) manet a nobis observantia summa, qui itinera hæc marittima, devia ea quidem aspera et plane ante hoc tempus inaccessa magnis suis sumptibus laboribus periculis majore regni et posteritatis commodo maxima suorum nominum gloria superarunt et apperuerunt quod si uti acceperit, successerit, ut eadem facilitate has regni et reipublica animas nervosque (sic enim a philosophis non inscite appellantur divitiæ) qua consequuti sumus, retineamus, nec priusquam accipiamus, ingratitudine nostra amittamus (quid enim aliud metuam non video) quid retribuamus Domino pro omnibus quæ tribuit nobis ab externis regibus nil speremus boni quia non volunt, nil metuamus mali, quia non audent. Verum non eo hæc dico quo quenquam in peccatorum sordibus delitescentem in utramque aurem dormire suadeam, sed ut amorem sigillatim singulos ab improbitate, et adigam ad res divinas universim omnes acrius celebrandas: quid enim dici cogitarive potest absurdius quam cum Deus optimus maximus sit ad dandum promptior et paratior nos ad promerendum simus tardiores. Attamen semper erit illa Dei veri, vera vox agnoscenda non quia tu dignus sed quia mitis ego. Si incantantium futilia fictiliaque præcaria, ceremoniæque inertes et ludinæ in morbis profligandis quicquam potuissent, hic profecto (dum in vivis esset) 'Calichoughe' (namque hoc ei nomen erat) eosdem pullulantes ut hydras amputasset et profligasset etenim hoc nemo fuit in hac arte excercitatior, nemo in ipsa superstitione (ni fallor) confidentior: qui tot incantationes usurpavit, quot dolorum termina emerserunt Mulieri laboranti tum e morbillis (qui postridie, quando hæc scripta sunt effloruerunt in cute frequentissimi) cadaver ostendi, unaque (meo suasu) ad sepulturam (quam nulla solemnitate de industria peractam esse volui, ne qua horror ei injiceretur de hominibus apud nos sacrificandis) licet invita, adducta est: ibique tamdiu detenta, donec terra undique coopertum esset cadaver: ossa humana ostendi effossa, fecique ut intelligeret omnes nos eodem modo esse inhumandos, quo omnem ex animo de humana carne comedenda (quæ in ipsis altas radices egerat) adimerem scrupum: ipsaque timorem deinceps deponere disceret. Sed ista aut prudentia et patientia homines omnes nostros exuperat mulier, aut ab ipsis brutis animantibus longe superatur humanitate, qua nihil omnino ejus morte commota est, neque eam (quantum ex vultu intelleximus) ægre tulit: ita ut hoc postremo ejus facto manefestius expresserit id quod longe antea conjecturis assequuti sumus, illum præ se mirum in modum contempsisse, et quamvis in uno eodemque lecto somnum capere solebat præter colloquia tamen nihil inter eos intercessisse, amplexus ejus abhorruisse. Vale. Bristolliæ 8° Novembris.

Tuus uti scis,

Edwardus Doddinge.

Si nihil infesti durus vidisset Ulisses
Felix Penelope, sed sine laude foret.

[*Colonial East Indies*, 57. *Domestic*, cxviii, No. 36.]

23RD OF NOVEMBER, 1577. FROM MR. MICH. LOCKE. WHAT THE GOODNESS OF THE GOLD ORE IS DECLARED.

Right honorable,—According to yor letter sent me by Mr. Watterhows, I have had conferens wth hym, and I have told hym my best opynyon playnly, and I wyll furder that matter the best I can and wyll joyne wth you therin for a part, yf it please you, wch I hartely pray you. The tryall of the ore brought home by Mr. Furbusher, moche paynes and labour hathe passed of late agayne, yet it is not brought to parfection, the iij workmasters cannot yet agree togethers, eche is jelous of other to be put out of the work and therby lothe to shew their conynge or to use effectuall conferens ; but amongest them all we doo very playnlye seę and fynd that the ure is very ryche, and the worst of all their doynges wyll yeld better then xl^{li} a ton, clere of charges. This is assuredly trew, wch may suffyse to embrase the enterpryse. And wth in few dayes yor honor shalbe better certyffyed of better matter herein when we have made better lykynge betwene the workmastars. And thus I commytt yor h. to the tuition of almighty God. From London, this xxiij November, 1577.

Yo^{r} h. most bounden,
Michael Lok.

[*Colonial* 58. *Domestic Eliz.*, cxviii. No. 39.]

NOV. 26TH, 1577. FROM SIR WILLIAM WYNTER, TOUCHYNGE THE GOLD ORE WHAT IT WILL YIELD.

Mr. Furbusher beyng bounde towardes Bristowe for the dischardginge of the maryners and takinge of order for the ships and ther furneture wch remayneth ther, hathe been enforsed to staie some what longer then willingly he would have don frome the doinge of these thinges before this tyme, because he hathe hade a desier that wth his travaile and others in comyssion tochinge this matter of the oore that he and they nowghte have understanded what sertaine accompt was to be made of the said oore to thend yor honur, and the rest of her Highnes cownsaile myght have been perfatly enformed. What hathe ben don hether to Mr. Furbusher will do yor honur to wyet. And albe hit the oore in reporte do not appiere to be of the vallew wch hathe ben looked for, yet yf the woorkmen be to belyved who offereth ther lieves to performe that wch they have set downe wth ther handes, the commodittie is suche as maye content resonable myendes, for my owne opynyon I beleve hit will fawll ow^t better than the woorkemen hathe set hit doune, and that it maye so come to pase I will use all the travaile I cane possiblie; not so muche for my pryvate gayne (trewly) as in respect of the Q. Matie that her highnes good hoepe be not made frustrate, and yet I dow^t not but you beleve I ame worse able to beare a lose than her Matie is. And thus most humbly takinge my leave, I rest, prainge God to kepe yow in helthe. Tower Hill, the xxv^th of November, 1577.

Yo^r honnors ever to comawnde,

W. Wynter.

[*Colonial*, 60. *Domestic Eliz.*, cxviii, No. 41.]

NOVEMBER 25TH, 1577. FROM JONAS SCHUTZ TOUCHING THE GOULD OORE.

Righte honnorable Mr. Secretarye Walsingham, my humble dutie premysed. These maye signifie unto youre honnoure that wheras I have bene by order from the Quenes Majestie and her most honnorable consail appointed to trye the ore brought into the realme by Maister Captaine Ffrobysher. Nowe, so it is that I have bene visited with sicknes and ame as yt weake, so that I have not bene able to accomplishe my dutie in tryall thereof, accordinge to my comyssion. And nowe, havinge recovered somewhat of my disease, I entende by the grace of God to ffinishe the profe therof. And whearas I dyde promyse before youre honnore halfe a nowce, I doo not mys dowte thereof. And yff the nexte doth fall any better which I ame in good hope then shall it be showed to youre honnore, and accordingelie one Saturdaye next to bringe a sample therof to the courte. Thus muche I thoughte it my dutie to signifie unto youre honnore. And so leavinge to trowble yo^u further, I reste prayinge the Almightie God to protecte yo^u. Ffrome the howsse of Johne Nighelson, scituat in Easte Smythfeilde, the xxv^th daye of November, A^o 1577.

Youre humble servaunte to commaunde,

Jonnas Schütz.

Too the righte honorable Mr. Secretarie Walsinghame, one of the Quenes Majestie previe consail.

[*Colonial* 62. *Domestic Eliz.*, cxviii, No. 43.]

NOV^R 26TH 1577. FROM DR BURCOT: WHAT HE THINCKETH THE GOOLD OORE WILL YEELD BY THE TONNE.

After most hartie and humble commendacions. Whereas the Queenes Matie (as I perceave by yor honors lettres) hathe required me to make an assaye and to shewe my judgm^t in that ooyre wch Captayne Ffurbysher brought into this lande, the truthe ys, I have so donne, and I desire yor hono^r to advertize her highnes that I have assayed and proved yt to the uttermost by dyvers and sundry assayes, and fynde not therein suche greate ryches as ys spoken and reported of. But the truthe ys, I have founde that in an hundred weight there ys half an onze of golde in the blacke ooyre that ys x ounces yn a tonne. Also, I fynde one other redd ooyre bearinge twoo ounces in an hundred weight, that ys ffortye ounces in a tonne. And yf the same be well husbanded by a skyllfull and expert man, that blacke oore will yelde in the great fyer half an ounce, and beare the charges of meltinge and puryfyenge of yt. I wolde therfore wyshe her highnes to allowe some yeerely consyderacion unto some expert and skyllfull man in the knowledge of myneralles that yf any suche roughe wyeld and forrayne ooyre at any tyme, hereafter happen to comme into this lande that he by his true assayes thereof may certyfie her highnes of the juste encrease of the same at his owne charges that thereby her Matie and subjectes may not (as heretofore they have bynne) be disceaved by suche vayne and untrue reports. And further that suche skyllfull man maye teache otheres in the same experyence of myneralles yf at any tyme hereafter the like vyage shalbe made for the like or other ooyre whose knoweledge of the travellers maye greatly encrese the commodytie of the viage, whoe by his instruccions in the same knoweledge may learne a brevyate and shorte assaye in the tryall thereof wth out any charges, ffurnys, or other instrument. And that in suche shorte tyme that they maye thereby make xij assayes in an houre, what goodnes ys in suche ooyre, and then to take the good and leave the badd. If age and sycknes did not so oppresse me, but that I were able to travell therein myself, I wolde willinglie bestowe my dyligence in that service. Thus I leave yor hono^r to Goddes direction, prayenge for her Maties longe and prosperous raigne. Dated this xxvj^th of November, 1577.

Yo^r hono^rs at commaundem^t,

Burchard Kraurych.

To the Right Woorshipfull Sr Fraunces Walsingham, Knight, Chief Secretary unto the Queenes Matie, my singuler good frende gyve these.

[*Colonial*, 63. *Domestic Eliz.*, cxviii, No. 54.]

30 NOVEMBER, 1577. FROM MICHAEL LOK, OF JONAS, NEW MANER OF TRYALLS OF THE GOOLD ORE.

Right honorable. I wrote you a letter vj daies past, wch I sent by Mr. Furbusher, myndinge at that tyme my sellf and Jonas to have byn wth yor honor at the court this daye. The onely cause of our staye was that Jonas is makinge triall of another order of meltinge to be used in the grosse worke wherby to melt the ore, wth halfe the charges and tyme of the ordinarie use of the grosse workes, and wth all to receave the fyne gold out of the fyer molten wth out any maner of yron or other matter of the ore to hynder the same, wch work is already done after one maner in grosse order, and found very good, and by Monday or Tewesday next wilbe also finished after another maner, wch is also hoped shall fawlle out as good or rather better. And therwthal we will repayre to yor honor wth full and parfect resolution of all matters to good lykynge by Gods grace, or on Wednesday to Hampton Court bycause it is nerer. And then, also, I wyll certyffy yor honor what I fynd in Sr L. Ducket and Sr R. Heyward, for the matter declared to me by Mr. Waterhowsse this mornyng, accordinge to yor letter received then, also of hym who sayethe that he also wilbe wth yor honor at the Court on Monday or Tewesday next wth answere therof.

That wch I wrote yor honor in my letter sent by Mr. Furbusher touchinge the rychesse of the ure, is very trew. Yt grevethe me to see so moche tyme lost before we begyn to gyve order for the makyng of the furnasses for the grosse work, wch of necessyty must be done before we shall have any goodnes when all is sayed and proved that can be done, I know not wherto to impute the fault, but to a *schisma* growen amonge us commissioners, throughe unbelefe, or I cannot tell what worse in summe of us, wch the tyme must open. And thus for this tyme humbly I take my leave and commytt yor honor to Godes protection. From London, this 30 November, 1577.

Yo^{r} honors most bounden,

Michael Lok.

I send this berer, my servant, purposely wth this letter, and yor honor maye command hym to retorne at yor plesure.

To the right honorable Mr. Francys Walsingham, Chyef Secretarie of the Queenes Matie, &c.

Delyvred at the Court.

[*Colonial*, 61. *Domestic Eliz.*, cxviii, No. 42.]

A NOTE OF ALL SUCH CHARGES AS SHALL AMOUNT IN MYLTYNGE DOWNE THE GOWLD OWRE, AND OTHER CHARGES AS FOLOWETH, BY MR. JONAS SHOWTES ACCOUNT.

Inprimis, every toune waight of owre to yeald the somme of .	xxx^{li}
Item, for the byldynge of the worke howse and furnysses .	iiij^{cli}
Item, for the charge of xij workemen wykely . .	v^{li}
Item, for cole, wood, fflowshe, and lead, to mylte it doune wykely	xv^{li}

Item, that the sayd Jonas shall trye wth the sayd charges afore say wykele ij tone wayte of owre wch shall yeald in fyne gowld the somme of . . . . xx ounces

Jonnas Schütz.

[*Colonial, East Indies*, 69. *Domestic Eliz.*, cxix, No. 15.]

AGNELLO ON THE TRIAL OF THE ORE.

Molto Magnifico et honorando Sr mio essellentissimo.

Non voici Sr mio che pensasti ch' io fussi cossi mal creuto che attanti che hora non fussi stato el debito mio ad responder alla vostra cortese lettera scrittame di 7 del passato ma la causa è stata ch' io desiderame di rispondemi con qualche sustantia, il che volendo fare son stato forciato di far molte esperienze et prove per conoscer la natura di questa minera portata in questo regno dal Sr Martino Furbisher; ultimamente di poi molte prove fatte ho trovato che bisogna separare la parte sulfurea combustile con conservatione dell' oro che in essa si sitruova, et più glie necessario separare quanto si puo la parte terrestra, et ancora el ferro che in essa si contiene, il che fatto detta minera sara più facile alla fusione et con maneo spesa, oltra di questo l'oro quale era disperso per minima in gran quantita di terrestreita et materia ferrea, sara redotto più unito ad diversi recuperare et unirsi insieme. El quale effetto penso haver trovato, la qual cosa se pensate possi suplire al desiderio di sua Mata me gli oferisco paratissimo. Ma perche intendo che vi sono molti che si proferiscono et fano proforte grande à sua Mata io sarsi di opinione che lei dettessi a cottare le loro proferte ch' io per me son di gia vechio et impotente a durar più faticha et tanto più che questa non è mia professione, per che veramente io non adoperai mai metalli, glie ben vero che essendo io stato sempre desideroso de intendere i secreti de natura ho pensato de intender la natura di questa minera come di sopra ho detto. Non voici pere Sr mio che pensasti ch' io habbi detto questo con intencione di non far piacer et servitio a sua Mata quando gli pincera comand . . . le ma l'ho solamente detto per le cause sopra detto. Et cosi facendo fine et pregando l'Altissimo vi conservi et prosperi, gli bascio l'honorata mano. Di Lon. adi 20 di Decemb. del 77.

Tutto al comando di V. S.,

Giovanni Battista Agnello.

Colonial East Indies, No. 64. *Domestic Eliz.*, cxix, 8.

A NOTE OF THE CHARGES REQUISITE FOR THE TRIALL OF ONE TONNE OF THE NORTH-WEST ORE.

Right honorable,—We have byn longe tyme about the second prooffe of the ure, and doo well parceave that this foinace is not great ynoughe for the common great workes of the meltinge and to bringe the work to that parfection wch is desyred, and so moche hathe Jonas said unto us from the begynnynge. And yet wee doo fynd by these two proffes made of c weight of ure apece that the ure hathe in it more then the valew of xl^li of money in gold in everie ton weight, after the rate that we fynd and sawe the said two proffes. Also in this meane tyme we have had triall of dyvers maner of workynges made by sundry men for the bettar preparynge of the ure that it might be easye in ye meltinge to avoyde great charges, and we have had dyvers small prooffes made by them wch have very good lykinge, but we are not able to say assuredlye what they wyll fawll out in the great worke untill the fornaces be made for the said workes. Thus is moche tyme passed awaye and money spent, and yet our expectation not satisfied. Wherfore we have thoughte good now to drawe this matter to the best end that we can. And uppon conferens had wth Jonas (whome we fynd very honest and trew in his doinges, and as our judgement wyll leade us the parfectest workmaster in this art of his profession). We have thus dealt wth hym as folowethe. He sayethe that this here new proved is poore in respect of the pece of ure brought last yere, and of sum other brought this yere, and of that wch he knowethe may be brought the next yere, and that it is of a strange nature, suche as he is not yet well acquaynted withall; but he doutethe not in the great work, he will learne to knowe it parfectlye. Neverthelesse, beinge, as it is or maye be, he dothe promysse to delyver halfe an once of fyne gold out of everye hundreth weight therof at the worst and least valew, and hopethe also assuredlie to delyver so moche more gold as shall pay all charges of the meltinge and workinge therof, wth an advantage wherein he will use his best endevour, as well for the servyce of her Matie as for his owne credite. And for the reward of his owne labour and industrye, he dothe reserve hym selfe to the good consideration of her Matie and yor honor wth the rest of the Lordes, desyrynge that whatsoever it be he maye be made sure therof duryng his lyffe by her Maties letters pattenttes before he begyn the great workes in the wch letters pattenttes he is contented ther shalbe a condicion sett downe, that yf he doo not parforme that wch before is declared that then he wyll take no benefite by that graunt. And he sayethe that he hathe no doubt that in the handelinge of the great workes his doynges shalbe suche as justelye shall deserve to augment his pencyon of lyvinge hereafter whatsoever it shalbe now. And touchinge the ordinarie charges of workinge the ure to parfection, he is not able yet to saye justelye what it wyll amount uppon the ton, untill he have made triall therof in the great work; but this much he sayethe that it wylbe under ten poundes the ton, exceptinge the charges of buyldynges of the workhowsse and fornaces, and so he wyll warrant it uppon forfeyture of his pençyon. And hereuppon he repayrethe to the Court to take sum end wth yor H., and the rest of the LL., wch beinge done, he wyll go presently to Bristow to Mr. Furbusher, wth yor good lykynge, to vew a place convenient to erect the workhowse and fornaces, and there to have conferens wth the workmasters that shall erect the same according to the plat, that he wyll gyve them the charges wherof he cannot judge wth out conferens had. wth them; but he thynkethe it wilbe under cc poundes. Thus moche we have thought good to certiffy yor honor, and in the rest therof hymsellffe shall satisfy you more at large. Robert[1] Denham is one suche as may be able to doo good sarvyces to the Q. Matie and stand the Company in great steade whatsoever should happen to Jonas, and therefore it were good he mought be remembered, wch we referre to yor honors good consyderation. And thus God preserve yor honor. From London, this vj December, 1577.

Yo^{r} honors to commaund,

W. Wynter,

Michael Loke.

To the right honorable Sr Ffrancys Walsingham, Knyght, one of her Maties principall Secretaries.

dct.

At the Court.

[*Colonial*, 65. *Domestic Eliz.*, cxix. No. 9.]

[2] The great proofe of the black oure of Alom and chayne.

[3] Thee third proofe of thee read sand of Alom and chayne.

[4] Thee second proofe of thee read sand of Alom and cheyne.

[5] Thee oure of Hynnesbury Gilles.

[*Colonial*, 66. *Domestic Eliz.*, cxix, 10.]

IIJ MELTING FURNACES WILL MELT IIJ TON OF URE IN A DAY AND NIGHT OR RATHER XIJ HOWARS VJ WORKMEN WYLL SARVE THESE IIJ FURNACES AND ALL OTHER WORKES THERTO BELONGING.

For vj men men's wagys and meate a daye . .	x^{s}
And for other ydle tyme of work . .	x^{s}
W^{ch} is for iij tons xx^{s} wch is j ton for . .	vj^{s} viij^{d}

[1] Robert, originally written William, but altered.
[2] A small piece of metal fixed to the paper by wax.
[3] Mark of the seal wax only remains.
[4] Part of the wax only left.
[5] Wax only left.

For coles and wood to melt j ton j lode . .		xx^{s}
For leade j ton for j ton ure wch leade wilbe gotten agayne in the last almost iij quarters of it so is lost but j quarter of j ton worthe x^{li} the ton, wch is for j ton of ure	iij^{li}	
For fyar to rost j ton ure . . .		vj^{s} viij^{d}
For a man to attend the same j day . .		xx^{d}
For mixture to melt the ure for j ton . .		x^{s}

Sum £5 5 0 a ton ure charges.

[*Colonial, East Indies*, 67. *Domestic Eliz.*, cxix, No. 12.]

DOCTOR BURCOTS ARTICLES AND CONDITIONS TO SERVE IN FYNING OF THE NORTH-WEST ORE OR ANY OTHER MINERALS.

Right honorable,

We have vewed all the watter mylles neere London and doo fynd the most of them to be tyde mylles wch wyll not sarve to work the ure.

Also we have vewed the Temple myll wch Jonas dothe well lyke for the watter course, but the same hathe very little or no ground wheron to buyld the workhowsses needfull nor no place there for habitation of the workmen and offycers requysyt for the workes.

Also we have seene the mylles at Dartford, whose water course Jonas doth allso lyke well. And consideryng the commoditie of the towne for habitation of the offycers and workmen also the water passage from the Tames to the towne and the good store of fewell in Kent, we thynk that place good fo^{r} the purpose wherof this berar Mr. Furbusher can certyffy yor H. particularlye referring all to the consyderation of yor H. and the lords of Maties honorable privie councell.

Also I have delyvred to Mr. Burkot yor H. letter and theruppon I and Mr. Furbusher have had large talke wth hym, and in the end we fynd hym farre out of reasson, and from that wch he wrote to yor H. as you shall parceave by the writinges herewthall sent of his demandes ; also Jonas is not wyllynge to joyne wth hym, and by our conferens had we doo see that Burcot wold doo in the workes no more but the same wch Jonas would doo and wyll doo and in sum poynttes not so moche nor so well as at yor H. commyng to London you shall more largely understand. The first thing that now is to be done for erection of the workehowsses for the ure is this : to wryte yor H. letters to Mr. Bartye, husband of the Duchesse of Suffolk to send hether Sebastian, a dockeman who now makethe certayne mylleworke for hym at Grymsthorp, wch workman must make the bellowes wheeles and all other tymber-work. Also yor letter to sum fryn[d] to send hether Hendrick the dockeman brykler or mason who is now in work at the glassehows in Sussex at a place called Lokwood, these ij men wth Jonas must presently vew and measure the plat of ground for erection of the mylle and furnaces and ordeyne for the plat of the work and for the stuffe to work w^th^all and buyld w^th^all. Also uppon yor H. resolution what place you think most meete to erect the workehowsses. The ownar therof must be agreed w^th^all presentlye for the same before we can begynne the workes of buyldynges. All other matters appertaynynge to the premysses may staye untyll yor H. come to London.

And thus I commytt yor honor to Almighty God.

Ffrom London the xiij of December, 1577.

Yo^r^ honors most bounden

Michael Lok.

To the right honorable Sr Francis Walsingham, knight, one of her Maties principall Secretaries

At the Court.

[Inclosure I.]

The 9th December, 1577.

Mr. Doctor Burcot shall doo as folowethe :—

1. He shalbe chyef master of the workes of provynge and meltinge the ures here at home yerelye and in his owne parson shall see and ordayne and command the same.

2. He shall delyver halfe an once at the least of fyne gold for every hundred weight of the ure, free and clere of all charges of ffyer and additions for the melting and mens labour for the workyng and all other charges except the charges of buyldyng and instrumenttes or workyng tooles.

And this shall he parforme or ells shall loose his pencion of cc^li^ and all other intertaynement.

Mr. Doctor Burcot shall have as folowethe :—

1. A pencion of cc^li^ yerely during his lyffe.

2. And xx^s^ day for his dyat when he or his deputye workethe.

3. And a better reward when the mynes prove bettar and l^li^ beforehand besydes his pencion. And this shalbe parformed to him by sufficient bondes.

[Inclosure II.]

Artycles off Burkard Kraynghe off the meltine and ffyninge of that ooyre that ys brought into this land and that wch here after shall come.

Inprimis that he will be a master teacher and instructer of Inglyshemen how they shall melte this presente blacke ooyre or any that comythe here after to puryffie and fyne yt and bringe yt to parffyte gold.

Item he will also have suche men as he will chuse and apounte they shalbe bound to the hole fellowshipe and unto him not to departe frome this busynes w^th^out the masters lycense and good will havinge ther wages reasonablye appounted unto them.

Item he will also erecte and buyld a meltene house wth vj fornaces axiltres, fyninge ovene vj pare off bellous wth all other instrumentes apperteyninge to suche a house of his owne device and knowlage profytable and mete for suche meltine at the fellowes cost and charges.

Item he will have too hundred pound ayeare duringe his naturall lyffe quarterlȳ to be payd and one hole quarter[1] in hand, and the next pament at oure Lady-day next followinge, and xx^s a day for his charges holy day and workie day as ofte as he ys in and aboute that busynes and yf yt fortune him to be charged wth bodylie syckenes and be not able to travile in the same arte and be present himselfe that he may have a sufficiente man ther in his place in the meane tyme and the same accountes and the xx^s to be payd monthly.

Item he will also have by that same meltine house sufficiente rostine house, coyle house, wth plentye of wood and coile.

Item the same Burkard hath takine upon him wth his affore appounted workemen and melters to bring out of the blacke oorye that ys present alredye in this lande halfe an ounce of a hundred weight gold and besydes that yt shall beare reasonable charges so that he may have the same ooyre cleanely delyvered unto him w^thout earthe drosse or stones havinge wood and coile wth workemen at ye quenes pryce.

Item will gyve a note what maner of bellowes and other instrumentes nessessary appertayninge to the same mayd here in London and carryed to suche a place as the Mr. and fellowes thinke mete to be buylden.

Item, he will instructe and teache to make proves and sayes to one man that will go suche a vioage agayne to bringe over treasure and ryches to pay for all and leave suche pooer and wyld ooryes behind yf ther be suche ryches in the land.

Item he ys also content to travill his old body in the fellowshipes cost and charge to vew se and fynd out in this land a place for buyldine suche a house bothe mete and profytable for the beste cheape of meltine and bringine in of the oorye.

Item, he will also make sayes of this ooryes that is in this land advertyce the comyssioners of the ryches of the same of his owne cost and charge, and in his owne house and showe and teache how yt shall be brought oute in the greate fyer because he hathe his pension for ye same.

Item, he will also have tow notable men in the fellowshipe that shalbe bound unto him in a pare of indentures and he to them for the hole fellowshipe wch one of them shalbe appounted to pay him at altymes for him and his men ther wages an his pension and xx^s a day.

Item, yf ther shall here after any more suche ooyre come into this land wch shall beare the charges and be more profytable then thys ys that where he hathe now xx^s a day then he shall have xl^s a day.

[1] "hole quarter" erased, and altered to ffyfti pound."

Item, that yf he do not performe the afforesayd artyckles then he shall losse his pittane and therto I have sette my hand.

Item, he will not have that his pension nor his xx^{s} shalbe accounted in the charges of the meltyng because yt is neyther for labourer nor workmenes wages.

Item, will also have that alwayes ther shall remane a peace of mony in the masters handes before hand in the buyldine and meltine to pay his men in dew season and he shall make acounte every sennet or xiiij dayes at ye leaste and send yt to him that payes the men to make his booke wth a trew accountes what is spent and payd.

Item, the M^{r} will also instructe and teache one of his secret and bounden sarvantes and prentyce durynge his lyffe as he hathe partely alredy done that yf yt happene that the same M^{r} dothe deseace or dye that the same his mane shall knowe suche secretes and mystories wch every worke man and laborrer ought not to knowe so that his service may be followed in his desseaces and after his deathe and to be joyned now wth him in patent.

[*Colonial* 113. *Domestic Eliz.*, cxxix, No. 2.]

JANUARY 2, 1578. FROM MR. EDWARD FENTON, WHAT SUCCESSE HE HATHE HAD IN TRAVELING TO GET OWRE IN THE WEST COUNTRIE.

My dutie to yor hono^{r} most humblie used. Makinge my L. of Bedforde acquainted wth her Maties commission and service I had in hande from yor ho: he presentlye directed his favorable letters unto Mr. Edgcombe (whose skill and indginete for that purpose and service his L. thought most sufficiente) to whom I repaired accordinglie. And making him acquainted therwth I desired his speedie good help and furtherance in the same and sheifest to be furnished of that oure or minerall (Mr. Burcott) affirmed to yor honor to have receved of him and gotten in his growndes wch he assured me by great othes was not true: for the same oare .... delivered unto (Burcott) by one of his bretheren who receavid the same of another man wch died longe time sithence, and where he had the same he knowes not neither can it be learned of any other. So that at my firste entraunce into the service I was voyde of that hoope and helpe I cheiflie exspected at his handes for the presente supplie of the same. Wherfore seinge the uncertentie of his help and that he sayde he had procured some other sortes of oare but not readie for me: I furthwth repaired into Cornewall to see what fruites I coulde reape, and fonde owt for that purpose by myne owne travaill: And coming amongest the mynes there (Christmas being at hand) and the myners being departed from their labours. Onlie in thende haping to one (Mr. Cosworth) receavo^{r} of her Maties revenew there, wth whom using some conferrence receavid bothe greate courtesie for my self and furtheraunce for the presente service I had in hande: he travailed with me into sondrie places and to divers gentlemen of that shier at whose handes and by whose meanes I was cheiflie to be holpen wth such mineralls as I serched for viz., Mr. Goodolphin, Mr. Arundell, and others wth whom after I had used some conferrence and given them some instructions towchinge thaction furthwth dispatched their letters to their servaunts best acquainted wth those cawses to make presente serch for all oares and mineralls remayninge in their workes from whom I have receavid such sortes of oare as I have sente to London (to Mr. Looke) putt in severall bagges marked wth figures accordinge to a kalendar herwith inclosed to yor honor.

But the oare (Mr. Burcott) had wherof Mr. Edgcombe delivered me a peice, I showed to divers tynners and others of skill in mineralls, but they never saw any suche in Cornwall or other places of their workinge.

Greator speede I could not make by reason thunfittnes of time as absence of all workmen from their workes, neither a greator quantitie of oare wch sorte will best serve the purpose it is gotten for, wch I coulde not do having no skill therin my self muche lesse here acquainted wth any that could do the same. And therfore thought it not good to entre into any further charges therin till I receaved yor honors further pleasurs and certificatt wch sorte or sortes therof will best aggree wth thaction it is provided for, wch I will most dutifullie and readelie followe accordinge to suche orders as yor honors shall direct me for the same. Humblie beseching yor ho: to direct yor favorable letters of thanks to (Mr. Coswarth) for the greate courtesie he hath shewed me in this service craving pardon for my boldness I beseche God to blesse yor honors with good success in all yor actions. Ffrom Mount Edgcombe the ij^{de} of Januarie, 1578.

Yo^{r} honors most humblie to commaunde,

Edward Fenton.

To the right honorable the Lords and others of her Maties most honorable Privie Counsaill.

haste.

[*Colonial*, 113. *State Papers. Domestic*, *Elizabeth*, Vol. 129, 2, i.]

THE KALLENDER OF SUCHE SORTS OF OARE AS I HAVE SENT IN SOUNDRIE BAGGS, VIZ.:

The first sort or kynd being liek copper called myndick growethe in St. Awstell Clives 3 milles from the haven of Foye.

There is liek to be good stoare therof.

2. The second sort comonly called by the tynners calle, there is great stoare and dyvers kynds therof growing in St. Tew and other places 3 milles from the sea syde: and from the haven of Foye vij milles.

3. The third sort lyke unto tynne or lead,[1] groweth in St. Awstell in the severall grownd of Hughe Collyns of Tregonie, ij milles from the see and vj from Foye.

4. The iiijth sort growethe in the parishe of Piryn in the grownd of (Mr. John Nance) and was one of the mynes (Mr. Burcot) wrought for silver: he gave to the honnor yerely v oz. of silver, it lyethe wthin 2 milles of New Kaie a littell harbor now dekayed, the work standethe xxij fethomes deape of water and the loade therof a foate broade.

5. The fift sort was gottin by me and Mr. Coswarthe in a silver work of Bircotts, at New Kaie, hard by the see side and in the parishe of St Collom (the lower, the loade scant a foat broade), I fownd also in a howse hard by the same, certayn slage wch he used to melt downe the same oare wth all, of what substaunce or from whence it came, I could not learne; it is amongest the oore in this bagge.

6. The vjth contayneth 4 sorts of oore received from Mr. Barnard Penrose dwelling nigh Helston.

7. The vijth sort was gotten in the parishe of St Tannesse, her Mats ;land, hard uppon the see side, the loade not above a handfull broad.

8. The viijth bagge contaynethe 7 sorts of oare wth their loads. Received of Mr. Edgcombe.

Fower sorts of oare in 4 severall baggs, marked wth the letter M., from Mr. Michell, of Trewroo.

Indorsed. The sortes of mynerالls received from C. Fenton, from Cornwall, the 8 Januarie, 1578.

[*Colonial*, 131. *State Papers. Domestic Eliz.*, Vol. 129, No. 43.]

THE XVIJ DAYE OE FEBROWARY IN ANO 1578, OF Xc OF ORE MELTYD AT DARTFORDE. A COWNT MADE OF Xc OF ORE MELTYD WCH CAME OUT OF THE JUDETH, AND 13c OF ORE WCH CAME OUT OF THE NORTHE, AND 3c OF LECTAGE WCH CAME FROME TOWER HIGHTT—26c IN ALL.

Where of came iijc ¼ of ryche leade, and that beyinge fynde downe there came viij oz. of selver, lackynge ijd weyght, where of beynge partyd, came of gowlde one q 3 qr oz. and xviij grains.

Where of came out of the leade ore and the lytarge, wch was xvijc ½ oz. ½ q 3, wch is x oz.

[1] In the toppe of this bagge you shall fynd ij peces of oare yello coller gotten at New Kaie.

Then meltyd the lytarge wth the slags where out is come ij^c^ of leade, wch ij^c^ of leade howldeth v oz.

All so there dothe remayne in stone iij^c^ ½, wch howldyth all v oz.

There remayns iij^c^ of lead at 30^s^

Where of all is xviij oz. of selver wth gowlde.

The gowlde wch is there in is ½ oz. 40 grains, wch is 35^s^ in valew.

There remayns xvij oz. j qr. iij^d^ weyght, ½ of selver, where of we take out x oz. for the xvj^c^ ore and lectarge. Reste in selver of owre owne ore 7 oz. j qr. 3^d^ weyt ½.

(On dors.)

Howe mych the x^c^ dothe make.

Furste, in sylver 17 oz. j qr. 3½^d^ weyte, at . .	4	7	0
Then the gowlde ½ oz. 40 grains, at . . .	1	15	0
Then 3^s^ lead lefte, at	1	10	0
Where of abate for x oz. wch came oute of the ore and lettarge of the northe	2	10	0
The reste clyer, wch is come out of ow^r^ ½ tunne of ow^r^ .	5	2	0

[*Colonial*, 134. *Domestic, Eliz.*, cxxx, No. 15.]

A NOTE OF THE VALUE OF 200 W^EIT^ OF OARE GOTTEN IN THE COUNTESS OF WARWICKS ILANDE IN (META INCOGNITA) AND PUTT DOWNE BY ME, JONAS SHUTE, AT THE TOWER HILL, THE XXIIIJ^TH^ DAIE OF MARCHE, 1578, AND PUTT OF ON III SEUERALL TESTES CONTEYINGE GOLD AND SILVER, AS FOLLOW^TH^, VIZ.

The prooffe of the first test.

The first prooffe waighed in gould and silver, vnrefyned . . .	11 oz. 4 pennye w^eit^ 16 graines.
Being refyned, in gould and silver .	11 oz. 1 penny w^eit^ and 11 gr.
In gould, beinge parted . .	20 graines and 3 quarters.

The prooffe of the second test.

The seconde waighed in gould and silver, vnrefyned . . .	1 oz. 3 qrters and 14 gr.
Being refyned, in gould and silver .	1 oz. 7 penny w^eit^ 14 grs.
In gould, being parted . .	1 penny w^eit^ 4 gr. 3 qrters.

The prooffe of the third test.

The thirde waighed in gould and silver, vnrefyned . . .	10 oz. 14 penny w^eit^ 18 gr.
Wherof there is a litle sample kept of the same for a sutle prooffe, if need require.	

Being refyned in gould . .	1 oz. 13 penny w^{eit}	
In gould, being parted . .	1 penny w^{eit} 8 gr. 1 qrter.	

The quantetie of gould and silver refyned in the iii tests.

The whole weight of the gould refyned amounteth to . . .	3 penny w^{eit} 10 gr. d.	x^{s} iii^{d}	xxxv^{s}
The whole w^{eit} of the silver refyned cometh to . . .	4 oz. 19 penny w^{eit} 3 grs. d.	xxv^{s} vii^{d}	

The quantetie and rate of thaditamets use in thies prooffes.

In litarg 400 w^{eit} held in silver . . .	2 oz. d.
In leade 56 pownds w^{eit} held in silver . .	1 qrter. of an oz.

All wch cometh to xiiii^{s} iiij^{d}, w^{ch} (I knowe) remayinth yet it the litarg and leade, and so will allowe for the same.

So that after this rate it cometh in the toone towards all chargs	xvii^{li} xviij^{s} ix^{d}
Wherof, I the said Jonas descireth allowance for waste	lvii^{s} ix^{d}
And so I, the said Jonas Shute, promisseth to make of euyre ton towards all chargs . . .	xv^{li}

STATE PAPERS RELATIVE TO THE OUTFIT FOR THE THIRD VOYAGE.

STATE PAPERS RELATIVE TO THE OUTFIT FOR THE THIRD VOYAGE.

[*Colonial*, 88. *Domestic Eliz.*, cxxiv, No. 1.]

A PROPORTION OF THE CHARGES FOR A THYRD VIAGE TO THE NORTHWEST TO FITCH 2000 TOONES OF OORE AND TO VITTAL AND KEEPE THERE 100 MEN 18 MONETHES.

Twoo thowsande toones of oure to be brought home at xxx^{li} le toone amounteth to . . .	lm^{l} *li*
Wherof	
I demaunde to furnishe the *Ayde* and *Gabriell* in presente and readie monie . . .	m^{li}
More for the wages of 80 men for v monethes at xxvj*s.* viij*d.* le monthe le man to be paid at per reatorne	dxxxiij^{li} vj$_{s}$ viij^{d}
Two shipps to be procured more of her Matie, viz., one of 400 toones and thother of 200 toones throughlie furnished wth tackle and munition wch maie amounte to	mmmdc^{li}
Marioners to saile the same ij shipps 150, at xxvj*s.* viiij*d.* le monthe le man in prest . .	cc^{li}
More for vittelling of thies 150 sailors at xx*s.* le monthe le man for vij monthes. . .	m^{l}*li*
More in prest for 120 pyoners to be convoyde in thies shipps for ij monthes wages le man at xx*s.* le monthe	ccxl^{li}
More for the vittelling of thies 120 pyoneers at xx*s.* le man le monthe for vij monthes . .	dccc,xl^{li}
More for iiij monthes wages for the pyoners to be paide at their reatorne	cccciij^{xx}*li*
More for v monthes wages to be paide 150 marioners at their reatorne	m^{li}
More for weapon and armo^{r} for thies 120 men .	cxx^{li}
More for soldio^{rs} and pyoners being 250 to be bestowed in shipps to be fraighted at xxvj*s.* viij*d.* le man le monthe for ij monthes in preste .	dclxvj^{li} xiij^{s} iiij^{d}
More for the vittellinge of thies 250 men for vij monthes at xx*s.* le man for every month . .	mdccl^{li}

P

More x halls or tentes for their harbor^{r} . . ccxl^{li}
More for armo^{r} and weapon for theis 250 men at xx*s*. le man . . . . . ccl^{li}
More for yronworke for tooles for the same pyoners and for viij smithes, their fourdges and bellowes . c^{li}
Ffor powder for their defence one laste . . c^{li}
More to be paide in wages at their reatorne for iiij monthes . . . . mmdclxvj^{li} xiij^{s} iiij^{d}
More for the fraight of 1200 toones at c*s*. le toone . vj^{mt}*li*.

Sum of all the charges to be disboursede as appereth by this particular. . xx^{m}d,ccc,xxxvj^{li} xiij^{s} iiij^{d}
And so remains cleare . . xxixmclxiij^{li} vj^{s} viij^{d}

M^{d} that there is in readie monie to be disboursed for the fetching of theis 2000 toones but . vj^{mt}dlxvj^{li} xiij^{s} iiij^{d}
Besides the ij shipps of her Matie wch maie come to . mmmdc*li*

A proportion for 100 men for victuall and wages to inhabit the Northwest.

More for the vittelling of 100 men to remayne there at xx^{li} le man for the yere and the proporcion to aunswere xviij monthes . . . mmm^{li}
More for their wages at xx^{s} le monthe le man . mdccc^{li} } mmmmdccc^{li}.

The Comoditie to be gayned by them.

Thies 100 men being laborers shall gett in this xviij monthes towardes their charges 2000 toones of oare wch shall yelde xx^{li} le toone cleare amounting to the some of . . . . xl^{m}*li*.

M^{d} that to fortefie and provide dwelling for thies 100 men wth munition for their defence is further to [be] provided and considered of.

[*Colonial* 89. *Domestic Eliz.*, cxxiii, No. 50.]

THE NAMES OF SUCHE GENTLEMEN AND OTHERS AS WENTE THE FIRST AND SECONDE VOYAGES WTH MARTIN FFROBISHER INTO THE LANDS NOW CALLED "META INCOGNITA," LATLIE DISCOVERED BY HIM TO THE NORTHWEST AND NOW IN CONSIDERACION OF THEIR SERVICE TO BE RECEAVID IN AS ADVENTURERS GRATIS, FOR SUCHE SEVERALL SOMES OF MONIES AS FOLLOWTH, VIZ.

The Names of the Gentlemen.

Edwarde Ffenton his lieutenaunte, by lande and sea in those partes. . . . . . . c^{li}
Gilberte Yorke his vice-admirall to go and reatorne wth the fleete l^{li}

George Best	l*li*
Richarde Philpott	l*li*
Henrie Carew	xxv*li*
Edmonde Stafforde	xxv*li*
Fraunces Brakenburie	xxv*li*
John Lee	xxv*li*
William Tanflde	xxv*li*
Edwarde Harvie	xxv*li*
Mathew Kindersley	xxv*li*
Thomas Chamberlaine, Abraham Linche, Dennys Sotle	xxv*li*
Roberte Kindersley, Henrie Kirkman	xxv*li*
Lucke Girido, vice-admirall at *Meta Incognita*	xxv*li*
The Maisters of Shipps and others.	
Christofer Hall, Mr in thadmirall	l*li*
Charles Jackman, Mr of the vice-admirall	xxv*li*
James Beare, Mr of the Reare-admirall	xxv*li*
Andro Dyer, Mr of the shipp that staies in the countrey	xxv*li*
Nicholas Chauncelor havinge been bothe the voyages and to remayne there	xxv*li*
Richarde Coxe Mr gonner of thadmirall	xxv*li*
Nicholas Counzer that tooke the man Thomas Boydell	xxv*li*
James Wallis, hurte and maymed by the countrey people	xxv*li*

[*Colonial*, 91. *Domestic Eliz.*, cxxiii. No. 51.]

INTERTAYNMENT OF GENTLEMEN AND OTHERS IN THE VOYAGE UNDER MR. FENTON, TO INHABITE IN THE NEWLAND META INCOGNITA.

Mr. Captayne Fenton	*li*10	0	0
George Beste	*li*5	0	0
Richard Philpot	*li*5	0	0
Luke Ward	*li*5	0	0
For ij lewtenanttes, eche	*li*2	10	0
For ij enseignes, eche	*li*2	0	0
And all the rest of the gentlemen	*li*1	10	0
And all others, soldyars, marynars, &c.	*li*1	6	8

By the monthe.

[*Colonial* 87. *Domestic Eliz.* cxix. No. 46.]

That Jonas may have l^{c}*li.* pension.

Shippes to be sent for v^{mt} ton weight.

A mynor to dig half ton adaye, in 28 days—a month.

i^{c} mynors for a month to digg . . .	j^{mt} iii^{c} ton.
ii^{c} mynors	ij^{mt} viij^{c}
iii^{c} mynons	iij^{mt} cc ton.
	Wages for ye mynors.
Y^{e} freight at iij^{li} ye ton . . .	ix^{mt} poundes.

Edm. Hogan, Sr Wm Wynter, Humfrey Lock, Rich. Ydys, Furbisher.

Dce.

Palmer to be allowed as an officer.

Wm Umfrey to be used. Humfrey Cole. Burchard to make a prooff of j^{c} weight of ye ure in ye towre.

[*Colonial*, 93. *Conway Papers.*]

INSTRUCTIONES GEVEN TO O^{R} LOVINGE FRIND MARTINE FFROBISER, ESQUIER, FOR THE ORDER TO BE OBSERVED IN HIS VOYAGE NOWE RECOMMENDED TO HIM FOR THE LANDE NOW CALLED BY HIR MA^{TIE} META INCOGNITA TO THE NORTHWEST PARTES AND CATHAYE.

Ffyrst, you shall enter as captain generall into the charge and government of theis shippes and vessells, viz., the *Ayde*, the *Gabriell*, *Michaell*, *Judethe*, the *Thomas Alline*, *Anne Fraunces*, the *Hoppewell*, the *Mone*, the *Ffeaunces of Ffoy*, the *Thomas*.

Item, you shall appoynte for the furnishinge of the *Ayde*, *Gabriell*, *Michaell*, and *Judith*, ffour-skore and ten hable and sufficient marinores and 130 pyoners and 50 soldiars, for the sarvyce and ladinge of all suche shippes and vessells as shall go under yor charge. and be appoynted to retourne againe wth yo^{u} for that purpose, and of the sayd shippes or vessells, and maryners, pyoners, and soldiors, you shall leave to remayne and to inhabite in the lande nowe called *Meta Incognita*, under the charg and government of Edward Ffenton, gent, your Lieutenaunte Generall, the *Gabriell*, the *Michaell*, and the *Judethe*, wch fortie hable marioners, gonners, shipwrights, and carpentars, 30 soldiors and 30 pyoners, wth sufficient vittalle for xviij monthes for their provisione, releife, and mayntenance, and also munition and armoure for their deefence, wch nomber of persones befor specified you shall not exced to carrie nor leve their.

Item, that the vittalls for vij monthes wch you deliver into the *Ayde* for provisone of 90 persones goinge, and to retorne in the said shippe, you shall carefulye see the same preserved and used in sarvyce wth out spoyle or hurte takinge by necligence. Item, you shall make a juste inventorie of every shippe to the companie belonginge of all the takell, munitione, and furnitur, to them belonginge at their settinge fourth from hens and the coppie therof under yor hand to be delivered to Michaell Lok, Treasorer of the Company. And the like to be done at yor retourne home, of all thinges then remaynyng in the said ships. And the like care you and yor Lieutenaunte Generall shall have of the victualls that shalbe by you delivered into any shippes or vessells whatsoever, for the provision of the 100 men appoynted to inhabite their.

Item, you shall not receve under yor charge and government any disordred or mutinous persone wch shall be appointed to goo or remayne their, but upon knowledge had to remove him before you departe hence, or ells by the way assone as you can avoyd hym.

Item, you shall use all dilligence possible to departe, wth yor said ships and vesselles frome the portes where they now remayne, before the firste of May next cominge, and to make your course eather by the northe or the west, as the winde will best serve yo^u.

Item, when you shall passe the landes of England, Scotlande, or Irelande, you shall direct yor course wth all yor shippes and vessells to the lande now called *Meta Incognita*, and to an iland and sounde[1] there called the Countess of Warwickes Iland and Sounde, being w^thin the supposed straight, wch we name Ffrobisers Straight, discovered by yor selfe 2 yeres past, and in yor voyage thither wardes you shall have speciall regarde so to order your course as yor shippes and vesselles do not losse the Companye one of an other, but may kepe company together. And the lyke also in yor retorne homewards. And yf any wilfulnes or negligence in this behalfe shall appeare in any persone or persons that shall have charge of any of the shippes aforesaide, or yf they or any other shall doo otherwyse then to them apperteyneth, you shall punishe suche offendor sharplye to the example of others.

Item, that at yor arryvall at the Countesse of Warwikes Iland and Sounde, you shall theron saffitee harbour yor shipps and vesselles, and frome thence you shall repayre to the mynes and mynerالls of the same iland wher you wrought this laste yeare wth myners and other men and furnyture necessarie, and ther shall place the myners and other men to worke and gather the oare, foreseinge they may be placed as well frome dainger and malyce of the people as frome anye other extremitye that maye happen.

Item, whyles these mynars are workyng in Warwyke Sound, you shall cause serche to be made for other mynes in other places, and yf uppon good proofe made, you shall happen to fynde other mynes to be richer then theis frome whence you had yor laste yeares ladinge, then you shall

[1] *Another hand.* Not for ye Isle of Foyzlin in the wey.

presentlie remove the shippes and myners to the same place of mynerall, and to lade of the same yf that may be done convenientlye.

Item, to searche and consider of an apte place wher you maie best plante and fortefye theise c men wch you shall leave to inhabite there aswell against the dainger and force of the natyve[1] people of ye countrey and any other y^t^ shall seke to arryve ther from any other part of Christendom,[2] as also to prevent and fore see as neare (as you cane) all other extremities and perills that maye happen, and necessaries to be considered of for them.

Item, you shall leave wth Captan Fenton, yor Lieuetenaunte Generall, the government of those 100 persons to remayne in that countrie wth instructions howe he maye best observe the nature of the ayre, and may discover and knowe the state of the countrie from tyme to tyme as moche as may be, and what tyme of the yeare the Straight is most free frome eysse kepyng to ye end a journall wekly of all accountes, wth whome you shalle leve the *Gabriell*, the *Michaell*, and the *Judith*, wth suche proportion of victualls and other necessarie thinges as are alredye appoynted to him and his companye for that purpose suppliing his want wth able and skyllfull men for that purpose, and wth any other thinges necessarie wch you or any other of the shippes maye convenientlie spare at yor reatorne.

Item, we require that you shall instructe all yor people rather to muche then any thinge to littell, aswell for yor owne saffetye there as of suche as you shall leave behinde you, that when you or they shall happen to come to have conference wth the people of those partes wher you shall arive, that in all yor doynges and theirs you so behave yor selves and theyme, towardes the said people as maye rather procure their frindships and good lykings towardes you by courtesyes then move them to any offence or myslikinge.

Item, uppon yor arrivall at the place before specified, and after you have bothe harbored safflie yor ships, sett yor myners one worke, and also have taken sufficient order for plantinge of those men wch shall inhabite ther, and appoyntinge in yor absence governers for all theis causes. We will then, yf leasure and tyme wille permitt the same that you wth the ij barkes shall repaire towardes the place where the first yeare you lost yor men, aswell to searche for mynes there as to discover 50 or 100 leages further westwardes frome that place as ye oppening of ye Streight by water will lowe, as you may be certayne that you are entride into the Southe Sea commonly called Mare di Sun. And in your passage to learne all that you cane in all thinges, and take parfect notes therof, not tarringe longe frome your shippes and workemen, but that you maye be hable to retorne homewardes wth them in due tyme.

[1] [Natyve] added by Lord Burleigh.
[2] [Of Christendom] added by Lord Burleigh.

Item, you shall well consider what place may be most aptest further to fortifye upon hereafter (yf nede requier), bothe for defence of the myners and also for possessinge of the countrie and bringe home wth you a perfecte platt and parfecte notes therof to be kept in secreat, and so delyvred unto us.[1]

Item, you shall not suffer any shippe or shippes beinge laden wth oare to sett sayle or departe from the place of their ladinge till the daye fixed in their charter partye except you see good cause otherwyse. And beinge so laden and redy to retorne homeward you shall reetayne them in flete and in companie all togethers as muche as in you liethe, and as the wether wyll suffer untill your retorne into this realme of England and arrivall at the place appoynted in the River of Thammes for unladinge of the same.

Item, for the succession of the Generall Governour of this whole voiage (yf he should fortune to die) for avoydinge of stryffe and kepinge of peace and fryndship there be the names of iiij gentlemen privatlie sett downe to succeade him in his place[2] on after ye other which ar severally wrytten in paper included in balls of wax sealed wth hyr Maties signett and put into boxes locked wth severall keys wherof on in your custody.[3]

Item, for the better and more circumspecte executions and determinacion in any waightie causes incident on land, we will that you shall call unto you for assistantes your Lieutenaunt Generall, Captayne Yorke, Richard Philpott, George Beast, and Henry Carewe, gent., wth whome you shall consult and confere what is beste to be done in the said causes, matteres, and actions of ymportaunce touchinge this service undertaken. And in all suche matteres so handeled, argued, and debated upon the some to rest, to be allowed, or disallowed at yor owne ellection, and that alwaies to be executed wch you shall thinke meeteste wth assent of any ij of them in general consent.[4] And like wyse in matteres of weight concerninge all yor shippes good government, aswell at the sea as in harboure, o^r wille is that the forenamed gent. and Christofer Hawle, Charles Jackeman, James Beare, and Andrewe Dier, ministers, in certayne of o^r shippes, presentlie ymployed in this north-west service, shalbe assistaunte unto you and consentinge to all determynacones concernynge the same. And in casse that of suche conference and descoursinge the opiniones of the aforesaid assistaunce be founde in effecte any waye to differ then o^r will is that thexecution of all suche matteres

[1] After us, "here to the Treasorer of the Companye" written and expuncted.

[2] [On custody], Lord Burleigh's hand; also the note.

[3] Three keys, Furbisher, Fenton, a m^r of a shipp. *See last paragraph but two.*

[4] [Wh to consent] also written in margin by Lord Burleigh.

so argued upon shall rest to be put in execution in suche sorte as you shall thinke moste metest, having the assent of any ij of them.[1]

Item, because the temprature of those northe-west partes and boundes of seas and landes are not yet sufficientlie knowne (wch thinge we principallye desyere), and for as much as verye good opertunitie in soundrie respectes maye falle out in tyme of yor absence to purchaze or attayne to the same, we thinke yt verye necessarie and to your better desert worthelie apperteninge that you shall enforme, advise, and auctoryshe by yor owne hande writtinge, in the beste manner you cane devise howe anye further descoverye, understandinge, or knowledge of the foresaid landes or seas (confynynge, borderinge, or lyinge, wthin 200 leages of the place wher at this voyage the habitacone or fortification of or people shalbe setled or situated) maye be executed and acheved by yor aforesaid Lieuetenante Generall or by suche other parson as he or the most parte of such as hereafter shalbe named to be his assystance shall deme and judge most apte and sufficient for the accomplishinge of the service their unto apperteyninge.

Item, that you shall have speciall care and geve generall warninge that no persone of what cawlinge soever he be shall make an assaye of any mannor of mettalle matter or oore on the foresaid partes of *Meta Incognita*, but onlie he or they to whome the offyce or feate of assayes makinge is asigned or comitted (onlie yor selfe, yor Leutenaunte Generall, and yor substitutes before named, from this article to be excepted), nor any persone under yor government shall take uppe or keape to him selfe and his private use anye parte or parcell of oare, precious stone, or other matter of comoditie, to be hade or founde in that lande but he, the said person so seazed of suche oare, stone, or other matter of commoditie, shall wth all speade or so sone as he cane detecte the same and make deliverey therof to yor selfe or yor Lieutenaunte Generall upon payne to forfite for everye ounce therof the valewe trible of any wages he is to receave after the daye of suche offence committed, and further to receave suche punishement as to hir Matie[2] shall seme good.

Item, or will is that you shall cause a recorde dilligentlye to be kept in wryttyng of all suche oare, myneralls, stones and other matters of vallew gotten or founde in that countrie, aswell of the time and place and places when or whear all and everye suche oare, minerall and other matter of suche vallewe is or shalbe founde or gotten, as also some parte, portion or example of all and everye the said oares, myneralls and other matter of vallewe in apte and peculiar boxes cause to be reserved wth theire due titles and notificacones. And further cause dulye to be layed uppe in the said boxes the severall rates and tryed valuacions of all assayes ther made of any the foresaid oares and myneralles,

[1] [Havinge them], Lord Burleigh's hand.
[2] [Hir Matie] altered to us, but altered back again by Lord Burleigh.

and all those foresaid boxes so furnished and distinctlie noted at yor reatorne to the citie of London you shall deliver or cause to be delivered to the tresorer of the companye of adventurers for those northeweste affayres, as well for the better directione and dealinge heare after wth any the foresaid oares or mynerallles ther as for the better and speedie account and reckinge, makinge in grosse heare at home of the valewe of suche quantitie or masse as any of them shall hether be brought. And of these doinges make two bookes, to be kept in ij severall shyps.

A dooble of this book to be made, and brought home in an other shipp.

Item, that the marioners of all the hired shippes imployed in this service shall geve, joyntlye wth all the other companies of or owne shipps, iij or iiij dayes travall and labor towardes thintrenchinge and fortifiinge of the place, wher the leutenante generall wth his charge shall remayne to inhabite there.

Item, that you shall make yor directe course from hence as neare as you cane, wth all suche shippes as passe under yor government, to the land now called *Meta Incognita*, and their lade 800 toones, or so muche more as the shippes of retorne cane saffie carrie of suche oare as you alredie have founde ther this last yeare, or rather richer yf you cane fynd the same. And so havinge laden your shippes wth the said nomber of 800 tonnes or more, as is aforesaid, shall make yor direct course frome thence into this realme of England into the river of Thames, where the shippes be appoynted to be unladen of the same.

A book conteyninge ye quantitie laden in every shipp. [Lord Burleigh.]

Item, that everye capten and mr of every shippe appoynted in this voyage shall joyntlie under their handes writinge by indenture deliver unto you a note and estimacone of suche nomber of toones of oare or other matter of vallew as they shall receve into their shippes theire. And all the same indentures to be registred in one booke, wherof iij copies to be made, and to be put in iij severall shypes to be delyvred to the tresorer of the compe at retorne home of the shypps.

Ye book to be indented (Lord Burleigh.)

That a minister or twoo do go in this jorney to use ministration of devyne service and sacraments, accordyng to ye churche of England. (Lord Burleigh.)
Nota, yt the victalls, munitions and other thynges to be carryed to be equally distributed into ye shippes, for dout of miscarrying of some of theme.

Item, in yor waye outward bound, yf it wylbe no hynderans to the rest of yor voyage, you shall doo yor endevour to dyskover the new land, supposed to be Ffryzeland, and to gett the best knowledge that you can of the state and nature therof. And yf you cannot conveniently doo it in yor waye outward bound, then doo your attempt h.... in yor waye homeward bound at retorne yf the same may be done convenientlye.

Item, when you shall passe, etc.

Item, that yf there should happen any person or persons ymployed in this service, of what calling or condition he or they shall be, should conspire or attempte privatlie or publiklie any treason, mutanie or other desorder, either towchinge the takinge awaie of yor owne life or any other of aucthoritie under you, whereby her Maties service in this voyage might therby be over throwen and ympugned, We will therfore that upon juste prooffe made of any such treasons, mutanie or other desorders attempted as aforesaid, the same shalbe punished by you or yor lieutenant generall, etc.,[1]

wch are severally wrytten in paper included in bawles of wax, sealed wth her Maties signet, and put into two severall boxes, locked wth iij severall keys, wherof one key in yor custodie, and one in custodie of Edward Fenton, and another in custodie of Christofer Hawlle. And the same two boxes to be put in ij severall shyps, to saye, one boxe in the *Ayde*, and the other in the ship where yor lieutenaunt generall shall passe.

Item, for the succession of the lieutenant generall of those c men wch shall remayne and inhabite there, there be named iij parsons to succede in order and maner as is sett downe before in the Article for the succession of the generall.

Item, that there be made a doble of this Commyssion to remayne wth the lieutenant generall.

Indorsed. 1578. Commyssion instructions to Mr. Ffurbusher to goo to sea, No. 1578.

[*Colonial*, 127. *State Papers. Domestic Eliz.*, Vol. 129, No. 36.]

THE INVENTARIE OF THE SHYP AYDE.

(2d page.)

The Inventarie of the shippe *Ayde* made the 10th of Februarye, 1571.

In primis her furniture as she was bought of the Quenes Matie in Aprille, 1577.

In primis, the bowsprite wth ij double pullies and iij shevers of bras.

The Bolt Sprite.—Item the yarde ; the saylle (worne) : the hallyares wth ij pullies cocked wth brasse ; the lifts wth iiij pulles ; the braces, wth 2 pulles, ; the sheate, wth pendannts ; the ij shankes paynters wth chaynes ; a boult, a collar and chaynes of irone ; the mayne staye ; the davette wth a claspe of irone ; ij shevers of brasse to the davetts a grapnell wth chayne (lost) Catts a false tyre for the spritte saile; the clewlynes.

The Fore-maste.—Item the mast wth a shiver of brase in the heade ; the fore topp not ; the yard wth gromets ; a swifter one aside wth iiij pullyes, worne ; the saylle viz., corse and bonnet, iij parts worne ; ij pendants on a syde wth iiij pulles, one shevered, and one cocked ; ij takels one a syde wth iiij pules iij coked wth brasse ; vj shroudes on a

[1] There is *nothing* lost here. The person who drew up this draft has re-written this passage to make it more clear.

syde ; the staye ; the lyftes wth iiij pullies ; the tye, worne ; the hallyares wth one shever of brase in the rames head and ij cocked in the rame hedd ; the parell wth lanyers and brest ropes ; ij trusses wth ij pulles ; ij bowlines (worne), wth a doble blocke and ij shyvers of bras ; the braces wth iiij pulles (worne) ; the sheats wth ij pulles cocked wth bras ; the shivers of brasse in the shippes side (none) ; the tacks (one of them newe) ; the martenetts ; the botts tacle wth iij shyvers of bras.

The fore tope maste.—Item, the toppe maste wth a cocke of brasse in the heade ; the yarde ; the saylle (iij parts worne) ; j tacle on a side wth iiij pulles ; iiij shroudes on a side ; iiij puttocks on a syde ; the staye and backstaye ; the tye and halliers wth ij pullies one shevered and one cocked wth brasse ; the liftes wth iiij pulles ; the sheates ; the parell, broken, laniers and brest ropes ; the truse wth ij pulles ; the boulines wth one doble (polle) ; the braces wth iiij polles ; the clulines wth ij pulles ; j crane line, bage and one pendante pulle.

The mayne maste.—Item, the maste wth ij shivers of bras in the heade (the mayne topp nawght) ; the yard wth grometts and stapells (broken and nawght) ; the saylle, viz., corse and bonnette (good) ; the drabler (newe) ; j swifter on a side wth iiij pulles (iij parts worne) ; iij pendants one a side wth vj pulles on a shever of brasse, and ij shevers of brasse for the botts tacle (the tackles worne) ; iij tackells on a side wth xij pulles, iij cocked wth brasse ; viij shroudes on a side ; the staye ; the liftes wth iiij pulles : the sheates supplied wth ij pulles, one shyvered wth bras, and th' other cocked with bras (the sheates worne), and ij shivers of brasse in the shippes side ; the tacks ; the tye (halph worne) ; ij shevers of brasse in the knight ; the halliers wth iij shevers of brasse, in the knight and ram heade ; the parrell wth laniers and brest ropes ; the trusse wth iiij pules (nowght) ; the martinetts (worne), and vj pulles ; the garnette wth ij pulles wth iij shevers of brasse ; the braces wth ij pulles ; the bowlines ; the clulines.

The mayne tope maste.—Item the toppe and maste wth a shever of brasse in the heade ; the yarde ; the saylle (newe) ; j tacle one a side wth iij pulles ; 4 shroudes one a side ; v puttockes one a side ; the staye and the backe staye ; the liftes wth iiij pulles ; the sheatts wth iiij pulles ij shevers, one of them brasse and ij cocked wth brasse, and ij of brasse in the bubbridge heade (none of brasse) ; the tye and halliers wth ij pulles one shevered and one cocked wth brasse ; the bowlines wth one doble pulle ; the braces wth iiij pulles ; the clulines wth ij pulles ; the the culine a rane bagge and one pendante pulle.

The myson maste.—Item the maste, wth a shevere of brasse in the hedd ; the yarde ; the saylle, viz., corse and bonet, nawght ; a swifter on a side wth iiij pulles, the swifter's nawght ; v shrouds one a side ; the staye ; the tye and halliers wth a shyver of bras, and brest ropes ; the trusse wth ij pulles ; the lyfts wth ij pulles ; the boulines—non ; the smitinge line—non ; the parell ; the myzon martinetts.

The mison tope maste.—Item the tope and maste; iij shroudes on a side; iiij puttocks on a side; the staye.

The botes maste.—Item, a shever of brasse in the heade; a paynter cheyne; a davett wth a shever of irone; a windlesse; a maste wth a sayll; a rother wth spindell and capps—(lost).

The skyffe.—Item, a skyffe; xij ores; a rother wth yrone worke—(none).

Implements.—Item, a mayne capstaine wth collor and paull of yrone and iiij bares; paule non nor bars; the fore capstene wth a paull of yrone and 2 barres; a state pompe wth a bracke; a bed sted and a table in the captaines cabbine, the table broken; a payre of bilbowes wth vj shakells; a grinstone wth spindle and winche of irone; a coper kettell; ij meate kettells, one very smalle; a barre wth a chayne and iij hockes in the cooke rome to hange the kettell one; but iij ankers, ankers great, iiij; cables of xij ynches that the shipe is mored by, ij; cables of xj ynches—iij, ij of them newe, one of the ij of 13 inches, one of them a juncke and cut; cables of x ynches, j halfe worne; cables of viij ynches for a botte rope, j halfe worn; cables of vj ynches, j newe, spent and gone; geste ropes of v ynches, j halfe worne; condinge hausers of v ynches, j; hausers of v ynches, j; fats (fathoms ?) of a hauser of v ynches, x fathom; a britton tackell wth iiiij shevers of brasse and one of irone coked wth iij blocks and j pendante tacle, j; boye ropes, j; catte ropes, ij worne; facks of coylle of iij ynches, x fathem; peces of coyles of ynches and ynches and halfe, iij peces; flaggs of Sainte George, j worne: compasses, ij; runnynge glasses, j nawght; soundinge lynes, ij; soundinge leades, iij, ij; bucketts, ij; boules, iiij; shovelles, iiij; skoppes, ij; spare pulles great and small, vj, ij coked wth brasse; marlienes, ij bundells; ratline, shyves; twine, x^{li}; item, boults of middremaxe, iiij; calappes, v; piche pottes, j nawght; fishe hokes, ij; leache hokes, ij, j; loffe hokes, iiij; balleste basketts, ij; canne hokes, j pare; fides, ij; boyes, iiij, iij; catte hokes shevered wth brasse, ij.

Summa of all, wch coste . . . . viij^{c} l^{li}

And the ordenans and munition aperinge hereafter, wch coste . . . . . iij^{c} xlv^{li}

Summa of all this shipp as ytt cost, amounteth . xj^{c} iiij^{xx} xv^{li}

We doe thincke that the foresaid ship, wth her masts, yards, sayles, anckers, cables, and other taikle and apparell conteined in particulers before sett downe in this book, so as the saume may be delyvered accordinglie, to be worthe . . . . vij^{c} ^{li}

Item, more for v peces of brasse in this book after specyfied amonge the ordenance and munitions, beinge ij mynions, and iij^{s} fawcons, weyinge iiij^{cwt} v^{c} xviij^{li} waight, at iij^{li} p^{r} c^{t}, cxxxv^{li} x^{s} And more for v cariadges perteineinge to the saide peces p^{r} estimation, iij^{li} vj^{s} viij^{d} . . . . cxxxviij^{li} xvj^{s} viij^{d}

Summa totalis . . viij^{c} xxxviij^{li} xvj^{s} viij^{d}

(Signed)

W. Wynter. Willm Holstok.

The rest of th' ordenance and munitions, in this inventorye we thinck them nott mete, for the Quenes Matie.

(The above letter is crossed off in the original.)

The xxiij^{d} of Februarie, 1578.

We doe thinck y^{t} the foresaid ship, wth her masts, yards, sailes, anckers, cables, and other taikle and apparell contened in particulars, before sett downe in this book, so as the same may be delivered accordinglie, to be worthe . . . . dcc^{li}

We doe also thinck y^{t} the v peces of brasse in this book afte specefied, amonge the ordenance, to be worthe the monye they are rated at, and mete for her Matie wth ther v cariadges, pertaining to them, wch dothe amount unto the some of . . . cxxxij^{li} ij^{s} xj^{d}

Totalis . . dcccxxxij^{li} ij^{s} xj^{d}

And as touchinge th' other ordenance, and munitions conteyned in this inventorye, we doe not thinck them mete for her highness.

(Signed)

W. Wynter. Willm Holstok.

More the ordenances and munition put into the shippe, after she was brought wch dide cost as followeth :—

Ordenans of brasse

Mynyones, ij wainge 22$^{cwt.}$ 2$^{qrs.}$ 4$^{lb.}$ at 3^{li} per cwt. .	£67	13	4
Fawcons, i wainge 7$^{cwt.}$ 2$^{qrs.}$ 14$^{lb.}$ at 3^{li} per cwt. .	£22	17	6
Faucons, ij wainge 15$^{cwt.}$ at 46^{s} 8^{d} per cwt. . .	£35	0	0
And for the carriages of all 5 peces . . .	£6	13	4
	£132	4	2

Of cast yrone.

Sacres, viij wainge . . .	5^{ton}	12$^{cwt.}$			
Mynyons, j wainge . . .		11$^{cwt.}$			
Fawcons, v wainge . . .	2^{ton}	2$^{cwt.}$			
Summa . .	8^{ton}	5$^{cwt.}$			
At £12 a ton . . .			£99	0	0
And for 14 carriages, all			17	13	4
Fowles, vj . . } Chambers to them, xij . } stoked at *li*5 pece . .			30	0	0

Munition, as followeth :—

	cwt.	qrs.	lbs.
Sacre shot, round, ij^{c}vij . . .	10	1	0
Fawcone shote, rounde, lxix . .	1	2	4

Colveringe shot, rounde, xix	1	2	4			
Mynion shot, rounde, xvij	0	2	14			
Summe waing	13	3	18			
At 10 shillings the cwt.				*li*6	18	0
Crosbar shotte.						
	cwt.	qrs.	lbs.			
For sacres, xlix	3	1	0			
For fawcone, xxvij	1	1	0			
For mynione, xj	0	2	14			
Summe	5	0	14			
At xxiij^{s} per cwt.				*li*5	18	0
Chayne shotte.						
	cwt.	qrs.	lbs.			
For sacres, 14	1	0	0			
For fawcone, 7	0	1	0			
Summe	1	1	0			
At xxiij^{s}				*li*1	8	6
Stone shote.						
For fowlers, liij, at xij^{d}				*li*1	13	0
Ladells wth staves for sacres and mynion, 15, at xij^{d}				*li*0	15	0
Sponges and staves for sacre, mynion, and faucon, 12, at xij^{d}				*li*0	12	0
Rammer staves, 20, at 8^{d}				*li*0	13	4
Formers for sacre, mynion, and fawcone, 3, at 6^{d}				*li*0	1	6
Armor, and weapon, and munitione.						
Calivers, 38, wherof 6 wthout stoks.						
Flaskes, 16; Toche boxes, 10; Moldes, 20 } at 13^{sh} 4^{d}				*li*25	6	8
Matche skines, weyinge cc$^{lbs.}$ at 16 shillings the c				*li*1	12	0
Bowes of ewe, 25, at 3^{sh} 8^{d}				*li*4	11	8
Shefes of arowes, xlv, at 2^{sh}				*li*4	10	0
Bow stringes, dossen, vij, at 8^{d}				*li*0	4	8
Partezans, iiij, at 13^{sh} 4^{d}				*li*2	13	4
Blacke bylles, xvj, at xij^{d}				*li*0	16	0
Pykes, 5, at 2^{sh}				*li*0	10	0
Crowes of yrone, 9, at 4^{sh}				*li*1	16	0
Trunkes of wylde fyer, ij, at 5^{sh}				*li*0	10	0
Balles, wilde fyer, 15, at 3^{sh}				*li*2	5	0
Arowes, wilde fyer, 11, at 1^{sh}				*li*0	11	0
Pykes, wilde fyer, 5, at 5^{sh}				*li*1	5	0
A drylle, j, at				*li*0	1	0
Tampyons, 29, at 1^{d} pece				*li*0	2	6

A gowge, j, at	*li*0	0	6
Chyssells, iiij, at 6d	*li*0	2	0
Peckers for stone shot, j, at	*li*0	0	6
A sledge, j, at	*li*0	2	0
Spare trockells, ix, at 12d a pare	*li*0	4	0
Summe this syde	*li*49	5	8
The last syde	*li*295	15	0
Summe of all this, which coste	*li*345	0	8

[*Colonial*, 133. *Domestic Eliz.*, cxxx, No. 10.]

MARCH 20TH, 1578. FROM MR. THO. ALLAN. THE "GABRIELL" PRICED AT LXXX£. MANYE THYNGS IN LOCKES HANDES TO BE SOULD AND TO BEE CALLED TO HIS REARE ACCOMPT.

My dewtie remembred: hit maye please youre honor to undarstande that I have receved ij letters this daye from youre honor, the one towchinge the ordenance to be solde, the other for the *Gabryell*, wch letters I ansure. The ordenance ys solde by the bryngar hereof to one Clement Draper for xij*li* the towne, redye monye, as he saythe to me, he havinge a lysence to transeporte the same, wich lysence yt maye please your honor to sende, and then the monye shalbe receved by me, and pade owte agayne to these men. Havinge some asystance wth me of the comyssioners at the payment there of wiche I desire to have ffor my dyscharge acordinge to ordar.

Ande for the *Gabryell*, she was prased by Mr. Locke and others at one hundrethe and li pounds. I sawe yt to moche. I browght hire downe to c*li*; yet no mane wyll by here at that pryce, so I have offered hire iiiixx*li*, and that I do here that Mr. Furbusher haythe byden for hire; but I thingke redye monye ys owte of the waye wth hyme, so I sent Clynton to knowe whether he wolde have hire or no, or else I wolde yt myght please youre honor that Sr Nycolas Malbe maye have hire, and paye vs this monye I thingke well of yt.

Sr, I wolde these men weare pade; I ame sore trobled wth them; youre honor sende them to me they saye, and youre honor knowythe I have no monye to paye them. I have had iij fytes of an agoo; Gode sende me to be quyte of yt. This is the gayne I do posese by folloyng of this besynes.

Sr, there ys manye things to sell yett, wiche do lye and are in Mr. Lockes kepinge, and there ys no mane taketh charge or care of them, and what he can sell he doythe, but paythe no mane a penye. It weare very goode that youre honor wolde commande that all things myght be solde owte of hande by a daie, and that Mr. Locke myght be dyscharged, and that he myght then bryngke in his reare acounte, and so to deale wth hym, for that he doythe owe to the companye, that men myght be pade, and that youre honor myght certenly knowe what ys yet owynge to men, and agayn what ys owinge to vs to dyscharge them, for yt doythe lyngar to longe for oure proffyt. This I take my leave of youre honor, wryten this xx[ti] of Marche, 1578.

Your honars to comavnde,

Thomas Allen.

To the Right Honorable S[r] Frances Walsingham,
Knyght and prensepall Secretorye to the
Quenes Ma[tie].
20 March, 1578.

[*Colonial*, 95. *Domestic Eliz.*, cxxiv, No. 2.]

THESE HAVE NOT PAYD THE 3 MAY 1578.

	Stok		Buildinges.
My Lord Admirall	£135		£20
My Lord Tresorer	£35		
My Lord Chamberlan	£135		£20
My Lord Leycester	£202	10	£30
My Lady Warrwyk	£32	10	
Mr. Secretarie Walsingham	£62	13	
My Lady Anne Talbot	£38	15	£5
S[r] John Brockett	£43	15	
S[r] William Wyntar	£250		£40
S[r] Leonell Duckett	£67	10	
Mr. William Pellham	£67	10	£10
Mr. Thomas Randolphe	£67	10	
Mr. Edward Dyar	£33	15	£5
Mr. Somers			£10
Mr. Coyar	£33	15	£5
Anthony Jenkynson	£67	10	£10
Jeffrey Turvyle	£67	10	£10
William Paintar	£67	10	£10
Richard Cowland	£67	10	£10
Mathew Fild	£32	10	
Thomas Allyn	£67	10	£10
Robert Martin	£33	15	£5
Christofer Androwes	£33	15	£5
S[r] Thomas Gresham	£70		£40
Martin Furbusher	£67	10	£10
My Lord Camberlan, Whaterton	£67	10	£10
	£1876		£260
Thomas Owen	£33	15	£5

THE THIRDE VOYAGE OF CAPTAINE FROBISHER, PRETENDED FOR THE DISCOVERIE OF CATAYA, BY META INCOGNITA. ANNO DO. 1578.*

THE Generall being returned from the second voyage, immediatlye after hys arrival in Englande repayred with all haste to the Court, being then at Windsore, to advertise hir Majestie of his prosperous proceeding, and good successe in this laste voyage, and of the plenty of gold ore, with other matters of importance which he hadde in these Septentrionall partes discovered. He was courteously enterteyned, and hartily welcomed of many noble men, but especially for his great adventure commended of hir Majestie, at whose hands he receyved great thanks, and most gratious countenance, according to his deserts. Hir Highnesse also greatly commended the rest of the gentlemen in this service, for their great forwardnes in this so dangerous toyling and painefull attempte: but especiallye she praysed and rejoiced, that among them there was so good order of governement, so good agreement, everye man so ready in his calling, to do whatsoever the Generall should commande, which due commendation gratiously of hir Majestie remembred, gave so greate encouragement to all the captaines and gentlemen, that they, to continue hir highnesse so good and honorable opinion of them, have since neither spared laboure, limme, nor life, to bring this matter (so well begon) to a happie and prosperous ende. And finding, that the matter of the gold ore had appearance and made shew of great riches and profite, and the hope of ye passage to Cataya, by this last voyage

Frobisher commended of hir Majestie.

The gentl men commended.

* Another account of this voyage was written by Thomas Ellis.

Commissioners appointed to examine the goodness of the ore.

greatly encreased, hir Majestie appointed speciall commissioners, chosen for this purpose, gentlemen of great judgement, art, and skill, to looke thorowly into ye cause, for ye true trial and due examination therof, and for the full handling of al matters thereunto appertaining. And bicause that place and countrey, hathe never heretofore bin discovered, and therefore had no speciall name, by which it might be called and known, hir Majestie named it very properly *Meta Incognita,* as a mark and bounds utterly hitherto unknown. The commissioners after sufficient triall and prooffe made of ye ore, and having understood by sundrie reasons, and substanciall grounds, the possibilitie and likelihoode of ye passage, advertised hir highnesse, that the cause was of importance, and ye voyage gretly worthy to be advanced again. Whereupon preparation was made of ships and al other things necessary, with such expedition, as ye time of the year then required. And bycause it was assuredly made accompt of, that the commoditie of mines, there already discovered, would at ye least countervaile in all respects, the adventurers charge, and give further hope and likelihood of greter matters to follow: it was thought needful, both for the better guard of those parts already found, and for further discovery of the inland and secreats of those countries, and also for further search of ye passage to Cataya (wherof the hope continually more and more encreaseth) that certain numbers of chosen soldiers and discreete men for those purposes should be assigned to inhabite there. Whereupon there was a strong forte or house of timber, artificially framed, and cunningly devised by a notable learned man here at home, in ships to be carryed thither, wherby those men that were appointed there to winter and make their abode ye whole yeare, might as wel be defended from the danger of ye falling snow and colde ayre, as also to be fortified from the force or offence of those countrie people, which perhaps otherwise with too greate companyes and multitudes

A name given to the place newe discovered.

The hope of the passage to Cataya.

A forte to be built in Meta Incognita.

might oppresse them. And to this greate adventure and notable exploit, many well minded and forward yong gentlemen of our countrey willingly have offered themselves. And first Captaine Fenton, Lieutenant Generall for Captaine Frobisher, and in charge of the company with him there, Captaine Beste, and Captaine Filpot, unto whose good discretions the government of that service was chiefly commended, who, as men not regarding perill in respect of the profite and common wealth of their countrie, were willing to abide the firste brunte and adventure of those daungers among a savage and brutishe kinde of people, in a place hitherto ever thought for extreme cold not habitable. The whole number of men whiche had offered, and were appointed to inhabite Meta Incognita al the yeare, were one hundreth persons, wherof xl shoulde be marriners, for the use of ships, 30 miners for gatheringe the golde ore togyther for the nexte yeare, and 30 souldiers for the better guarde of the reste, within which last number are included the gentlemen, goldfiners, bakers, carpenters and all necessarye persons. To eche of ye captaines was assigned one ship, as well for the further searching of the coast and countrie there, as for to returne and bring backe their companies againe, if the necessitie of the place so urged, or by miscarying of the fleete in the yeare following, they mighte be disappointed of their further provision.

A hundreth men appointed to inhabit there.

Being therefore thus furnished with all necessaries, there were ready to depart upon the said voyage xv sayle of good shippes, whereof the whole number was to returne agayne with their loadinge of gold ore in the end of the sommer, except those three ships, which should be left for the use of those captaynes whiche should inhabite there the whole yeare. And being in so good readynesse, the Generall, with all the captaynes came to the court, then lying at Greenewich, to take their leave of hir Majestie, at whose hands they all receyved greate encouragemente and gracious coun-

Fifteene sayle.

tenance. Hir Highnesse, besides other good giftes, and greater promises, bestowed on the Generall a faire cheyne of gold, and the rest of the captaines kissed hir hande, tooke their leave, and departed every man towardes their charge.

A cheyne of gold given to Frobisher.

THE NAMES OF THE SHIPPES, WITH THEIR SEVERALL CAPTAYNES.

1. In the *Ayde* being Admirall was the Generall - -	Captayne Frobisher.
2. In the *Tho. Allen,* Viceadmirall	Ca. Yorke.
3. In the *Judith,* Lieutenant Generall	Ca. Fenton.
4. In the *Anne Frances* - -	Ca. Best.
5. In the *Hopewell* - -	Ca. Carew.
6. In the *Beare* - - -	Ca. Filpot.
7. In the *Thomas,* of Ipswich -	Ca. Tanfield.
8. In the *Emanuell,* of Exceter -	Ca. Courtney.
9. In the *Frances,* of Foy -	Ca. Moyles.
10. In the *Moone* - - -	Ca. Upcot.
11. In the *Ema,* of Bridgewater -	Ca. Newton
12. In the *Salomon,* of Weymouth	Ca. Randal.
13. In the barke *Dennis* - -	Ca. Kendall.
14. In the *Gabriell* - -	Ca. Harvey.
15. In the *Michaell* - -	Ca. Kinnersley.

The sayd xv sayle of shippes arrived and mette togyther at Harwitch, the seauen and twentith day of May, anno 1578, where the Generall and the other Captaynes made view, and mustered theyr companyes. And every severall Captaine receyved from the Generall certayne articles of direction, for the better keeping of order and company togither in the way, which articles are as followeth:—

Articles and orders to be observed for the fleete, set downe by Captayne Frobisher, Generall, and delivered in writing to every Captayne, as well for keeping company as for the course, the 31 of May.

1. *Inprimis,* to banishe swearinge, dice, and card-playing,

and filthy communication, and to serve God twice a day, with the ordinarie service, usuall in churches of England, and to cleare the glasse,* according to the old order of England.

2. The Admirall shall carrie the light, and after his light be once put out, no man to go a head of him, but every man to fitte his sayles to follow as neere as they may, without dangering one another.

3. That no man shall by day or by night depart further from the Admirall than the distance of one English mile and as neere as they may withoute daunger one of another.

4. If it chance to growe thicke and the wind contrarie, eyther by daye or by night, that the Admirall be forced to cast aboute, before hir casting aboute, she shal gyve warning by shooting off a peece, and to him shall answere the Vize-admirall and the Rere-admirall with every one of them a peece, if it be by nighte or in a fogge, and that the Vize-admirall shall aunswere firste and the Rere-admirall last.

5. That no man in the fleete descrying any sayle or sayles, give uppon anye occasion anye chace, before he have spoken with the Admirall.

6. That everye evening all the fleete come uppe and speake with the Admirall at seaven of the clocke, or betweene that and eyght, and if weather will not serve them all to speake with the Admirall, then some shall come to the Vize-admirall, and receyve your order of your course of Maister Hall, chiefe pylot of the fleete, as he shall direct you.

7. If to any man in the fleete there happen any mischance, they shall presently shoote off two peeces by day, and if it be by night two peeces, and shew two lightes.

8. If any man in y[e] fleete come up in y[e] night, and hale his fellow, knowing him not, he shall give him this watch-

* In Earl Essex's expedition to Cadiz, Dr. Marbeck records that "to inculcate discipline and subordination, and to impress the sacredness of their cause, the Lord Admiral had service performed three times a day, in the morning, in the evening, and at bed-time, at the clearing of the glasse.

In Blanckley's *Naval Expositor*, 1750, under "Glasses Watch," will be found "Being fourhours governs them at sea for changing the watch."

word, Before the world was God. The other shall aunswere him, if he be one of our fleete, After God, came Christ, his Sonne. So that if anye be founde amongst us, not of oure owne company, he that firste descryeth anye suche sayle or sayles shall give warning to the Admirall by himselfe or any other that he can speake to that sailes better than he, being neerest unto him.

9. That every ship in the fleete in the time of fogges, whiche continually happen with little winds and most parte calmes, shall keepe a reasonable noyse with trumpet, drumme, or otherwise to keepe themselves cleere one of another.

10. If it fall out thicke or misty that we lay it to hull, the Admirall shall give warning by a peece, and putting out three lightes one over another, to the ende that every man may take in his sayles, and at his setting of sayles agayne do the like, if it be not cleare.

11. If any man discover land by nighte, that he give the like warning that he doth for mischances, two lightes and two peeces, if it be by day one peece, and putte out hys flagge and strike all his sayles he hath aboorde.

12. If any shyppe shall happen to lose company by force of weather, then anye suche shippe or shippes shall gette hir into the latitude of and so keep that latitude, untyll they gette Freeselande. And after they be past the west partes of Freeselande, they shall gette them into the latitude of and and not to the northwarde of and beeing once entred within the straytes, all suche shippes shall everye watche shoote off a good peece, and looke out well for smoke and fire, whych those that gette in first shall make every night, untill all the fleete bee come togither.

13. That uppon the sighte of an ensigne in the mast of the Admirall, a peece shotte of the whole fleete shall repaire to the Admirall, to understande such conference as the Generall is to have with them.

14. If we chance to meet with any enemies, that foure shippes shall attend uppon the Admirall, viz.—the *Frances*

of Foy, the *Moone*, the barke *Dennis*, and the *Gabriell*: and foure upon my Lieutenant Generall in the *Judith*, viz. —the *Hopewell*, the *Armenall*, the *Beare*, and the *Salomon*: and the other foure upon the Vize-admirall—the *Anne Frances*, the *Thomas of Ipswich*, the *Emanuell*, and the *Michaell*.

15. If there happen any disordered person in the fleete, that he be taken and kept in safe custodie until he may conveniently be brought aboorde the Admirall, and there to receive such punishment as his or their offences shal deserve.

By me, MARTINE FROBISHER.

OURE DEPARTURE FROM ENGLAND.

Having received these articles of direction, we departed from Harwich the one and thirtith of May; and, sayling alongst the south partes of England westward, wee at length came by the coast of Ireland, at Cape Cleare, the sixth of June, and gave chace there to a small barke, which was supposed to be a pyrat or rover on the seas; but it fell out in deede that they were poore men of Bristowe, who hadde mette with suche company of Frenchmen as had spoyled and slayne manye of them, and left the rest so sore wounded that they were lyke to perishe in the sea, havyng neyther hande nor foote hole, to helpe themselves withall, nor victuals to susteyne theyr hungrie bodyes. Oure Generall, who well understandeth the office of a souldioure and an Englishman, and knoweth well what the necessity of the sea meaneth, pitying much ye miserie of the poore men, releived them with surgerie and salves, to heale their hurtes, and with meate and drinke to comfort their pining hartes. Some of them having neither eate nor drinke more than olives and stinking water in many days before (as they reported). And after this good deed done, havinge a large winde, we kept our course uppon our sayde voyage withoute staying for the taking in of freshe water or any other pro-

A charitable deede.

vision, whereof many of the fleete were not thoroughly furnished (and sayling towardes the north-west partes from Ireland, we mette with a great currante from oute of the south-west, which carryed us (by our reckoning) one point to the north-estwardes of our said course, which currant seemed to us to continue itselfe towards Norway and other the north-east partes of the world, whereby we may be induced to believe that this is the same whiche the Portugalles meete at Capo d'Buona Speranza, where, striking over from thence to the Straytes of Magellanes, and finding no passage there for the narrownesse of the sayde Straytes, runneth alongst into the greate Bay of Mexico, where, also having a let of lande it is forced to strike backe agayne towardes the north-east, as we not only heere, but in another place also, further to the northwardes, by good experience this yeare have founde, as shall be heereafter in his place more at large declared.

Marke thys curiant.

Nowe had wee sayled aboute foureteene dayes withoute sight of any land or any other living thing, except certayne fowles, as wylmots, nodies, gulles, etc., whiche there seeme only to live by sea.

Weast England.

The twentith of June, at two of the clocke in the morning, the Generall descryed land and found it to be Weast Freeselande, now named Weast England. Heere the Generall and other gentlemen wente ashoare, being the fyrste knowen Christians that we have true notice of, that ever set foote upon that ground; and therefore the Generall toke possession thereof to the use of our Soveraigne Lady the Queenes Majestie, and discovered heere a goodly harborough for the shippes, where were also certaine little boats of that countrey. And being there landed, they espied certayne tents and people of that countrey which were (as they judge) in all sortes, very like those of Meta Incognita, as by theyr apparell and other things whych we found in theyr tentes appeared.

The savage and simple people, so soone as they perceyved

our men comming towardes them (supposing there had bin no other worlde but theirs), fledde fearfully away, as men muche amazed at so strange a sight, and creatures of humane shape, so farre in apparell, complexion, and other things differente from themselves. They left in their tents all their furniture for haste behinde them, where, amongst other things, were founde a boxe of small nayles and certaine redde hearings, boordes of firre tree well cutte, with dyvers other things artificially wroughte, whereby it appeareth that they have trade with some civill people, or else are in deede themselves artificiall workemen.

Oure menne broughte awaye wyth them onelye two of theyr dogges, leaving in recompence belles, looking glasses, and dyvers of oure countrey toyes behynde them.

This countrie no doubte promiseth good hope of great commoditie and riches, if it maye be well discovered. The description whereof you shall finde more at large in my seconde booke, page 5.

Some are of opinion that this Weast Englande is firme land with the north-east partes of Meta Incognita, or else with Gronelande. And their reason is, bicause the people, apparell, boates, and other things are so like to theirs; and another reason is, the multitude of islands of ise whyche lay betweene it and Meta Incognita, doeth argue, that on the north side there is a bay, whych cannot be but by conjoining of these two landes togither.

And havinge a fayre and large winde, wee departed from thence towardes Frobyshers Straites, the three and twentith of June. But fyrste we gave name to a hyghe cliffe in Weast England, the laste that was in oure sight, and for a certaine similitude we called it Charinge Crosse. Then we bare southerly towardes the sea, bycause to the northwardes of this coaste wee mette wyth muche driving ise, whiche by reason of the thicke mistes and weather might have bin some trouble unto us.

Charing Crosse.

On Monday, the laste of June, wee mette with manye greate whales, as they hadde beene porposes.

This same day the *Salamander* being under both hir corses and bonets, hapned to strike a greate whale with hir full stemme, wyth such a blow, that the ship stoode stil and stirred neither forwarde nor backward. The whale thereat made a great and ugly noise, and caste up his body and tayle, and so went under water, and within two dayes after there was founde a greate whale dead, swimming above water, which we supposed was that the *Salamander* stroke.

A whale stroke a ship.

The seconde daye of July, early in the morning, wee had sight of the Queenes Forelande, and bare in with the lande all the daye, and passing thorow great quantitie of ise by nighte, were entered somewhat within the straites, percieving no waye to passe further in, the whole place being frosen over from the one side to the other, and as it were with many walles, mountaines, and bulwarkes of yse, choaked uppe the passage, and denied us entrance. And yet do I not thinke that this passage or the sea hereaboutes, is frosen over at any time of the yeare; albeit it seemed so unto us by the abundance of ise gathered together, whyche occupyed the whole place. But I do rather suppose these ise to be bredde in the hollowe soundes and freshets thereaboutes, whyche, by the heate of the sommers sunne beeing loosed, doe emptie themselves wyth the ebbes into the sea, and so gather in great abundance there togither.

Frobishers straites choked up with ice.

And to speake somewhat here of the auncient opinion of the frosen sea in these partes, I doe thinke it to be rather a bare conjecture of menne, than that ever anye manne hathe made experience of anye such sea. And that whiche they speake of Mare Glaciale may be truly thought to be spoken of these parts; for this maye well be called indeede the Ysie Sea, but not the Frosen Sea, for no sea consisting of salte water can be frosen, as I have more at large herein shewed my opinion in my seconde booke, page 6; for it seemeth im-

Salte water cannot freese.

possible for any sea to be frozen which hath his course of ebbing and flowing, especiallye in those places where the tides doe ebbe and flowe above tenne fadome. And also all these aforesaide ise, which we sometime met a hundreth mile from lande, being gathered out of the salt sea, are in taste fresh, and being dissolved, become sweet and holesome water.

The cause why thys yeare we have beene more combred with ise (than at other times before) may be by reason of the easterly and southerly windes, whyche brought us more timely thither now than we looked for. Whiche blowing from the sea directlye uppon the place of our straites, hath kept in the ise, and not suffered them to be caryed out by the ebbe to the maine sea, where they woulde in more shorte time have been dissolved. And all these fleeting ise are not onelye so daungerous, in that they winde and gather so neare togither that a man may passe sometimes tenne or twelve myles as it were uppon one firme ilande of ise, but also for that they open and shutte togither againe in suche sorte wyth the tydes and sea-gate, that whilest one shippe followeth the other wyth full sayles, the ise whyche was open unto the foremoste will joyne and close togyther before the latter can come to followe the fyrste, whereby manye tymes oure shippes were broughte into greate danger as beeing not able so sodainely to take in oure sayles, or staye the swifte way of oure shippes.

We were forced manye tymes to stemme and strike great rockes of ise, and so as it were make way through mightye mountaines, by which means some of the fleete, where they founde the yse to open, entred in, and passed so farre within the daunger thereof, with continuall desire to recover their post, that it was the greatest wonder of the world that they ever escaped safe, or were ever heard of againe. For even at this present, we missed two of the fleete, that is, the *Judith*, wherein was the Lieutenant Generall Captaine Fenton, and the *Michaell*, whome both we supposed hadde bene utterlye lost, having not heard any tydings of them in moe than twentie dayes afore.

Bark *Dennys* sunk.

And one of our fleete named the barke *Dennys*, being of an hundreth tunne burden, seeking way in amongst these ise, received such a blowe with a rocke of ise, that she sunke downe therewith, in the sighte of the whole fleete. Howbeit, having signified hir daunger by shooting of a peece of great ordinaunce, newe succour of other shippes came so readily unto them, that the men were al saved with boates.

Part of the house lost.

Within this shippe that was drowned there was parcel of our house, whiche was to be erected for them that shoulde staye all the winter in Meta Incognita.

Thys was a more fearefull spectacle for the fleete to beholde, for that the outragious storme, whiche presentlye followed, threatened them the like fortune and daunger. For the fleete being thus compassed (as aforesayde) on every side with ise, having left muche behynde them, through which they had passed, and finding more before them, through which it was not possible to passe, there arose a sodaine and terrible tempest at the southeast, which blowing from the mayne sea directlye upon the place of the straytes, brought togither all the yse aseaborde of us upon our backs, and thereby debarde us of turninge backe to recover sea roome againe: so that being thus compassed with danger on every side, sundrie men with sundrie devises sought the best way to save themselves. Some of the ships, where they could find a place more cleare of ise, and get a little berth of sea roome, did take in their sayles, and there lay adrift. Other some fastened and mored ancker uppon a great iland of ise, and roade under the lee thereof, supposing to be better garded thereby from the outrageous windes and the daunger of the lesser fleeting ise. And againe some were so fast shut up and compassed in amongst an infinite number of great countreys and ilands of ise, that they were fayne to submit themselves and their ships to the mercie of the unmercifull ise, and strengthened the sides of their ships with junckes of cables, beds, masts, planckes, and such like, whiche being hanged overboord, on the sides of their shippes, mighte the better defende them from the outrageous sway and strokes of the said ise. But as in greatest distresse, men of best value are best to be discerned, so it is greatly worthy commendation and noting with what invincible mind every captayne encouraged his company, and with what incredible labour the paynefull mariners and poore miners (unacquainted with suche extremities), to the everlasting renoune of our nation, dyd overcome the brunt of these so greate and extreame daungers; for some, even without boorde uppon the ise, and some within boorde, uppon the sides of their shippes, having poles, pikes, peeces of timber, and ores in their hands, stoode almost day and night, withoute anye rest, bearing off the force, and breaking the sway of the ise, with suche incredible payne and perill that it was wonderfull to behold, which otherwise no doubt had striken quite through and through the sides of their shippes, notwithstanding our former provision; for plancks of timber, of more than three ynches thick, and other things of greater force and bignesse, by the surging of the sea and billow, with the ise were shevered and cutte in sunder at the sides of oure shippes, that it will seeme more than credible to be reported of. And yet (that which is more) it is faythfully and playnely to be proved, and that by many substantiall witnesses, that our shippes, even those of greatest burdens, with the meeting of contrary waves of the sea, were heaved up betweene ilandes of ise a foote welneere out of the sea above their watermarke, having their knees and timbers within boorde both bowed and broken therewith.

And amidst these extremes, whilest some laboured for defence of the shippes and sought to save their bodyes, othersome of more mylder spirit soughte to save the soule by devoute prayer and mediation to the Almightie, thinking indeede by no other meanes possible than by a divine miracle to have their deliverance; so that there was none that were eyther ydle or not well occupied, and hee that helde himselfe in best securitie had (God knoweth) but only bare hope remayning for his best safetie.

Thus all the gallant fleete and miserable men, without hope of ever getting forth agayne, distressed with these extremities, remayned heere all the whole night and parte of the next day, excepting foure shippes, that is, the *Anne Frances*, the *Moone*, the *Frances of Foy*, and the *Gabriell*, which being somewhat a seaboorde of the fleete, and beeing fast ships, by a winde, having a more scope of cleere, tryed it out all the time of the storme under sayle, beeyng hardly able to beare a coast of each.

And albeit, by reason of the fleeting ise, whych were dispersed heere almost the whole sea over, they were broughte manye times to the extreamest poynte of perill, mountaynes of ise tenne thousande tymes scaping them scarce one ynch, whiche to have stricken, had bin theyr presente destruction, considering the swifte course and way of the shippes, and the unwildynesse of them to stay and turne as a man would wish. Yet they esteemed it their better safetie, with such perill to seeke searoome, than without hope of ever getting libertie, to lie striving against y[e] streame, and beating amongst the isie mountaines, whose hugenesse, and monstrous greatnesse was suche, that no man woulde credite, but such as to their paynes sawe and felt it. And these foure shippes by the next day at noone, gote out to sea, and were fyrste cleere of the ise, who nowe enjoying theyr own libertie, beganne anew to sorrowe and feare for their fellowes safeties. And devoutely kneeling aboute theyr mayne mast, gave unto God humble thanks, not only for themselves, but besought him lykewise highly for theyr friends deliverance. And even nowe, whilest amiddest these extremities, thys gallant fleete and valiant men were altogither over laboured, and forewatched, with the long and fearefull continuance of the foresayde dangers, it pleased God with his eyes of mercie to looke downe from heaven, to sende them help in good time, giving them the next daye a more favourable wind at the west northwest, whiche did not only disperse and drive forthe the ise before them, but also gave them libertie of more scope and searoome, and were by night of the daye following perceyved of the other foure shippes, where to their greatest comfort they enjoyed agayne the fellowship of one another. Some in mending the sides of theyr shippes, some in setting up their toppe mastes, and mending theyr sayles and tacklings. Agayne, some complayning of theyr false stemme borne away, some in stopping their leakes, some in recounting their dangers past, spent no small time and labour, that I dare well avouche, there were never men more daungerously distressed, nor more mercifully by God's Providence delivered. And heere of both the torn shippes, and the forweeryed bodyes of the men arrived, doe beare most evidente marke and witness. And now the whole fleete plyed off to seaward, resolving there to abide, untill the sunne might consume (or the force of wind disperse) these ise from the place of theyr passage: and beeing a good berth off the shore, they took in their sayles, and lay adrift.

The seaventh of July, as men nothing yet dismayed, we cast about towards the inward, and had sighte of lande, which rose in forme like the northerlande of the straytes, which some of the fleete, and those not the worst marriners, judged to be the north forlande: howbeit, other some were of contrary opinion. But the matter was not well to be discerned, by reason of the thicke fogge, whiche a long time hung uppon the coast, and the newe falling snowe which yearely altereth the shape of the land, and taketh away oftentimes the marriners markes. And by reason of the

Another assault.

Fog, snow, and mists hinder the mariners marks.

darke mists, whiche continued by the space of twenty dayes togither, this doubt grew the greater and the longer perillous. For wheras indeede we thought our selves to be upon the northeast side of Frobishers straytes, we were now carried to the southwestwards of the Queenes forlande, and being deceyved by a swift currant comming from the northeast, were brought to the southwestwards of our sayd course, many miles more than we dyd thinke possible could come to passe. The cause whereof we have since found, and shall be at large hereafter declared.

Here we made a poynt of land, which some mistooke for a place in the straytes, called Mount Warwicke: but howe we shoulde be so farre shotte up so suddaynely within the sayde straytes, the expertest mariners began to marvell, thinking it a thing impossible, that they coulde be so farre overtaken in their accompts, or that any currant coulde so deceyve them heere, whiche they had not by former experience proved and found out. Howbeit, many confessed, that they founde a swifter course of floud than before time they had observed. And truly it was wonderfull to heare and see the rushling and noyse that the tydes do make in thys place, with so violente a force that our shippes lying a hull, were turned sometimes rounde aboute even in a momente, after the manner of a whirlpool, and the noyse of the streame no lesse to be hearde a farre off, than the waterfall of London Bridge.

A currant.

But whilest the fleete lay thus doubtfull amongst greate store of ise in a place they knewe not, withoute sighte of sunne, whereby to take the height, and so to know the true elevation of the pole, and withoute any cleare of lighte to make perfite the coast, the Generall with the captaynes and maysters of his shippes beganne doubtfully to question of the matter, and sent his pinnesse aboorde to heare eache mans opinion, and specially of James Beare, mayster of the *Anne Frances*, who was knowen to be a sufficient and skilfull

James Beare a good mariner.

mariner, and having bin there the yeare before, had well observed the place, and drawne out cardes of the coast. But the rather this matter grew the more doubtful, for that Christopher Hall, chiefe pylot of the voyage, delivered a plaine and publike opinion in the hearinge of the whole fleete, that he had never seene the foresayd coast before, and that he could not make it for any place of Frobishers straites, as some of the fleete supposed, and yet the lands do lye and trend so like, that the best mariners therin may be deceived.

Christopher Hall chief pylot.

The tenth of July, the weather still continuing thicke and darke, some of the shippes in the fogge loste sighte of the Admirall and the rest of the fleete, and, wandering too and fro with doubtful opinion whether it were best to seeke backe againe to seaward through great store of ise, or to follow on a doubtfull course in a sea, bay, or straytes, they knew not, or alongst a coast, whereof by reason of the darke mists they could not discern the daungers, if by chance any rock or broken ground should lye of the place as commonly in these partes it doth.

The Vize-admirall, Captayne Yorke, considering the foresayd opinion of the pylot, Hall, who was with him in the *Thomas Allen*, having lost sight of the fleete, turned back to sea agayne, having two other shippes in company with him.

Also the Captaine of the *Anne Fraunces* having likewise lost companye of the fleete, and being all alone, helde it for best to turne it out to sea agayne, untyll they mighte have cleere weather to take the sunnes altitude, and with incredible payne and perill got out of the doubtfull place into the open sea agayne, being so narrowly distressed by the way by meanes of continuall fogge and ise, that they were many times ready to leape upon the iland of ise to avoyde the present daunger, and so hopyng to prolong life awhile, meante rather to dye a pining death.

Some hoped to save themselves on chestes, and some determined to tye the hatches of the shippes fast togyther and to bynde themselves wyth theyr furniture fast thereunto, and so to be towed with the shipboat ashore, whyche otherwise could not receyve halfe of the companye; by whiche means, if happilie they hadde arrived, they shoulde eyther have perished for lacke of foode to eate, or else shoulde themselves have bene eaten of those ravenous, bloudye, and man-eating people.

Hard shifts to save mens lives.

The rest of the fleete following the course of the Generall, whyche ledde them the way, passed up above 60 leagues within the sayd doubtfull and supposed straytes, havyng alwayes a fayre continente uppon their starreboorde syde, and a continuance still of an open sea before them.

The Generall, albeit with the fyrste perchance he found out the error, and that this was not the old straytes, yet he persuaded the fleete alwayes that they were in theyr righte course and knowne straytes. Howbeit, I suppose he rather dissembled his opinion therein than otherwyse, meaning by that policie (being hymself ledde with an honorable desire of further discoverie) to enduce ye fleete to follow him to see a further proofe of that place. And, as some of the company reported, he hath since confessed that, if it had not bin for the charge and care he had of ye fleete and fraughted shippes, he both would and could have gone through to the south sea, called Mare del Sur, and dissolved the long doubt of the passage which we seeke to find to the rich countrey of Cataya.

Mistaken straits.

Frobisher could have passed to Cataya.

1. Of which mistaken straytes, considering the circumstance, we have great cause to confirme our opinion to like and hope well of the passage in this place. For the foresaide bay or sea the further we sayled therein the wyder we found it, with great likelyhoode of endlesse continuance. And wherein other places we were muche troubled wyth ise, as in the entrance of the same, so after we had sayled 50

Fair open way.

or 60 leagues therein, we had no let of ise or other thing at all, as in other places we found. Reasons prove a passage here.

2. Also this place seemeth to have a marvellous greate indrafte, and draweth unto it most of the drift yse and other things which do fleete in the sea, eyther to the north or eastwards of the same, as by good experience we have founde. Great indrafts.

3. For heere also we mette with boordes, lathes, and divers other things driving in the sea, which was of the wracke of the shippe called the barke *Dennys*, which perished amongst the ise, as beforesaid, being lost at the first attempt of the entrance overthwart the Queens Foreland, in the mouth of Frobishers Straits, whiche coulde by no means have bin so brought thither neyther by winde nor tide, being lost so many leagues off, if by force of the sayde currant the same had not bin violently brought. For if the same hadde bin brought thither by the tyde of flodde, looke how farre in the said flodde had caried it, the ebbe woulde have recaryed it as farre backe agayne, and by the winde it could not so come to passe, bycause it was then sometime calme, and most times contrary. Currant.

And some marriners doe affyrme that they have diligently observed y[t] there runneth in this place nine houres flodde to three ebbe, which may thus come to passe by force of the saide currant: for whereas the sea in most places of the world doth more or lesse ordinarily ebbe and flow once every twelve houres, with sixe houres ebbe and sixe houres floud, so also would it doe there, were it not for the violence of this hastning currant, which forceth the floud to make appearance to beginne before his ordinary time one houre and a halfe, and also to continue longer than his natural course by an other houre and a halfe, until the force of the ebbe be so greate that it will no longer be resisted (according to the saying: *Naturam expellas furca licet tamen vsq. recurrit.* Although nature and natural courses be forced Nine hours floude to nine hours ebbe.

and resisted never so muche, yet at laste it will have their owne sway againe).

Moreover, it is not possible that so great course of flouds and currant, so highe swelling tides with continuance of so deepe waters, can be digested here without unburdening themselves into some open sea beyonde this place, which argueth the more likelihood of the passage to be hereabouts. Also we suppose these great indrafts do growe and are made by the reverberation and reflection of that same currant, whiche at oure comming by Irelande mette and crossed us, of which in the firste parte of this discourse I spake, whyche comming from the bay of Mexico, passing by, and washing the south weast parts of Ireland, reboundeth over to the northest parts of the world, as Norway, Islande, etc., where, not finding any passage to an open sea, but rather is there encreased by a new accesse, and another currant meeting with it from ye Scythian Sea, passing the bay of Saint Nicholas westwarde, doeth once againe rebounde backe by the coasts of Groenland, and from thence uppon Frobishers straites being to the southwestwardes of the same.

The sea moveth from east to west continually.

5. And if that principle of philosophy be true, that *Inferiora corpora regunter à superioribus,* that is, if inferior bodies be governed, ruled and caried after the maner and course of the superiors, then the water being an inferior element, must needes be governed after the superior Heaven, and so to follow the course of *Primum mobile* from east to weast.

Authoritie.

6. But everye man that hathe written or considered anye thing of this passage, hath more doubted the retourne by the same waye, by reason of a greate downefall of water, whyche they imagine to be thereabouts (which we also by experience partly find) than anye mistruste they have of the same passage at all. For we find (as it were) a great downfall in this place, but yet not muche, but that we may return, although with suche adoe. For we were easilyer caried in in

one houre than we coulde gette forth againe in three. Also by an other experience at another time we founde thys currant to deceive us in this sort:—That, whereas we supposed to bee 15 leagues off, and lying a hull, we were brought within 2 leagues of the shoare, contrarie to al expectation. Hard, but yet possible, turning back again.

Oure menne that sayled furthest in the same mistaken straites (having the maine lande uppon their starbord side), affyrme that they mette with the outlet or passage of water whiche commeth throwe Frobyshers straites, and followeth as all one into this passage.

Some of our companye also affyrme that they had sighte of a continent upon their larbord side, being 60 leagues within the supposed straites: howbeit excepte certaine ilandes in the entraunce hereof, we could make no parte perfect thereof. All the foresaid tract of land seemeth to be more fruitful and better stored of grasse. Deere, wilde foule, as partridges, larkes, seamews, guls, wilmots, falcons, and tassell gentils, ravens, beares, hares, foxes, and other things, than any other parte we have yet discovered, and is more populous. And here Luke Ward, a gentleman of y[e] company, traded merchandize, and did exchange knives, bells, looking-glasses, &c., with those countrey people who brought him foule, fishe, beares-skinnes, and suche like, as their countrey yeeldeth for the same. Here also they saw of those greater boates of the country with twentie persons in apeece. Traffic.

Nowe, after the Generall hadde bestowed these manye dayes here, not without many daungers, he returned backe againe. And by the way sayling alongst this coaste (being the backside of the supposed continent of America), and the Queenes Forelande, he perceived a great sounde to goe thorowe into Frobyshers Straits. Whereuppon he sente the *Gabriell* the one and twentith of July, to prove whether they mighte go thorough and meete agayne with him in the Return out of the mistaken straits.

straites, which they did, and as we imagined before, so the Queenes Forelande proved an iland, as I thinke most of these supposed continentes will. And so he departed towardes the straites, thinking it were highe time nowe to recover hys porte and to provide the fleete of their lading, whereof he was not a little carefull, as shall by the processe and his resolute attempts appeare. And in his returne with the rest of the fleete, he was so entangled by reason of the darke fogge, amongst a number of ilandes and broken ground that lyeth of this coast, that many of the ships came over the top of rocks, which presently after they might perceive to ly a drie, havyng not halfe a foote water more than some of their ships did draw. And by reason they coulde not with a small gale of wind stem the force of y[e] floud, wherby to go cleare of y[e] rocks, they were faine to let an ancker fall with twoo bent of cable togither, at a c and odde fadome deapth, where otherwise they had bin by the force of the tides carried upon y[e] rocks again, and perished: so that if God in these fortunes, as a merciful guyde, (beyond y[e] expectation of man) had not carried us thorow, we had surely more than x. m. times perished amiddest these dangers. For being many times driven harde aboorde the shoare withoute any sighte of lande, untill we were readye to make shipwracke thereon, beeing forced commonly with oure boates, to sound before oure shippes, least we might light thereon before we could discerne the same. It pleased God to give us a cleare of sunne and light, for a short time, to see and avoide thereby the daunger, having been continually darke before, and presently after. Manye times also by means of fogge and currants, being driven neare uppon the coaste, God lent us even at the very pintch one prosperous breath of winde or other, whereby to double the land and avoyde the perill, and when that we were all withoute hope of helpe, every man recommending himselfe to death, and crying out, Lorde now helpe or never: nowe Lorde look downe from Heaven and save us sinners, or else oure safetie commeth too

Great dangers.

late: even then the mightie maker of Heaven, and oure mercifull God, did deliver us: so that they who have bin partakers of those daungers, do even in their souls confesse, that God even by miracle hath sought to save them, whose name be praised evermore.

Long tyme nowe the *Anne Frances* had layne beating off and on all alone, before the Queenes forelande, not beeing able to recover their porte for ise, albeit many times they daungerously attempted it, for yet the ise choaked up the passage, and woulde not suffer them to enter. And havyng never seen any of the fleete since twentie dayes past, when by reason of the thicke mistes they were severed in the mistaken straites, did nowe this present three and twentith of July overthwart a place in the straites called Hattons Hedland, where they met with seven shippes of the fleete again, which good happe did not only rejoice them, for themselves, in respect of the comforte whiche they received by suche good companye, but especially, that by this means they were put out of doubt of their freendes, whose safeties long time they did not a little suspect and feare.

Anne Frances met with some of the fleet.

At their meeting they haled the Admirall after the manner of the sea, and with great joy welcomed one another with a thundring voly of shot. And now every man declared at large the fortunes and dangers which they hadde passed.

The foure and twentith of July we mette with the *Frances of Foy*, who with much adoe soughte way back againe thorowe the yse from out of the mistaken straites, where to their greate perill, they proved to recover their porte. They broughte the first newes of the Vizeadmirall Capitaine Yorke, who many dayes with themselves, and the *Busse* of Bridgewater was missing. They reported that they left the Vizeadmirall reasonably cleare of the ise, but the other shippe they greatly feared, whom they coulde not come to helpe, being themselves so hardly distressed, as never men more. Also they told us of the *Gabriel*, who having got

Frances of Foy.

Bridgewater ship.

thorow from the backside, and wester point of the Queens forelande, into Frobyshers Strates, fell into their companye about the Cape of Good Hope.

And uppon the seaven and twentith of Julye, the ship of Bridgewater gote oute of the ise, and met with the fleete whiche laye off and on under Hattons Hedland. They reported of their marvellous accidents and daungers, declaringe their shyppe to be so leaky, that they must of necessitie seeke harborow, having their stem beaten within theyr huddings, that they hadde muche adoe to keepe themselves above water. They had (as they say) five hundreth strokes at the poupe in lesse than half a watche, being scarce two houres. Their menne being so over-wearied therewith, and with the former dangers, that they desired helpe of menne from the other shippes. Moreover, they declared, that there was nothing but ise and daunger, where they hadde bin, and that the straites within was frosen uppe. And that it was the moste impossible thyng in the world, to passe up unto the Countesse of Warwicks sounde whiche was the place of our porte.

Straits frosen over.

The reporte of these daungers by these shyppes thus published amongst the fleete, wyth the remembraunce of the perills past, and those present before their face, brought no small feare and terror into the hartes of many considerate men. So that some beganne privily to murmur against the Generall for this wilfull manner of proceeding. Some desired to discover some harborowe thereaboutes, to refreshe themselves, and reforme their broken vesselles for a while, untill the north and northwest winds might disperse the ise, and make the place more free to passe. Other some forgetting themselves, spake more undutifully in this behalfe, saying: that they hadde as leeve be hanged when they came home, as without hope of safetie, to seeke to passe, and so to perishe amongst the ise.

The Generall not opening his eares to the peevishe passion

of anye private person, but chiefly caryng for the publicke profite of his countries cause, and nothing at all regardyng hys owne ease, lyfe, or safetie, but especiallye respecting the accomplishment of the cause he had undertaken, (wherein the chiefe reputation and fame of a Generall and Capitaine consisteth), and calling to his remembrance the shorte time he hadde in hande, to provide so great number of shyppes their loading, determined with this resolution, to passe and recover his porte, or else there to bury himselfe with hys attempte, and if suche extremitie so befell him, that he muste needes perish amongst the ise, when all hope shoulde be past, and all hope of safetie set aside, having all the ordinaunce within boorde well charged, resolved wyth pouder to burne and bury himselfe and all togither with hir Majesties shyppes. And with this peal of ordinance, to receive an honourable knell, instead of a better burial, esteeming it more happy so to end hys life, rather than himself, or any of his company or anye one of hir Majesties shyppes shoulde become a praye or spectacle to those base bloudye and man eating people.

A valiant mind of Frobisher.

Notwithstanding, somewhat to appease the feeble passions of the fearefuller sorte, and the better to entertaine time for a season, whilest the ise might the better be dissolved, hee haled on the fleete, wyth beleefe, that he would put into harborowe: thereuppon whilest the sheppes laye off and on, under Hattons Hedlande, he soughte in wyth his pynnesses amongest the islands there, as thoughe hee meant to searche for harborow, where indeede he meant nothinge less, but rather sought if any ore mighte be found in that place, as by the sequel appeared.

In the mean time, whilest the fleete laye thus doubtfull withoute anye certaine resolution what to do, being harde aboorde the leeshore, there arose a sodaine and terrible tempest at the southsouthest, whereby the ise began marvellously to gather about us.

Whereuppon everye manne, as in such case of extremitie he thoughte beste, soughte the wisest waye for his owne safetie. The most parte of the fleete whych were further shotte uppe within the straites, and so farre to the leewarde, as that they coulde not double the lande, following the course of the General, who led them the way, tooke in their sailes, and laide it a hull amongst the ise, and so passed over the storme, and hadde no extreamitie at all, but for a short time in the same place.

Howbeit the other shyppes whiche plyed oute to seawarde, hadde an extreame storme for a longer season. And the nature of the place is suche, that it is subject diversely to divers winds according to the sundrie situation of the great alps and mountaynes there, every mountayne causing a severall blaste, and pirrie, after the manner of the Levant.

Snow in July.

In this storme being the sixe and twentith of July, there fell so much snow, with such bitter cold air, that we could scarce see one another for the same, nor open our eyes to handle our ropes and sayles, the snow being above halfe a foote deepe uppon the hatches of oure shippe, which did so wette thorow oure poore marriners clothes, that he that hadde five or sixe shifte of apparell, had scarce one drie threede to his backe, whiche kinde of wette and coldnesse, togither with the over labouring of the poore menne amiddest the ise, breed no small sicknesse amongest the fleete, which somewhat discouraged some of the poor men, who had not experience of the like before, everye man perswading himselfe, that the wynter there must needs be extreme, where they be found so unseasonable a sommer.

Extreme winter.

Great heat in Meta Incognita.

And yet notwythstandyng this cold ayre, the sunne many times hathe a marvellous force of heate amongst those mountains, insomuche, that when ther is no breth of wind to bring ye cold ayre from the dispersed ise uppon us, we shall be weary of the blominge heate, and then sodainly with a perry of wind whiche commeth down from ye hollownes of

Unconstant weather.

ye hilles, we shal have such a breth of heate brought upon our face, as though we were entred some bastow or hote-house, and when the first of the pirry and blast is past, we shall have the winde sodainly anew blow cold againe.

In this storme the *Anne Fraunces*, the *Moone*, and the *Thomas of Ipswich*, who founde themselves able to holde it up with a sayle, and could double aboute the Cape of the Queens forelande, plyed oute to seawarde, holding it for better policie and safetie, to seeke sea roome, than to hazard the continuance of the storme, the daunger of the ise and the leeshore.

And being uncertaine at this time of the Generalls private determinations, the weather being so darke, that they coulde not discerne one another, nor perceive which waye he wrought, betooke themselves to this course for best and safest.

The Generall notwithstanding the greate storme, following his owne former resolution, soughte by all meanes possible, by a shorter way, to recover his port, and where he saw the ise never so little open, he gat in at one gappe, and out at another, and so himself valiantly ledde the way through before, to induce ye fleete to followe after, and with incredible payne and perill, at length gat through the ise, and uppon the one and thirtith of July, he recovered his long wished porte after many attempts, and sundry times being put backe, and came to anker in the Countesse of Warwicke's sound, in the entrance whereof, when he thoughte all perill past, he encountred a great iland of ise, whyche gave the *Ayde* suche a blow, having a little before wayed hir anker a cocke bill, that it stroke the anker flouke through the shippes bowes under the water, whych caused so greate a leake, that with muche adoe they preserved the shippe from sinking.

The General recovereth his port.

At theyr arivall heere, they perceived two shippes at anker within the harborough, whereat they began muche to marvel, and greatelye to rejoice, for those they knew to be the

Michaell, wherein was the Lieutenant generall Captayne Fenton, and the small barke called the *Gabriell,* who so long tyme were missing, and never hearde of before, whome every man made the last reckning, never to heare of agayne.

Heere every man greately rejoysed of their happie meeting, and welcomed one another after the sea manner, with their great ordinance, and when eache partie hadde reaped up their sundrie fortunes and perils past, they highlye praised God, and altogither uppon their knees gave hym due, humble and harty thanckes, and Mayster Wolfall, a learned man, appoynted by hir Majesties Councell to be theyr minister and preacher, made unto them a godly sermon, exhorting them especially to be thankefull to God for theyr strange and miraculous deliverance in those so dangerous places, and putting them in mynde of the uncertainetie of mans life, willed them to make themselves alwayes ready as resolute men to enjoy and accept thankefully whatsoever adventure his divine Providence should appoynt. This Mayster Wolfall being well seated and setled at home in his owne countrey, with a good and large living, having a good honest woman to wife, and very towardly children, being of good reputation among the best, refused not to take in hand this paynefull voyage, for the only care he had to save souls, and to reform those infidels if it were possible to Christianitie; and also partly for the great desire he hadde that this notable voyage, so well begun, might be brought to perfection, and therefore he was contented to stay there the whole yeare, if occasion had served, being in every necessary action as forward as the resolutest men of all. Wherfor in this behalfe he may rightly be called a true pastor and minister of Gods word, which for the profite of his flocke spared not to venture his owne life.

Mayster Wolfall preacher.

But to retorne agayne to Captayne Fentons company, and to speake somewhat of their dangers (albeit they bee more

The adventures of Captain Fenton and his company.

than by writing can be expressed). They reported, that from the night of the first storme, whiche was aboute the first day of July, untill seaven dayes before the Generalls arrivall, which was the sixe and twentith of the same, they never saw any one day or houre wherein they were not troubled with continuall daunger and feare of death, and were twentie dayes almost togither fast amongst the ise. They had their shippe stricken through and through on both sides, their false stem borne quite away, and could go from their shippes in some places uppon the ise very many miles, and might easily have passed from one iland of ise to another, even to the shore, and if God had not wonderfully provided for them and theyr necessitie, and time had not made them more cunning and wise to seeke strange remedies for strange kinds of dangers, it had been impossible for them ever to have escaped: for among other devises, wheresoever they founde any iland of ise of greater bignesse than the rest (as there be some of more than halfe a mile compasse about, and almost 400 fadome high), they commonly coveted to recover the same, and thereof to make a bulwarke for their defence, whereon having mored ancker, they roade under the lee thereof for a time, beeyng therby garded from ye danger of the lesser driving ise. But when they must needes forgoe this newe founde forte, by meanes of other ise, whiche at length woulde undermine and compasse them round aboute, and when that by heaving of the billow they were therwith like to be brused in peces, they used to make fast the ship unto the most firme and broad peece of ise they could find, and binding hir nose fast thereunto, would fill all theyr sayles, whereon the winde havinge great power, would force forward the ship, and so the ship bearing before hir the yse, and so one yse driving forward another, should at length get scope and searoom. And having by this means at length put their enemies to flight, occupied the cleere place for a prettie season, among

Extremitie causeth men to devise new arts and remedies.

Hard shifts.

sundry mountaynes and Alpes of ise. One there was founde by measure to be sixty-five fadome above water, which for a kind of similitude was called Salomons porch. Some think those ilands eight times so muche under water as they are above, by-cause of their monstrous weight. But now I remember, I saw very strange wonders, men walking, running, leaping, and shoting upon the maine seas forty miles from any land, withoute any shippe or other vessell under them. Also I saw fresh rivers running amidst the salt sea a hundred myle from land, which if any man will not beleeve, let him know that many of our company lept out of their shippe uppon ilandes of ise, and running there uppe and downe, did shoote at buttes uppon the ise, and with their calivers did kill great ceales, whiche use to lye and sleepe upon the ise, and this ise melting at the top by reflexion of the sun, came down in sundrye streams, whyche, uniting togither, made a prettie brooke able to drive a mill.

Strange wonders.

The sayd Captayne Fenton recovered his porte tenne dayes before any man, and spent good time in searching for mine, and found good store thereof, which bycause it proved good, was after called Fentons Fortune. He also discovered about tenne miles up into the countrey, where he perceived neyther towne, village, nor likelyhoode of habitation, but seemeth (as he sayeth) barrenous as the other parts which as yet we have entred upon; but their victuals and provision went so scant with them, that they had determined to returne homeward within seaven dayes after, if the fleete had not then arrived.

The Generall after his arrivall in the Countesses Sound spent no time in vayne, but immediately at his first landing called the chiefe captaynes of his councell togither, and consulted with them for the speedier execution of such things as then they had in hand. At first, for searching and finding out good minerall for the miners to be occupied on. Then to give good orders to be observed of the whole company on shore. And lastly, to consider for the erecting up the forte and house for the use of them which were to abide there the whole yeare. For the better handling of these and all other like important causes in this service, it was ordeined from hir Majestie and the Councell that the Generall should call unto him certayne of the chiefe captaynes and gentlemen in councell, to conferre, consult, and determine of all occurrets in this service, whose names are here as folow:

Captayne Fenton
Captayne Yorke
Captayne Best
Captayne Carew
Captayne Philpot

And in sea causes to have as assistants, Christopher Hal and Charles Jackman, being both very good pylots and sufficient mariners, whereof the one was chiefe pylot of the voyage, and the other for the discoverie. From the place of our habitation weastward, Maister Selman was appointed notarie, to register the whole manner of proceeding in these affaires, that true relation thereof might be made, if it pleased hir Majestie to require it.

The first of August every captaine, by order from the Generall and his counsell, was commanded to bring ashore unto the Countesses Iland al such gentlemen, souldiers, and myners, as were under their charge, with such provision as they had of victuals, tents, and things necessarye for the speedie getting togither of mine, and fraught for the shippes.

The muster of the men being taken, and the victuals with all other things viewed and considered, every man was set to his charge, as his place and office required. The myners were appointed where to worke, and the mariners discharged their shippes.

Uppon the seconde of August was published and proclaymed, uppon the Countesse of Waricks Iland, with sound of trumpet, certain orders by the General and his counsel appointed to be observed of the companye during the time of their abiding there. The copie whereof here followeth:

Orders sett down by M. Frobisher, Esquire, Captaine Generall for the voyage to "Cataya," to be observed of the companie during the time of their abode in "Meta Incognita." Published the second day of August, 1578.

1. *Inprimis*, the Generall, in hir Majesties name, straightly chargeth and commandeth that no person or persons, with boate nor pinnesse, shall go ashoare, for any cause, but to the Countesse of Warwickes Ilande and Winters Fornace, without licence of the General, or his deputies. And if they fortune at anye time, having licence, to meet with any of the countrey people, that they shall not enter into any conference or armes wyth them, untill they have given intelligence thereof to the Generall or hys lieutenant.

2. Item, that no person, of what calling soever he be, shal make an assay of any maner of mettall, matter, or ore, in ye partes nowe called Meta Incognita, but only suche as shal be appointed by the General, or in his absence by his lieutenant, to do the same: nor that anye person shall take up and keepe to his private use anye parte or parcel of ore, pretious stone, or other matter of commoditie to be had or founde in that lande, but he the sayde person so seased of such ore, stone, or other matter of commoditie shall with all speede, as soon as he can, defect the same, and make deliverie thereof to the generall, or his lieutenant generall, uppon paine to forfaite for everye such ounce thereof, the value treble of anye wages he is to receive after the daye of such offence committed: and further, to receyve suche punishment as to hyr Majestie shall seem good.

3. Item, that no shippe or shippes shall take uppon them

to loade any manner of ore without licence of the general, or he that shal be appointed deputie for him, for y[e] view of the same.

4. Item that all the maisters of everye shippe or shippes within the fleete shal upon Mundaye next comming, by foure of the clocke in the morning, wyth all the most parte of theyr companies, make theyr repayre to the Countesses Ilande aforesaide, there to view and make such places, for loading and unloading of ore and other thyngs, as shall be most commodious and meete for that purpose.

5. Item, that no person or persons within this service, by sea or lande, shall use anye discovered[1] speeches, swearing, brauling, or cursing upon payne of imprisonmente.

6. Item, that no person or persons, eyther by sea or lande, shal draw his or theyr weapons in quarrellyng manner, to the intente to offende or disturbe the quiete of anye person or persons wythin thys service, uppon paine that being so taken, he or they whatsoever immediately to loose his right hande.

7. Item, that no person or persons shall washe their handes or anye other things, in the spring, uppon the Countesses Iland, where the water is used, and preserved for the dressing of their victuals, upon paine to receive such punishment as shall be thought good, by the Generall or his Lieutenant, for the same. And for the better preservation and health of everye manne, that no person or persons shall doe his easement but under the cliffes where the sea may washe the same awaye, upon paine that everye one so offending, for the first time shall be imprisoned in the billowe fourteene houres, and for the second time being so taken by the provost Martiall, to pay twelve pense.

8. Item, that no person or persons, of what nature or condition soever, shall cast out of their shippe or shippes, anye ballast or rubbish, into the roade, where these shippes now rydeth, or may conveniently ride, within this sounde, that

[1] Sic in original.

therby the same sounde or roade steade may be impaired, but shall carrie the same, and lay it where it may not offend. Uppon paine that every man so offending, the owner of such shippe or shippes, shall forfaite the fraught of one tunne.

By me MARTYN FROBISHER.

In the meane time, whylest the mariners plyed their work, ye Captains sought out new mynes, the goldfinders made tryall of the ore, the mariners discharged their shippes, the gentlemen for example sake laboured hartily, and honestlye encouraged the inferiour sorte to worke. So that small time of that little leasure, that was left to tarrie, was spent in vaine.

The second of August the *Gabriel* arrived, who came from the Vizeadmirall, and being distressed sore with yse, put into harborrow neere unto Mount Oxford. And now was the whole fleete arrived safely at their port, excepting foure, besides the shippe that was loste, that is, the *Thomas Allen*, the *Anne Frances*, the *Thomas* of Ipswich, and the *Moone*, whose absence was some let unto the works and other proceedings, as well for that these shippes were furnished with the better sort of myners and other provision for the habitation.

The ninth of August, the Generall with the Captaynes of his counsell assembled togither, beganne to consider and take order for the erecting up of the house or forte, for them that were to inhabit there the whole yeare, and that presently the masons and carpenters might go in hande therewith. First therefore they perused the bills of ladyng what every man received into his shippe, and found that there was arrived only the east side, and the south side of ye house, and yet not that perfect and intier, for many peeces thereof were used for fenders in many shippes, and so broken in peeces, whyles they were distressed in the ise. Also after due examination had, and true accompt taken, there was founde want of drinke and fuel, to serve one hundreth men, which was the

Consultation for inhabiting Meta Incognita.

number appointed firste to inhabite there, bycause their greatest store was in the ships which were not yet arrived. Then Captaine Fenton seeing the scarcity of ye necessary things aforesaid, was contented, and offred himselfe to inhabite there, with sixtie men. Wherupon they caused the carpenters and masons to come before them, and demaunded in what time they woulde take upon them to erect up a lesse house for sixtie men. They required eight or nine weeks, if there were tymber sufficient, whereas now they had but six and twentie dayes in all to remayne in that countrey. Wherefore it was fully agreed upon, and resolved by the General and his counsell, that no habitation shoulde be there this yeare. And therefore they willed Maister Selman the Register, to set down this decree, with all their consents, for the better satisfying of hir Majestie, the Lords of the Counsel, and the adventurers.

No habitation this yere.

The *Anne Frances*, since she was parted from the fleete, in the last storme before spoken of, could never recover above five leagues within the straights, the wind being sometyme contrarie, and moste times the ise compassing them round about. And from that time, being aboute the seaven and twentith of July, coulde neyther heare nor have sight of any of the fleete, untill the third of August, when they descried a sayle near to Mount Oxford, with whome when they had spoken, they could understande no newes of anye of the fleete at all. And this was the *Thomas of Ipswich*, who hadde layne beating off and on at sea, with very foule weather, and contrarye winds, ever since that foresaide storme, without sight of any man. They kept company not long togyther, but were forced to lose one another again, the *Moone* being consort always with the *Anne Fraunces*, and keeping verie good companye plyed up togither into the straites, with great desire to recover their long wished port; and attempted as often, and passed as far as possible the winde, weather, and ise, gave leave, whyche commonly they found very contrary. For when the weather was cleare, and withoute fogge, then commonly yᵉ wind was contrarie. And when it was eyther easterly or southerly, which woulde serve their turnes, then had they so great a fogge, and darke miste therewith, that eyther they could not discerne way throw the ise, or else the ise laye so thicke togither, that it was impossible for them to passe. And on the other side, when it was calme, the tydes hadde force to bryng the ise so sodaynlye about them, that commonlye then they were moste therewith destressed, having no winde to carry them from the daunger therof.

And by the sixte of August, being with much adoe got up as high as Leicester point, they had good hope to find the souther shore cleare, and so to passe uppe towardes their porte. But being there becalmed, and lying a hull openly upon the greate bay whiche commethe oute of the mystaken straites before spoken of, they were so sodainely compassed with ise rounde about, by means of the swifte tydes whiche runne in that place, that they were never afore so hardly beset as nowe. And in seeking to avoyde these dangers in the darke weather, the *Anne Frances* lost sighte of the other two ships, who being likewise hardly distressed, signified their daunger, as they since reported, by shooting off their ordinaunce, which the other coulde not heare, nor if they had hearde, could have given them no remedie, being so busily occupied to winde themselves out of their owne troubles.

The Moone. The fleeboate called the *Moone,* was here heaved above the water with the force of the ise, and received a great leake therby. Likewise the *Thomas of Ipswich,* and the *Anne Frances* were sore brused at that instant, having their false stemme borne away, and their shippe sides stroken quite through.

Now considering the continuall daungers and contraries, and the little leasure that they had lefte to tarrie in these partes, besides that every night the ropes of theyr shippes were so frosen, that a man coulde not handle them without cutting his handes, togither with the great doubt they had of the fleetes safety, thinking it an impossibility for them to passe unto their port, as well for that they saw themselves, as for that they harde by the former reporte of the shippes which had proved before, who affirmed that the straites were all frosen over within. They thought it now very hie time to consider of their estates and safeties that were yet left togither. And hereuppon the Captaines and maisters of these shippes desired the Captaine of the *Anne Frances* to enter into consideration with them of these matters, wherefore Captaine Tanfield of the *Thomas of Ipswich*, with his pylot Richard Coxe, and Captaine Upcote of the *Moone*, with his maister John Lakes came aboorde the *Anne Frances* the eight of August to consult of these causes. And being assembled togither in the Captayne's cabin sundrie doubtes were then alleaged. For the fearefuller sorte of mariners being overtyred with the continuall labour of the former daungers, coveted to returne homewarde, sayinge that they woulde not againe tempt God so much, who had given them so many warnings, and delivered them from so wonderfull daungers; that they rather desired to loose wages fraughte and all, than to continue and follow such desperate fortunes. Again their shippes were so leake, and the men so wearie, that to amende the one, and refreshe the other, they muste of necessitie seeke into harborow.

The *Anne Frances*, the *Thomas of Ipswich* and the *Moone* consult.

But on the other side, it was argued againe to the contrarie, that to seeke into harborowe thereaboutes was but to subject themselves to double daungers, for if happilye they escape the daungers of rockes in their entring, yet being in, they were nevertheless subject there to the daunger of the ise, which with the swift tydes and currents is carried in and out in most harborows thereaboutes, and may thereby gaule their cables asunder, drive them uppon the shoare, and bring them to much trouble. Also the coast is so much subject to broken ground and rockes, especially in the mouth and entraunce of every harborow, that albeit the channell be sounded over and over againe, yet are you never the neare to discerne the daungers. For the bottome of the sea, holding like shape and forme as the lande, beyng full of hilles, dales, and ragged rockes, suffereth you not, by your soundings, to knowe and keepe a true gesse of ye depth, for you shall sounde upon the side or hollownesse of one hil or rocke under water, and have a hundreth, fiftie, or fortie fadome depth ; and before the next cast, ere you shall bee able to have your lead againe, you shall be uppon the toppe thereof, and come aground to your utter confusion.

Another reason against going to harborow, was, that the colde ayre did threaten a sodaine freezing uppe of the sounds, seeing y^t^ every night there was new congealed ise, even of that water which remained within their shippes. And therefore it should seeme to be more safe to lye off and on at sea, than for lacke of winde to bring them forth of harborow, to hazard by sodaine frostes to be shut up the whole yeare.

Captaine Bests resolution.

After many such daungers and reasons alleaged, and large debating of these causes on both sides, the Captaine of the *Anne Frances* delivered his opinion unto the company to this effect. First, concerning the question of returning home, he thought it so much dishonorable, as not to grow in any further question: and, agayne, to returne home at length (as at length they must needes), and not to be able to bring a certayne report of the fleet, whether they were living or lost, or whether any of them had recovered their port or not in the Countesses Sounde (as it was to be thoughte the most part would if they were living), he sayd that it would be so great an argument, eyther of wante of courage or discretion in them, as he resolved rather to fall into any danger, than so shamefully to consent to retourne home, protesting that it should never be spoken of him, that he woulde ever returne withoute doing his endeavour to finde the fleete, and knowe the certaynetie of the Generals safetie. He put his companie in remembrance of a pinasse of five tunne burthen, which he hadde within his ship, which was caryed in peeces, and unmade up for the use of those which shoulde inhabite there the whole yeare, the which if they coulde fynde meanes to joyne togither, hee offered himselfe to prove before therewith, whether it were possible for any boate to passe for ice, whereby the shipps myghte bee broughte in after, and mighte also thereby gyve true notice, if any of the fleete were arrived at theyr porte or not.

But, notwithstanding, for that he well perceyved that the most parte of hys companye were addicted to put into harborow, he was willing the rather for these causes somewhat to encline thereunto. As first, to search alongst the same coast, and the soundes thereaboutes, he thoughte it to be to good purpose, for that it was likely to fynd some of the fleete there, whiche being leake, and sore brused with the ise, was the rather thoughte lykely to be put into an yll harborough, beying distressed with foule weather in the last storme, than to hazard theyr uncertayne safeties amongst the ise; for about this place they lost them and lefte the fleete then doubtfully questioning of harborow.

It was lykely also, that they might fynde some fitte harborow thereaboutes, whyche myghte be hovefull for them against another tyme. It was not likewise impossible to fynde some ore or myne thereaboutes, wherewithall to fraughte theyr shyppes, whiche woulde bee more commodious in this place, for the neerenesse to seawarde, and for a better outlette, than further within the straytes, beyng lykely heere alwayes to loade in a shorter time, howsoever the strayte shoulde be pestered wyth ise within; so that if it myghte come to passe that thereby they mighte eyther fynd the fleete, mine, or convenient harborough, any of these three would well serve theyr presente turnes, and gyve some hope and comforte unto theyr companyes whiche nowe were altogyther comfortlesse. But if that all fortune shoulde fall out so contrarye that they coulde neyther recover theyr porte nor anye of these aforesaide helpes, that yet they would not yet departe the coast, as long as it was possible for them to tarrie there, but would lye off and on at sea athwart the place. Therefore hys final conclusion was sette downe thus:—Firste, that the *Thomas of Ipswiche* and the *Moone* shoulde consorte and keepe companye togyther carefully with the *Anne Frances* as neere as they could, and as true Englishmen and faythful friends should supplye one anothers want in all fortunes and dangers. In the morning following every shippe to sende of hys boate with a sufficient pylot to searche out and sounde the harboroughs for the safe bringing in of theyr shippes. And beeyng arrived in harborough where they mighte finde convenient place for the purpose, they resolved forthwith to joine and set togyther the pinasse, wherewythall the Captayne of the *Anne Frances* might, according to his former determination, discover up into the straytes.

After these determinations thus sette downe, the *Thomas of Ipswiche* the nyghte following lost company of the other shyppes, and afterwarde shaped a contrarye course homewarde, whyche fell oute, as it manyfestlie appeared, very much agaynst theyr Captayne, Mayster Tanfieldes, mynde, as by due examination before the Lordes of Hir Majesties most Honorable Privie Counsell, it hathe since been proved to the greate discredite of the Pilot Coxe, who specially persuaded his company againste the opinion of hys sayde Captayne to returne home.

And, as the Captayne of the *Anne Frances* dothe witnesse, even at theyr conference togither, Captayne Tanfield tolde hym that he did not a little suspect the said Pylot Coxe, saying, that he had neyther opinion in the man of honest duetie, manhoode, or constancie. Notwithstanding the sayde shippes departure, the Captayne of the *Anne Frances,* beeying desirous to putte in execution hys former resolutions, went with hys shyppeboate (beeyng accompanied also wyth the *Moones* skyffe) to prove amongst the ilandes which lye under Hattons Headland, if anye convenient harborough, or any knowledge of the fleete, or anye good ore was there to be found. The shyppes lying off and on at sea the whyle under sayle, and searching through many soundes, they saw them all full of manye dangers and broken grounde, yet one there was which seemed an indifferent place to harborow in, and whiche they did very diligentlye sounde over and searched agayne.

Heere the sayde Captayne founde a great blacke iland, whereunto he had good liking, and certifying the company therof they wer somewhat comforted, and with the good hope of his words, rowed cheerfully unto the place where, when they arrived, they founde such plentie of blacke ore of the same sorte whiche was broughte into Eyglande thys last yeare, that if the goodnesse myghte aunswere the greate plentye thereof, it was to be thoughte that it might reasonably suffise all the golde gluttons of the worlde. Thys ilande the Captayne, for cause of his good happe, called after his owne name, Bestes blessing, and wyth these good tydings returning aboorde hys shippe the ninth of August, about tenne of the clocke at night, he was joyfully welcomed of hys companye who before were discomforted and greatelie expected some better fortune at hys handes.

Best's Blessing.

The next daye beeyng the tenth of August, the weather reasonably fayre, they put into the foresayde harborough, having their boate for theyr better securitie sounding before theyr ship. But for all the care and diligence that coulde be taken, in soundyng the Channell over and over agayne, the *Anne Frances* came aground uppon a sunken rocke within the harborough, and lay thereon more than halfe drye untill the next flood, when, by Gods Almighty Provi-

The *Anne Frances* in danger.

dence, contrarye almost to all expectation, they came afloate agayne, beeyng forced all that tyme to undersette theyr shippe wyth their mayne yarde, whyche otherwyse was lykely to oversette and put thereby in daunger the whole company. They hadde above two thousande strokes togyther at the pumpe, before they coulde make theyr shyppe free of the water agayne, so sore shee was brused by lying uppon the rockes. The *Moone* came safely, and roade at ancker by the *Anne Frances*, whose helpe in theyr necessitie they coulde not well have missed.

The *Moone* in harborow.

Now, whilest the marriners were romaging theyr shyppes and mending that whiche was amisse, the miners followed their laboure, for getting togither of sufficient quantitie of ore, and the carpenters endeavoured to do theyr best for the making uppe of the boate or pinnesse, whiche to bring to passe, they wanted two speciall and moste necessary things; that is, certaine principal timbers that are called knees, which are the chiefest strength of any boate, and also nayles wherewithall to joine the plancks togither. Whereupon, having by chance a smyth amongst them (and yet unfurnished of his necessarie tooles to worke and make nayles withall), they were faine of a gunne chamber to make an anvil to worke upon, and to use a pickaxe instead of a sledge to beat withall, and also to occupy two small bellows insteede of one payre of greater smyths bellows. And for lack of small iron, for the easier making of the nayles, were forced to breake their tongs, grydiern, and fiershovell in peeces.

The eleventh of August, the Captaine of the *Anne Frances* taking the maister of hys ship with hym, went up to the toppe of Hattons Hedland, which is the highest lande of all the straites, to the ende to descry the situation of the country underneath, and to take a true plot of the place, whereby also to see what store of the ise was yet lefte in the straites, as also to searche what mine, matter, or fruite that

Hattons Hedland.

soyle myght yeelde. And the rather for the honor ye said Captaine doth owe to that honorable name which himselfe gave thereunto the last yeare in the highest parte of this hedlande, he caused his companye to make a columne or crosse of stone, in token of Christian possession. In this place there is plenty of blacke ore and divers preatie stones.

The seaventeenth of Auguste, the Captaines wyth their companies chased and killed a greate white beare, whiche adventured and gave a fierce assaulte upon twentie men being weaponed. And he served them for good meat many dayes after.

The eighteenth of August, the pinnesse with muche adoe being set togyther, the saide Captaine Beste determined to depart upon the straites to prove and make trial, as before was pretended, some of his companye greatlye persuading him to the contrarie, and specially the carpenter that set the same togither, who saide that he would not adventure himselfe therein for five hundreth poundes, for that the boate hung togither but onelye by the strength of the nayles, and lacked some of her principall knees and tymbers.

A pinnesse there built.

These words somewhat discouraged some of the company which should have gone therein. Whereupon the Captaine, as one not altogither addicted to his owne selfe will, but somewhat foreseeing how it might be afterwards spoken, if contrarye fortune should happen him (lo, he hathe followed his owne opinion and desperate resolutions, and so thereafter it has befallen him), calling the maister marriners of beste judgement togyther, declared unto them howe much the cause imported him in his credite to seeke out the Generall, as well to conferre with him of some causes of waight as otherwise to make due examination and triall of the goodnesse of the ore, whereof they had no assurance but by guesse of the eye, and was wel like the other: which, so to carry home, not knowing the goodnesse thereof, might be as much as if they should bring so many stones. And, therefore, hee desired them to delyver their plaine and honest opinion, whether the pinnasse were sufficient for him so to adventure in or no. It was answered, that by carefull heede taking thereunto amongst the ise and the foule weather, the pinnesse might suffice. And hereuppon the maisters mate of the *Anne Frances*, called John Gray, manfully and honestly offering himself unto his Captain in this adventure and service, gave cause to others of hys marriners to follow the attempt.

And upon the nineteenth of August the said Captain being accompanied with Captaine Upcote of the *Moone*, and xviii persons in the small pinnesse, having convenient portion of victualles and things necessary, departed upon the said pretended voyage, leaving their shippe at ancker in a good readinesse for the taking in of their fraight. And having little winde to saile withall, they plyed alongest the souther shoare, and passed above 30 leagues, having the onely helpe of mans labour with ores, and so entendyng to keepe that shoare aboorde untill they were gote up to the farthest and narrowest of ye straites, minded there to crosse over and to search likewise alongest the northerland unto the Countesses Sound, and from thence to passe all that coaste along, whereby if any of the fleete hadde been distressed by wracke of rocke or ise, by that meanes they might be perceived of them, and so they thereby to give them such helpe and reliefe as they could. They did greatly feare and ever suspecte that some of the fleete were surely caste awaye and driven to seeke sowre sallets amongest the colde cliffes.

And being shot up about 40 leagues within ye straites, they put over towards ye norther shore, which was not a little daungerous for theyr small boate. And by meanes of sodaine flawe were driven and faine to seek harborow in the Gabriels Ilands. night amongst all the rockes and broken grounde of Gabriells Ilandes, a place so named within the straites above the Countesse of Warwicks Sounde. And by the way where they landed they did find certaine great stones sette uppe by the countrie people, as it seemed for markes, where they also made manye crosses of stone in token that Christians had bin there. The xxii of August they hadde sighte of the Countesse Sounde, and made the place perfecte from the toppe of a hill, and keepyng along the norther shoare perceived the smoake of a fyre under a hylles side, whereof they diverslye deemed when they came nearer the place, they perceyved people whiche wafted unto them, as it seemed, with a flagge or auncient. And bycause the canniballes and countrie people had used to doe the lyke when they perceived any of our boats to passe by, they suspected them to be the same. And coming somewhat nearer they might perceive certayne tents and discerne this auncient to be of mingled colours, black and white, after the English fashion. But bycause they could see no shippe nor likelihoode of harborow within five or sixe leagues aboute, and knewe that none of oure men were wonte to frequent those partes, they coulde not tell what to judge thereof, but imagined that some of the shyppes being caried so highe wyth the storme and mistes, had made shipwracke amongest the ise or the broken ilandes there, and were spoyled by the country people, who might use the sundrie coloured flagge for a policie to bring them likewise within their daunger. Whereupon the saide Captaine, wyth his companies, resolved to recover the same auncient, if it were so, from those base, cruell, and man-eating people, or else to lose their lives, and all togither. One promised himselfe a payre of garters, another a scarffe, the third a lace to tye hys whistle withal of the same. In the ende, they discerned them to be their countreymen, and then they deemed them to have loste theyr shyppes, and so to be gathered togyther for theyr better strength. On the other side, the companye a shoare feared that the Captayne having loste his shippe, came to seeke forth the fleete for his reliefe in hys poore pinnesse, so that their extremities caused eache parte to suspect the worste.

The captaine nowe with his pinnesse being come neere the shoare, commanded his boate carefully to be kepte afloat, least in their necessitie, they might winne the same from hym, and seeke first to save themselves (for everye manne in that cause is nexte himselfe). They haled one another according to the manner of the sea, and demanded what cheare; and either partie answered y[e] other, that all was well; whereuppon there was a sodaine and joyfull oute-shoote, with greate flinging up of cappes, and a brave voly of shotte to welcome one another. And truelye it was a moste straunge case, to see howe joyfull and gladde everye partie was to see themselves meete in safetie againe, after so strange and incredible daungers; yet to be shorte, as theyr daungers were greate, so their God was greater.

Proximus sum egomet mihi.

And here the company were workyng uppon newe mines, which Captayn Yorke being here arrived not long before, hadde founde out in this place, and it is named the Countesse of Sussex Mine.

Captaine York arrived.

After some conference wyth oure friends here, the Captaine of the *Anne Frances* departed towardes the Countesse of Warwickes Sounde to speake with the Generall, and to have triall made of suche mettall as he hadde broughte thither, by the goldfinders. And so determined to dispatche againe towards his shippe. And having spoken wyth the Generall, he received order for all causes, and direction as well for the bringing uppe of his shippe to the Countesses Sounde, as also to fraight his shippe with the same ore he himselfe hadde found, which upon triall made, proved to be very good.

The thirteenth of Auguste, the saide Capitaine mette togither with the other Capitaines (Commissioners in counsell with the Generall) aboorde the *Ayde*, where they considered and consulted of sundrie causes, which, particularly registred by the notarie, were appointed, where and howe to be done againste an other yeare.

The fourteenth of August the Generall with two pinnesses and good numbers of men, wente to Beare's Sounde, commanding the said capitaine with his pinnesse to attend the service, to see if he could encounter or apprehend any of the caniballes, for sundry tymes they showed themselves busy thereabouts, sometimes with seven or eight boates in one company, as though they minded to encounter with oure companye, whiche were working there at the mines, in no greate numbers. But when they perceived anye of oure shippes to ride in that roade (being belike more amazed at the countenance of a shippe, and a more number of men) didde never shewe themselves againe there at all. Wherefore oure men soughte with their pinnesses to compasse aboute the iland, where they did use, supposing there sodainely to intercept some of them. But before oure men coulde come neare, having belike some watch in the toppe of the mountaines, they conveyed themselves privily away, and lefte (as it shoulde seeme) one of their great dartes behinde them for haste, whiche we founde neare to a place of their caves and housing. Therefore, though our Generall were very desirous to have taken some of them to have broughte into Englande, they being nowe growen more wary by their former losses, would not at any time come within our daungers. About midnight of the same day the captaine of the *Anne Frances* departed thence and set his course over the straites towards Hattons Hedland, being about fifteene leagues, and returned aboord his ship over, the five and twentithe of Auguste, to the greate comforte of his company, who long expected his comming, where he founde hys shyppes ready rigged and loaden. Wherefore he departed from thence agayne the next morning towardes the Countesses Sounde, where he arrived the eight and twentith of the

None of the people will be taken.

same. By the waye he sette hys miners ashoare at Beares Sounde, for the better dispatche and gathering the ore togither, for that some of the ships were behinde with their fraighte, the time of the yeare passing speedily away.

The thirtith of August the *Anne Frances* was brought aground, and had viij great leakes mended, whiche she had received by means of the rocks and ise. This daye the masons finished a house whiche Captaine Fenton caused to be made of lyme and stone upon the Countesse of Warwickes Ilande, to the ende we mighte prove against the nexte yeare, whether the snow coulde overwhelm it, the frosts break uppe, or the people dismember the same. And the better to allure those brutish and uncivill people to courtesie, againste other times of our comming, we lefte therein dyvers of our countrie toyes, as bells, and knives, wherein they specially delight, one for the necessarie use, and the other for the great pleasure thereof. Also pictures of men and women in lead, men a horsebacke, lookinglasses, whistles, and pipes. Also in the house was made an oven, and breade left baked therein, for them to see and taste.

A house builded and left there.

We buried the timber of our pretended forte, with manye barrels of meale, pease, griste, and sundrie other good things, which was of the provision of those whych should inhabite, if occasion served. And insteede therof we fraight oure ships full of ore, whiche we holde of farre greater price. Also here we sowed pease, corne, and other graine, to prove the fruitfulnesse of the soyle against the next yeare.

Maister Wolfall on Winters Fornace preached a godly sermon, which being ended, he celebrated also a communion upon the lande, at the partaking whereof was the capitaine of the *Anne Frances*, and manye other gentlemen and soldiours, marriners and miners wyth hym. The celebration of divine mistery was ye first signe, seale, and confirmation of Christes name, death and passion ever knowen in all these quarters. The said M. Wolfall made sermons, and celebrated the communion at sundrie other times, in severall and sundrie ships, bicause the whole company could never meet togither at any one place. The fleet now being in some good readinesse for their lading, ye General calling togither the gentlemen and captains to consult, told them that he was very desirous yt some further discovery should be attempted, and yt he woulde not only by Gods help bring home his shippes laden with golde ore, but also meant to bring some certificat of a further discoverie of ye countrie, which thing to bring to passe (having sometime therein consulted) they founde verye harde, and almost invincible. And considering that already they hadde spente some time in searching out the trending and fashion of the mistaken straites, and had entred verye farre therein, therefore it coulde not be saide but that by thys voyage they have notice of a further discovery, and that the hope of the passage thereby is much furthered and encreased, as appeared before in the discourse thereof. Yet notwythstanding, if anye meanes mighte be further devised, the captaynes were contented and willing, as the Generall shoulde appointe and commande, to take any enterprise in hande. Whiche, after long debating, was found a thing verye impossible, and that rather consultation was to bee had of returning homewarde, especiallye for these causes following. First, the darke foggy mistes, the continuall fallyng snowe and stormy weather which they commonly were vexed with, and nowe daylye ever more and more encreased, have no small argument of the winters drawing neare. And also the froste everye nighte was so harde congealed within the sounde, that if by evill happe they shoulde be long kepte in wyth contrarye windes, it was greatly to be feared that they should be shutte uppe there faste the whole yeare, whych being utterly unprovided, would be their utter destruction. Againe, drincke was so scant throughout all the fleete, by means of the great leakage, that not onely the

Consultation for a further discoverie.

provision whyche was layde in for the habitation was wanting and wasted, but also eache shyppes severall provision spent and lost, which many of oure companye, to their great griefe, founde in their returne since, for al the way homewards they dranke nothing but water. And the great cause of this lekage and wasting was, for that ye great timber and seacole, which lay so waighty upon ye barrels, brake, brused, and rotted ye hoopes in sunder. Yet notwithstanding these reasons alledged, ye Generall himselfe (willing the rest of the gentlemen and captaines every man to looke to his severall charge and lading, that against a day appointed they shoulde be all in a readinesse to sette homeward) himselfe went in his pinnesse and discovered further northward in the straytes, and found that by Beares Sound and Halles Iland the land was not firme, as it was first supposed, but all broken ilandes in manner of an archipelagus; and so, with other secret intelligence to himselfe, he returned to the fleete. Where presentlye, upon his arrivall at the Countesses Sound, he began to take order for their returning homeward, and first caused certayne Articles to be proclaymed, for the better keeping orders and courses in their returne, which Articles were delivered to every captayne, and are these that follow :—

ARTICLES SETTE DOWNE BY MARTIN FROBISHER, ESQUIER, CAPTAYNE GENERALL OF THE WHOLE FLEETE, APPOYNTED FOR THE NORTHWEAST DISCOVERIES OF CATAYA, PUBLISHED AND MADE KNOWEN TO THE FLEETE FOR THE BETTER OBSERVING CERTAYNE ORDERS AND COURSE IN THEIR RETURNE HOMEWARDE.

1. Firste and principallie he doth straytely charge and commaunde, by vertue of hir Majesties commission which he hath, and in hir Majesties name, that every captayne and captaynes, master and masters of the sayde fleete do vigilently and carefully keepe company with the Admirall, and

by no manner of meanes breake companye willingly now in our returne homewards, uppon peyne of forfeture his or their whole freyte, that shall be found culpable therein, and further to receyve suche punishment, as to hir Majestie shal seeme good therein, and also to answere all such damages or losses as may happen or growe by dispersing and breaking from the fleete. And therefore for the better keeping of companye, the Generall straytely chargeth and commaundeth all the maysters of these shippes, and every of them, that they repayre to speake with the Admirall once every daye, if he or they may convenientlye doe it, uppon payne of forfeting of one tunne fraighte to hir Majestie, for every daye neglecting the same.

2. Item, that every mayster in the sayde fleete observe and keepe orderly and vigilantly all such articles as were outwards bounde, drawen, and published by the Generall in hyr Majesties name, whereof there was delyvered to every shippe a copie.

3. Item, that all captaynes and maysters of everye ship and shippes doe proclaime and make it knowen to their companye, that no person or persons within the sayde fleete, of what condition soever, doe take or keepe to theyr use or uses any ore or stones, of what quantitie so ever it be, but forthwith upon publication hereof, to delyver them and yeelde them to the custodie of the captayne to deliver unto the Generall his officers, that shall be appointed to call for them upon payne or losse of his or their wages, and treble the value of them or him that shall be founde giltie, the one halfe thereof to be given unto him that shal apprehend any suche person, and the other halfe at hir Majesties appoyntment, and the partie founde guiltie therein to be apprehended as a fellon.

4. Item, that no person or persons convey or carrie out of any ship or shippes any ore or stone or other commoditie whatsoever were had or found in the land called Meta In-

cognita, before they came in the place appoynted, which is against Dartford Creeke in yᵉ River of Thames, and then and there to deliver none to anye person or persons, but such as shall be appoynted by hir Highnesse most honorable Privie Counsell, upon the payne and danger abovesaid.

5. Item, forasmuche as in my voyage hither bounde, I landed upon Freseland, and divers other of the said fleete, which land I named West England, from which land some brought stones, ore, and other commodities, whereby hereafter they might use coulorable means, to convey as well ore, stones, and other things found in the abovesayd land, I do therefore charge every person and persons in the sayd fleete to deliver, or cause to be delivered, al maner of ore, stones, and other commodities founde as well there as here, to the captaynes of every shippe or shippes, to be redelivered by him or them to the Generall, upon payne and danger aforesayd.

6. Item, that if any shippe or shippes by force of weather shall be separated from the Admirall, and afterwards happen to fall, or shall be in danger to fall into the handes of their enimies, that then all and everye suche shippe or shippes shall have speciall regard before his falling into theyr handes, to convey away and cast into the seas all suche plattes or cardes, as shall be in any suche shippe or shippes of the abovesaide discovered lande, and all other knowledge thereof.

7. Item, that if any such shippe or shippes by force of weather shall be separated from the fleete or Admirall, and shall afterwardes arrive at any port in England, that then in such case he shall not depart from that porte, but shall give order and advertisement to Michaell Locke, treasourer of the companye, by whom hee or they shall have order from the Lordes of the Privie Councell what they shall do.

8. Item, forasmuch as sundry of the fleets companies have had lent them crowes of iron, sledges, pixeaxes, shovels,

spades, hatchets, axes, and divers other instruments for mines and mining used. And also dyvers of the sayde kind of instruments above named, was lefte at the Countesse of Sussex mine by the *Aydes* companye, and are yet kept from their knowledge by such as wrought at the sayd myne, which instruments do apperteyne to the righte honorable and worshipfull company of the abovesaide discoverie. I do therefore charge all captaynes and maysters of every shippe or shippes to make it knowen to his or their companies, to the end that all such instruments, as well those lent, as those that are otherwise deteyned and kept away, may be agayne restored, and broughte aboord the Admirall upon payne and danger expressed in the third article.

By me, MARTIN FROBISHER.

THE FLEETES RETURNING HOMEWARD.

Having nowe receyved articles and direction for oure returne homewardes, all other things being in forwardnesse and in good order, the last day of August the whole fleete departed from the Countesse Sound, excepting the *Judith* and the *Anne Frances*, who stayed for the taking in of fresh water, and came forth the next daye and mette the fleete lying off and on, athwart Beares Sounde, who stayed for the Generall, which then was gone ashore to dispatch the two barkes and the *Busse*, of Bridgewater, for their loading, whereby to get the companyes and other things aboorde. The captayne of the *Anne Frances* having most part of his company ashore the first of September, went also to Beares Sound in his pinnesse to fetch hys men aboorde, but the winde grew so great immediately uppon their landing that the shippes at sea were in great danger, and some of them hardly put from their ankers, and greatly feared to be utterly lost, as the *Hopewell*, wherein was Captayne Carew and others, who could not tell on which side their danger was most, for having mightie rockes threatening on

Return homeward.

the one side, and driving ilands of cutting ise on the other side, they greatly feared to make shipwrack, ye ise driving so neare them that it touched their borde sprete. And by meanes of ye sea that was growen so hie, they were not able to put to seas with their smal pinnesses, to recover their shippes. And, againe, the ships were not able to tarrie or lye athwarte for them by meanes of the outrageous windes and swelling seas. The General willed the captaine of the *Anne Frances* with his companye for that nighte to lodge aboorde the *Busse* of Bridgewater, and went himself with the rest of his men aborde the barkes. But their numbers were so great and the provision of the barkes so scant that they pestered one another exceedingly. They had good hope that the next morning the weather woulde be faire, wherby they might recover their shippes. But in the morning following it was farre worse, for the storme continued greater, the sea being more swollen and the fleete gone quite out of sighte. So that now their doubts began to growe great, for the ship of Bridgewater which was of greatest receit, and wherof they had best hope and made most accompt, roade so far to leewarde of the harborow mouth, that they were not able for the rockes (that lay betweene the winde and them) to leade it out to sea with a sayle. And the barkes were so already pestered with men and so slenderly furnished of provision, that they had scarce meate for sixe dayes for such numbers.

The Generall in the morning departed to sea in the *Gabriell*, to seeke for the fleete, leaving the *Busse*, of Bridgewater, and the *Michael* behinde in Beares Sound. The *Busse* set sayle, and thought by turning in the narrowe channell within the harborow, to get to windewarde; but being put to leewarde more by that meanes was faine to come to ancker for hir better safetie amongst a number of rockes, and there left in great danger of ever getting forth againe. The *Michaell* set sayle to follow the Generall, and could give ye *Busse* no reliefe, although they earnestly desired the same. And the captaine of the *Anne Frances* was lefte in harde election of two evils: either to abide his fortune with the *Busse,* of Bridgewater, which was doubtfull of ever getting forth, or else to be towed in his smal pinnesse at the sterne of the *Michael* thorow the raging seas, for that the barke was not able to receive or releeve halfe his company, wherein his daunger was not a little perillous.

So, after resolved to committe himselfe, with all his company, unto that fortune of God and sea, hee was daungerously towed at the sterne of the barke for many myles, untill at length they espyed the *Anne Frances* under sayle, harde under their lee, which was no small comforte unto them. For no doubt both those and a great number moe had perished for lacke of victuals, and convenient roome in the barkes, without the helpe of the sayde ships. But the honest care that the maister of the *Anne Frances* had of his captaine and the good regarde of dutie towards his General, suffered him not to depart, but honestly abode to hazarde a daungerous roade all the night long, notwithstanding all the stormy weather, when all the fleete besides departed. And the pinnesse came no sooner aborde the shippe, and the men entered, but she presently sheavered and fel in peeces, and sunke at the ships sterne with al the poore mens furniture: so weake was the boate with towing, and so forcible was the sea to bruse hir in peeces. But (as God woulde) the men were all saved.

At this presente in this storme manye of the fleete were dangerously distressed, and were severed almost al asunder. And there were lost in the whole fleete well neere xx boates and pinnesses in this storme, and some men stroken over boorde into the sea, and utterly lost. Manye also spente their mayne yardes and mastes, and with the continuall frostes and deawe, the roapes of our shippes were nowe growen so rotten, that they went all asunder. Yet, thanks be to God, all the fleete arrived safely in Englande aboute the first of October, some in one place, and some in another. But among other, it was most marvellous how y^e^ *Busse*, of Bridgewater, got away, who being lefte behinde the fleete in great daunger of never getting forth, was forced to seeke a way northwarde, thorowe an unknowen channel full of rockes, upon the back side of Beares Sounde, and there by good hap found out a way into the north sea (a very daungerous attempte), save that necessitie, which hath no lawe, forced them to trie masteries. This foresaide north sea is the same which lyeth upon the backe side of all the northe lande of Frobishers Straits, where first ye Generall himself in his pinnesses, and some other of our company have discovered (as they affirme) a great forelande where they would have also a greate likelyhoode of the greatest passage towardes the South Sea, or *Mare del Sur*.

A fruitful new iland discovered.

The *Busse*, of Bridgewater, as she came homeward to ye southestwarde of Freseland, discovered a great ilande in the latitude of degrees which was never yet founde before, and sayled three dayes alongst the coast, the land seeming to be fruiteful, full of woods, and a champain countrie.

There dyed in the whole fleete in all this voyage not above fortie persons, whiche number is not great, considering howe manye ships were in the fleete, and how strange fortunes we passed.

A GENERALL BRIEFE DESCRIPTION OF THE COUNTREY, AND CONDITION OF THE PEOPLE, WHICH ARE FOUND IN "META INCOGNITA."

Having now sufficiently and truly set forth ye whole circumstance, and particular handling of every occurence in the three voyages of our worthy Generall, Captayne Frobisher, it shal not be from the purpose to speake somewhat in generall of the nature of this countrey called *Meta In-*

cognita, and the condition of the savage people there inhabiting.

First therefore concerning the topographicall description of the place. It is nowe founde in the last voyage that Queen Elizabeths Cape, being situate in latitude of degrees and a halfe, whiche before was supposed to be parte of the firme land of America. And also all the rest of the south side of Frobishers Straytes, are all severall ilands and broken land, and likewise so will all the north side of the said straytes fall out to be, as I thinke. And some of our company being entred above 60 leagues within the mistaken straytes, in the third booke mentioned, thought certaynely that they had descryed the firme lande of America towards the south, which I thinke will fall out so to bee. A topographicall description of Meta Incognita.

These broken landes and ilandes, being very many in number, do seeme to make there an archipelagus, which as they all differ in greatnesse, forme, and fashion one from another, so are they in goodnesse, couloure and soyle muche unlike. They all are very high lands, mountaynes, and in most parts covered with snow, even all the summer long. The norther lands have lesse store of snow, more grasse, and are more playne countreys; the cause may be, for that the souther ilands receive all the snow, y^t the cold winds and percing ayre bring out of the north. And contrarily the norther partes receive more warme blastes of milder aire from the south, whereupon may grow the cause why the people covet and inhabit more upon the north partes, than the south, as farre as we can yet by our experience perceive they doe. These people I judge to be a kinde of Tartar, or rather a kind of Samowey, of the same sort and condition of life y^t the Samoweides be to the northeastwards, beyond Moscovy, who are called Samoweyes, which is as much to say in the Moscovy tong, as eaters of themselves, and so the Russians their borderers doe name them. And by late conference with a friend of mine (with whome I dyd sometime travell in the parts of Moscovy) who hath great experience of those Somoweides and people of ye northeast, I finde, that in all their maner of living, those people of the northeast, and these of the northweast, are like. They are of the coloure of a ripe olive, which how it may come to passe, being borne in so cold a climate, I referre to ye judgement of others, for they are naturally borne children of the same couloure and complexion as all the Americans are, which dwell under the equinoctiall line.

They are men very active and nimble. They are a strong people, and very warlike, for in our sighte, uppon the toppes of the hilles, they would often muster themselves, and after the maner of a skirmish, trace their ground very nimbly, and mannage their bowes and dartes with great dexteritie. They goe clad in coates made of the skinnes of beastes, as of ceales, dere, beares, foxes, and hares. They have also some garments of feathers, being made of the cases of foules, finely sowed and compact togither. Of all which sortes, we broughte home some with us into England, which we founde in their tents. In sommer, they use to weare the hearie side of their coates outwarde, and sometime go naked for too much heate. And in winter (as by signes they have declared) they weare foure or five folde uppon their bodies with ye heare (for warmth) turned inward. Hereby it appeareth, that the ayre there is not indifferente, but eyther it is fervent hote, or else extreeme colde, and far more excessive in both qualities, than the reason of the clymate shoulde yeelde. For there it is colder, being under degrees in latitude than it is at Warhus in the voyage to Saint Nicholas in Moscovie, being at above 70 degrees in latitude. The reason hereof, perhaps, maye be, that thys Meta Incognita is much frequented and vexed with eastern and northeastern windes, whiche from the sea and ise bringeth often an intollerable cold ayre, whiche was also the cause that this yere our straites were so long shutte up. But there is great hope and likelyhoode, that further within the straights it will be more constant and temperate weather.

These people are in nature verye subtil, and sharpe witted, readye to conceive our meaning by signes, and to make answere, well to be understoode againe. As if they have not seene the thing whereof you aske them, they wyll winck, or cover their eyes with their hands, as who would say, it hath bene hyd from their sighte. If they understande you not, whereof you aske them, they will stoppe their eares. They will teach us the names of eache thing in their language, which we desire to learne, and are apt to learne any thing of us. They delight in musicke above measure, and will keep time and stroke to any tune which you shal sing, both wyth their voyce, heade, hande and feete, and wyll sing the same tune aptlye after you. They will rowe with our oares in our boates, and kepe a true stroke with oure mariners, and seeme to take great delight therein. They live in caves of the earth and hunte for their dinners or praye, even as the beare or other wilde beastes do. They eate rawe fleshe and fishe, and refuse no meate, howsoever it be stinking. They are desperate in their fight, sullen of nature, and ravenous in their manner of feedinge.

Their sullen and desperate nature doth herein manifestly appeare, that a companie of them being environed of our men, on the toppe of a high cliffe, so that they coulde by no meanes escape our handes, finding themselves in this case distressed, chose rather to cast themselves headlong downe the rockes into the sea, and so to be brused and drowned, rather than to yeeld themselves to our men's mercies.

For their weapons, to offende their enimies, or kill their pray withall, they have dartes, slings, bowes, and arrows headed with sharp stones, bones, and some with yron. They are exceedingly friendly and kinde harted, one to the other, and mourne greatly at the losse or harme of their fellowes, and expresse their griefe of minde, when they part one from an other, with a mournefull song, and Dirges. They are very shamefast in bewraying the secretes of nature, and verye chaste in ye maner of their living: for when the man which we brought from thence into England (ye last voyage) should put of his coat, or discover his whole body for change, he would not suffer the woman to be present, but put hir forth of hys cabin. And in all the space of two or three monethes, while the man lived in company of the woman, there was never any thing seene or perceived betweene them more than might have passed betweene brother and sister: but the woman was in all things very servicable for the man, attending him carefully, when he was sick, and he likewise in al the meates whiche they did eate togither, would carve unto her of the sweetest, fattest, and best morsels they had. They wondred muche at all our things, and were afraide of our horses, and other beastes, out of measure. They beganne to grow more civill, familiar, pleasant, and docible amongst us in a verye shorte time.

They have boates made of leather, and covered cleane over, saving one place in the middle to sit in, plancked within with timber, and they use to rowe therein with one ore, more swiftly a great deale, than we in our boates can doe with twentie. They have one sort of greater boates wherin they can carrie above twentie persons, and have a mast wyth a sayle thereon, whiche sayle is made of thinne skinnes or bladders, sowed together with the sinewes of fishes.

They are good fishermen, and in their small boates, beeing disguised with their coates of ceales skinnes, they deceyve the fishe, who take them rather for their fellowe ceales, than for deceyving men.

They are good marke men. With their darte or arrowe they will commonly kill ducke or any other foule in the head and commonly in the eye.

When they shoote at a greate fishe with anye of theyr

dartes, they use to tye a bladder thereunto, whereby they may the better finde them againe, and the fishe not able to carrie it so easily away, for that the bladder dothe boy the darte, will at length be weerie and dye therewith.

They use to traffike and exchange their commodities with some other people, of whome they have such things as their miserable country and ignorance of arte to make, denyeth them to have, as barres of iron, heads of iron for their dartes, needles made foure-square, certayne buttons of copper, whiche they use to weare uppon theyr forheads for ornaments, as oure ledyes in the Court of England do use great pearle.

Also they have made signes unto us that they have seen gold and such bright plates of mettals whiche are used for ornaments amongst some people with whome they have conference.

We found also in their tents a Guiney beane of redde couloure, the which dothe usually grow in the hote countreys: whereby it appereth they trade with other nations whiche dwell farre off, or else themselves are great travellers.

They have nothing in use among them to make fyre withall, saving a kind of heath and mosse which groweth there.

And they kindle their fyre with continuall rubbing and fretting one sticke againste another, as we do with flints. They drawe with dogges in sleads upon the ise, and remove their tents therwithal, wherein they dwel in sommer, when they goe a hunting for their praye and provision againste winter. They doe sometime parboyle their meate a little and seeth the same in kettles made of beasts skins: they have also pannes cutte and made of stone very artificially: they use preaty ginnes wherewith they take foule. The women carry their sucking children at their backs, and do feed them with raw flesh, which first they do a little chawe

How they make fyre.

The kettles and pannes.

in their owne mouths. The women have their faces marked or painted over with small blewe spots: they have blacke and long haire on their heads, and trimme the same in a decent order. The men have but little haire on their faces, and very thinne beardes. For their common drincke, they eate ise to quench their thirst withal. Their earth yeeldeth no graine or fruite of sustenance for man, or almost for beast to live uppon; and the people will eate grasse and shrubs of the grounde, even as our kine do. They have no woode growing in theyr countrey thereaboutes, and yet wee finde they have some timber among them, whiche we thinke doth grow farre off to the southwardes of this place, about Canada, or some other part of Newe Founde Land: for there belike, the trees standing on the cliffes of the sea side, by the waight of ise and snowe in winter overcharging them with waight when the sommers thawe commeth above, and also the sea underfretting them beneath, whiche winneth daylye of the lande, they are undermined and fall down from those cliffes into the sea, and with the tydes and currants are driven to and fro upon the coasts further off, and by conjecture are taken uppe here by these countrie people to serve them, to plancke and strengthen their boats withall, and to make dartes, bowes, and arrowes, and suche other things necessarie for their use. And of this kind of drift wood we finde all the seas over great store, which being cutte or sawed asunder, by reason of long driving in the sea, is eaten of wormes, and full of hoales, of whych sorte theirs is founde to be.

The people eat grasse shrubbes.

The moone maketh a revolution above ground.

We have not yet founde anye venemous serpent or other hurtefull thing in these partes, but there is a kinde of small fly or gnat that stingeth and offendeth sorelye, leaving manye red spots in the face, and other places where she stingeth. They have snowe and hayle in the beste time of their sommer, and the ground frosen three fadome deepe.

These people are great inchaunters, and use many charms

of witchcraft: for when their heads do ake, they tye a great stone with a string unto a sticke, and with certayne prayers and wordes done to the sticke, they lift up the stone from the ground, which sometimes wyth all a mans force they cannot stir, and sometime againe they lift as easily as a feather, and hope thereby with certayne ceremonious words to have ease and helpe. And they made us by signes to understand, lying groveling with their faces uppon the ground and making a noise downwarde, that they worshippe the devill under them.

They have great store of deere, beares, hares, foxes, and innumerable numbers of sundry sortes of wilde foule, as seamewes, gulles, wilmotes, duckes, &c., whereof our men killed in one day fifteene hundred.

They have also store of hawkes, as falcons, tassels, &c., whereof two alighted upon one of our shippes at theyr returne, and were brought into England, which some thinke will prove very good.

There are also greate store of ravens, larkes, and partridges, whereof the countrey people feede.

All the fowles are farre thicker clothed with downe and feathers, and have thicker skinnes than any in England have: for, as that country is colder, so Nature hathe provided a remedie thereunto.

Our men have eaten of their beares, hares, partriches, larkes, and of their wilde fowle, and find them reasonable good meate, but not so delectable as ours.

Their wilde fowle must be all fleyne, their skinnes are so thick: and they tast best fryed in pannes.

The countrie seemeth to be much subjecte to earthquakes.

The ayre is very subtile, piercing, and searching, so that if any corrupted or infected body, especially with the disease called *Morbus Gallicus,* come there, it will presentlye breake forth and shewe it selfe, and cannot there by anye kinde of salve or medicine be cured.

Their longest sommers day is of greate length, without any darke night, so that in July all the night long we might perfitely and easilie wright and reade whatsoever had pleased us, which lightsome nights were very beneficiall unto us, being so distressed with abundance of ise as wee were.

The length of their day.

The sunne setteth to them in the evening at a quarter of an houre after tenne of the clocke, and riseth agayne in the morning at three quarters of an houre after one of the clocke, so that in sommer theyr sunne shineth to them twentie houres and a halfe, and in the nighte is absent three houres and a halfe. And although the sunne be absent these 3½ houres, yet is it not darke that time, for that the sunne is never above three or foure degrees under the edge of the horizon: the cause is, that the tropicke cancer doth cutte their horizon at very uneaven and oblique angles. But the moone at any time of the yeare beeing in Cancer, having north latitude, doth make a full revolution above their horizon, so that sometimes they see the moone above 24 houres togither. Some of oure companie, of the more ignorant sort, thought we might continually have seene the sunne and the moone had it not bin for two or three high mountaynes.

The people are nowe become so warye and so circumspect, by reason of their former losses, that by no means we can apprehend any of them, although we attempted often in this last voyage. But to saye truth, we could not bestowe any great time in pursuing them, bycause of oure greate businesse in lading and other things.

Commodities of Meta Incognita.

To conclude, I finde all the countrie nothing that may be to delite in, either of pleasure or of accompte, only the shewe of mine, bothe of golde, silver, steele, yron, and blacke leade, with divers preaty stones, as blewe saphire very perfect, and others, whereof we founde great plentie, maye give encouragement for men to seeke thyther. And there is no doubt, but being well looked unto and thorowly discovered, it wyll make our countrie both rich and happye, and of these prosperous beginnings will growe hereafter (I hope) most happye endings. Which GOD of his goodnesse graunte, to whome be all prayse and glorie. Amen.

At London:
Printed by HENRY BYNNYMAN,
Anno Domini 1578,
Decembris 10.

ANOTHER ACCOUNT OF THE THIRD VOYAGE, BY EDWARD SELLMAN.

EDWARD SELLMANN wrote this booke; and he delivered yt to Michael Lok, the 2 of October 1578, in London:—

The 2 of May 1578, we departed from Bristoll with the *Ayde* and the *Gabriell,* Christopher Hall, and Robert Davis M[s].

The 6 said we arrived at Plymouth, where we stayed to take in our myners.

The 19 said we departed from Plymouth, with the *Ayde,* the *Fraunces,* and the *Moone of Foy,* the *Admirrell,* and the bark *Denis,* and arrived at the Downes the 24th said and the said at midnight we departed thence and arrived at Harwiche the 22 said to stay for the reste of the fleete, where we found the *Thomas* of Harwich.

The 27 said, there arrived at Harwich, the *Thomas Alin,* the *An Fraunces,* the *Hopewell,* the *Beare Lester,* the *Judith,* the *Gabriell,* and the *Michael,* the *Salomon* of Weymouth came to us to Harwich, and the *Emanuel* of

The 31 said the *Aide* with all the above named ships departed from Harwiche with the winde at N.E. making our passage towards the west coast and arrived at Plymouth the 3 of June.

The 3 of June, 1578, at night we departed from Plymouth, with the winde west hand at east, and to the westwards of the Cape 7 leags we had sight of a bark of Bristoll with whome after we had spoken, they declared that they came out of Spayne and were robbed by 2 French men-of-war, and five of their companye slayne, their lading was oyle and sack, they spoiled them of all their victuall allso and left them nothing to eate of but oyle beryes: The Generall gave them 3 sackes of bisket, and j barrell of butter, peas and chese to releve them withall, by which bark I wrote a letter, and sent it to Mr. Kitchen to be conveyed to my master, Mr. Michael Lock, advertising him of all the fleets arrivale uppon the coast of Zealand.

The 7 said we sailed N.W. and by W. the winde at S.E. a fyne bearing gale, with the winde sometimes at N.E. sometimes at E. sometimes at S.W. still keping our course (for the most part) N.W. and by W. and N.W. untill the 19 said at none, at which time we went in 60 degrees of latitude, and to the eastwards of Friseland, 30 leags, by the reckening of some 40, and 50 by others, and bearing N.W. and by W. and W.N.W. of us at the going down of the sunne, but at that time we had not made the land perfect, and so sayling untill mydnight, we came nerer unto yt and made yt perfectly. At which tyme we shot of a pece of ordonance to geve the flete warning thereof: I judge the voyage is better to be attempted, followed and used by the west parts, then by the north parts, as well for the avoyding of much cold within the north passage we had, as allso redyer windes to follow our said viadge, as by the falling out of this passage doth appere.

The 20 of June, 1578, earely in the morning, the Generall caused a small pynnas to be hoysed out of the *Ayde*, and with her he passed a boord the *Gabriell*, and did beare in with the land sayling alongst yt, untill he found a sound to enter in uppon the south side of the land, which sound after he was entred, called yt Luke's Sound, by reason of one Luke Ward that went with him a land; in which sound they found people and tents, but the people fled from them, and they entred their tents, finding thereby by all things therein that they are a people like the people of Meta Incognita with like boates of all sortes, but the Generall doth take them to be a more delicat people in lodging and feeding then the other: They found of their seals which they had taken sundry, and other victuaill which they could not tell what flesh or fish yt was: At their said tente they found allso 40 yong whelps, whereof 2 they brought away with them, they are allso like the dogs of the place afore named: Some of our men that were with the Generall aland did see in their tente nayles like scupper nayles, and a tryvet of yron, but the Generall toke order with the company, that none shold bring any of their things away: The Generall hath named this iland West England, and a certayn hedland uppon the south side, he hath called yt Furbushers foreland, with other names he hath geven to particular places which I know not.

The said at night we departed thens with the winde N.E. and sayled W.N.W. towards the Streits untill 9 or 10 a clock the 21 said.

The 21 said the winde N.W. we sayled N.E. and by E. towards the said West England to make better discovery of yt, bycause yt served not us to procede of our pretended viage, and so sayling till 3 a clock, yt fell caulme, being 16 leags from yt: About 6 a clock the winde at N.N.E. we sayled N.W. and by W. towards the straits.

The 22 the winde at E.S.E. we sayled N.W. and by W. untill none, and then we met with great store of yse, of broken ilands in great peeces, which we iudge to be the ilands dissolved, that were there seene the last yere and driven upon the N.W. coast, by reason of the easterly windes which we had comming hitherwards, and for that we coveted to discover more of the north west coast by reason of clere weather which we had, we were the rather put amongst them, and thereby to clere ourselves of them againe, to sayle south, S. and by W. and S.W. for the space of 3 or 4 howres with a great gale of winde: And we feared the coast to lye out more westerly then we could make yt by reason of foggy weather and thereby might have bin driven uppon a lee coast, but ofter we found ourselves clere of the yse we sayled agayne N.W. and by W. with the winde at S.E. untill the 25 said, and then the winde came W. and we sailed N.N.W. untill the 27 said, at which time we came amongst as well great ilands of yse, as allso great quantity of broken yse of both sides of us being shotte within channells' of them, whereby yt was iudged that we were open of the straits, and we made sundry foggy land to be the Queens foreland, and thereuppon did beare the bolder in amongst them, at which tyme we found our selves in the latitude of $62\frac{2}{3}$ and some $62\frac{1}{2}$. And the 28 said they observed the latitude by the sunne and found them in $62\frac{2}{3}$ of latitude and afterwards had sight of 2 ilands to the northwards of Warwicks foreland, and after had sight of the same foreland, we being to northwards of yt 14 or 16 leags. And the said day we lay to the ofwards south-east and south-south-east: And the 29 said south-west, the winde at W.N.W. untill the 30 said, and then we sayled south and by E. and S.S.E. untill we came in the latitude of $61\frac{2}{3}$ the first of July, at which tyme we had the winde at S.S.E. and then we sailed in W. And the 2 said we had sight of the Queens foreland and sometimes did beare in N.W. and by W. and N.W. finding stragling over all the straicts and after we did beare in further uppon the south side we found great quantity of yse driving together, yet we had sundry channels to pas betwene them, and after that we sent the pynnas from the ship to discover the best way our passage amongst them, and so we followed with divers other of the fleete after the pynnas, untill she could not pas any furder, fynding the yse all closed abowt us, and afterward sent our boate and pynnasses of divers of the flete to breake a small neck of yse for passage farder places that we did see clere: and at that tyme the winde began to blow vere boystrous at the S.S.E. and caused the sea to heave and set very cruell; at that instant we were divers of the flete in a great channell indifferent free of yse, in which channell we determined to spend the night with bearing small sayles, being environed with yse: The bark *Denis* at that tyme plying up and down, did strike uppon a great yse and there perished, so that the boates which were sent to breake the yse for passage, returned to her to save her men and presently after the ship did sink down right; divers of the flete, notwithstanding with small sayles did ply up and down in the same channell, and others as they could fynde all that night, but we in the *Ayde*, and the *Thomas Alin* did forsake yt, bearing no sayle, but lay adrift amongst the yse all the night, being terribly tormented therewith untill 11 of the clock, the 3 of July, occupying our men with oares, pikes and other powles to break the force of the yse from beating of the ship as much as we might, notwithstanding we had terrible blowes therewith, and were preserved by the mighty power of God from perrishing, contrary to our expectations. The winde afterwards comming to the S.W. and having the with us, we did drive out, sometyme setting sayle and sometimes a hull south east: And being allmost out of the danger of the yse, we did discrye the most of the flete, which rejoyced us very much: And the said 3 day about night, some of us talking with others did understand we were all in saffety, except onely the *Michael* of whome as yet we cannot understand where she ys, we did arme the bowe of the ship with sundry planks of 3 inches thick and with capstayne barrs and junks, for that the yse stroke terribly against that place of her in so much that some of the planks did perrish with the blowes: The rest of the flete except one or two more did not pas the like mysery, by reason they did kepe the channell betwene the yse with small sayles, which we could not do, for that our ship was long, and could not work with her as others did: And besides that yf we had kept that channell with sayle, where the rest did, we had burded one an other and thereby perished, as we had like to have don by the ship of Weymouth (owner Hugh Randall) in boording of us that night by drift and forcing uppon us by yse, the boystrous winde that then did blow did cause us to unrig and take down both our topmasts for the ease of the ship, the which topmasts we did hang over boord allso to save the ship from the yse.

The 4 said being in the morning clere withowt the streicts and the winde at west, we did sayle S.S.W. bearing alongst the coast of America, fynding yse driving from the coast as though yt were long hedges into the sea to the eastwards, we sayled as aforesaid untill 4 a clocke at afternone, and then we layde yt a hull, untill 6 aclock the 5 said, and then we sailed southwest alongst America coast, the winde at W. northwest untill the 5 said at night about 6 aclock at which tyme we had sight of the coast, and very huge ilands of yse, higher than ever we did see any, at which time we did cast about and did lye north of the land, the winde as before untill the 6 said at night, at which time we were within the streicts and did perfectly make the Queens forelande; to the southwards of the Queens foreland, we had sight of a head of a land, being from yt about 20 leags, which untill we had taken the latitude of yt we made yt to be the Queens foreland all but the Master Robert Davis onely, but he would not agree to yt, nor so allow yt, alledging sundry reasons to prove the contrary by his marks when he saw yt a few dayes before. The 6 said being as aforesaid shot within the Queens forelande on the south side, the winde came up to the S.S.E. and did blow a great gale, and the weather waxed thick and foggy, and therefore all the night we layde yt a hull. And uppon the 7 said following we had sight of the north shore as we toke yt: And the 8 and 9 we did beare wyth yt and alongst yt lying north and by west, but did not make yt perfectly; some imagining rather that yt was the S. side of the Queens foreland (as afterwards yt proved in dede), and Master Hall of the same opinion, but yf yt fall out so, they were deceyved with the setting of the tides. The Generall and our master could not be dissuaded, but doth still make yt to be the north shore, the Generall assuring himself thereof to this present (the 10 said) that yt is so, and Jame Beare allso, but being foggy and darkened with mystes, they cannot yet make yt perfectly, I pray God send yt clere, that we may make yt perfectly: Alongst the said shore in sight and out of sight by reason of fogs, we did runne in by the judgement of the master 35 leags bearing sayle and hulling, and there did remaine hulling being dark and foggy untill the 16 said, at which tyme we had yt somewhat clere, and thereuppon did beare towards the shore to make yt, at which tyme we did fall with the opening of a sound which we made the Counte's Sound and did beare in with yt, all men that had seene it the yere before (except two, called Stobern and Bert) allowed yt to be the same, which afterwards proved the contrary: The 17 said we toke the altitude of the sunne and found us but in the latitude of 62 and 10 minuts, and thereuppon found the error which we were in, then knowing that we were uppon the S. side of the S. shore called the Queens foreland, and with the winde at W. we did beare out agayne, and the 18 said being shot out so far as to the masters judgement that we had sight of the Queens foreland being E. from us and then running alongst till we brought yt thwart of us the weather being foggy, notwithstanding we did alter our course more northerly and brought us to be impatched with great quantity of yse and dark weather, being allso shot very nere the shore, still thinking that we had byn at the Queens foreland, and altering our course more northerly, did bring ourselves hard aboord the shore, at which tyme yt pleased God to geve us sight of yt, and thereby found yt did not lye as the Queens foreland did, fynding us deceyved and not so far shot as the said foreland, but being imbayed uppon a lee coast and in sight of divers ilands and rocks, not knowing how to escape with life, and in the depe of 50 faddoms of water, so that we could not well anker, but yet sometimes in mynde to anker yf we could have got a poynt of an iland which we made unto, and then fearing allso we shold have had byn put from our anker, or greatly impatched with yse which we were allso amongst and then caulme, and could not get of from the rocks or ilands which we did see, did strike all our sayles to anker, but before we were all ready to cast anker, the Eternall God (who delivereth all men being in perills) did send us a gale of winde to beare of from the said ilands, but afterward we wished that we had ankered there, for that when we were of a small way from yt we sounded and found us in but 7 faddoms of water and hard rocks, we lying under sayle towards the west which was our best way, for sure we were we could not dubble the land to the eastwards, the winde being at S.S.E. and the land lying E.S.E. and W.N.W. we after yt pleased God to send us 10 faddoms and then 17, and then 25, and so into 30 and 40, and allso did sende us the winde at W.S.W. so that we did lye S.S.W. of into the sea untill we came into 120 faddoms with our sayling and towing out with our boates, still having the eb with us untill night and then being caulme and little winde, we did strike our sayles and did lye a hulling, so that the flud did port us in towards the shore againe untill we came into 80 faddoms, and then we were forced to make a brude of cabells, and did anker untill the eb did come being the 19 day of July in the morning, at which tyme we did set sayle with a small gale of winde, the winde at S. and by E. and did sayle S.W. and by W. the weather still foggy. The 20 said the weather began to clere, the winde westerly, at which tyme we had sight of the ships that were before in our company, and towards the afternone we came to speake with some of them, and they declared that some of our company were in 2 faddoms of water uppon the lee shore, being in great danger amongst the rocks and broken grounds, and delivered by Gods allmighty power thus twise from perisshing, towards night yt waxed somwhat foggy agayn, and a little before night we having sight of a point of land, bearing E.S.E. of us making yt the Queens foreland, we did beare with it in such sort as we thought to go clere of yt, and the land lying out farder then we had sight of yt, we being not so far shot out of the streict that we were in by 20 leags which 20 leags we were in furder then we made account of, being entred within yt at the least 60 leags, fell agayn in danger of that land in the night, but kept us of from yt, by our sounding lead: And in the morning the 21 said yt waxed clerer, and then we made the land of the Queens foreland perfect and towards night opening a great bay at the wester end of the souther parte of that land, which we imagined to go through into the Streicts of Frobusher, which to make triall thereof, the *Gabriell* was sent to discover, and we bearing about with the easter end of yt the 22 towards night, had sight of the *Gabriell* comming into the streicts through that sound passage at the Cape of Good Hope, so that it is proved that the land of the Queens foreland to be an iland; the *Gabriell* having order to passe to the Countesses Sound, did beare in towards yt, and we followed untill we could not passe any farder for yse lying so thick, and the *Gabriell* being within the yse, did still beare up into the streicts, and we forced to retire outwards agayn, being very much impatched therewith all the whole night.

The 23 said we had sight of the *Anfraunces*, whose company we lost as before said, and when we came to the speeche of the captayne and master, they declared they had layn of and on open of the streicts 12 dayes and could not entre for fogs and yse, and was in danger before that uppon the lee shore of the S. side after she departed from us.

The 24 said the Generall being mynded to beare into the

streicts, bycause the *Gabriell* passed up in our sight, supposing allso the *Tho. Alin*, the *Fraunces of Foy*, the *Emanuell* of Bridgewater, the *Judith*, and the *Michael*, to be above in the sound; notwithstanding the great quantity of yse, we were impatched withall the 23 said, and the winde at S.W. a good and reasonable gale, did mynde to beare up into the streicts agayn this present, alledging that the said wynde had brought out all the yse, whereof great quantity we did see blown uppon the lee coast: but yt pleased God to send us a messinger out of the streicts called the *Fraunces of Foy*, who did kepe company with the *Tho. Alin*, and the *Emanuel* of Bridgewater, and did enter into the streicts the 19 said and the 20 said, being shut up as far as Jackmans Sound, did put over with the Countes Sound among very much yse and were environned therewith, frosen and shut up therein, being marveylously tormented therewith, not onely with yse comming down, but allso with yse carried up with the winde and tyde. This *Ffraunces of Foy* (I say) was a blessed messinger of God, sent to us to warn us of the daungers that she and the others passed, who still did leaye the *Tho. Alin*, the *Busse* or *Emanuel* of Bridgewater, and the *Gabriell*, last come unto them in great danger, being carried towards the coast lee in the frosen and thick yse as the winde did carry them. God deliver them for his mercyes sake and for his blessed sonne Jesus Christes sake. The Master Tho. Noris of the said ship the *Ffraunces of Foy*, before Master Hall, and he with others entred the streict was in a sound uppon the N. side of the Queens Foreland, where they were they found very good owr by our judgements to the sight, and therefore the Generall is gone this morning a land to seke the same, purposing to go into the said sound with all our 9 ships now in company untill tyme may serve us to go farder and other our ports of lading. The 25 at night we did beare into the streicts and then had sight of the *Emanuel* of Bridgewater. And, comming to the speche of them, the master of her affirmed that the yse did ly very thick over all the streicts, so that we could not attayn to the Countess Sound as yet; the Generall, notwithstanding, wold geve no credit thereunto, but did beare in with the streicts to make triall thereof the 26 in the morning, and finding great store of yse did retire back or out agayn with the winde at north and much yse following us. At that tyme (the 25 said) these ships did entre in with us, the *Emanuel*, the *Armonell*, the *Hopewell*, and the *Beare*, and 5 others of the flete did put to sea, having the winde then at the S. E. and east, which was a scant winde for them to dubble out the foreland, being nere the land.

The 26, at night, we came back to the sea againe and brought the foreland of us south-west.

The 27, towards night, the winde at west, we did beare in towards the foreland, and did lye of and on all the night.

The 28, in the morning, we did beare agayn into the straight, the winde westerly, bearing inwards still untill we were repulsed and forced to put out agayn by reason of much yse driving out, but the *Hopewell* finding some clerer slade then we could do, did still beare in. God send her good hap. And then we did seke to recover the wether shore which was the foreland, the winde at W.N.W., blowing somewhat boystrous.

The 29, in the morning, we did beare into the streict agayn with winde at W., a small leading gale, and sometimes at W.S.W., we lying up N.W., passing up amongst great quantity of yse, sometime thick, and sometime thinner, and so did still procede, bearing inwards untill the 30 said at none, keping about the middle of the streict. And in the morning the 30 said we were thwart of Yorks Sound, which I affirmed to the Generall to be so: but he denyed yt, saying that we were not shot up as high as Jackmans Sound by 16 leags, at which instant the Generall went up to the top and descried Gabriels Iland, making yt to be Penbroke

Iland, going into the Countess Sound; and so directing his course with yt, Christopher Jackson, the trumpetter, being in the top, did make yt playnly Gabriels Iland, and allso made the Countess Sound, to the which the Generall yelded, and then presently did allter his course, and embarked him self in a pynnas with sayles and oares, bycause yt did blow but little wynde for the ship, and gave us tokens to follow him, and so signifyed to us thereby that yt was the right place or sound as before is said. Into the which he entred with his pynnas, and being entred therein fownd there the *Judith* and the *Michael*, and cause them to shote of certayn peces of ordonance, to geve knowledge there were certayn of our flete which comforted us very muche; but we imagined those ships to be the *Tho. Alin* and the *Gabriel*; for we did think verily the *Judith* and the *Michael* could not have escaped the dangers that they were in, being not of our company a month or more.

The 30 of July, at night, we entred into the mowth of the Cowntesse Sound, and there came to us sent from the generall, Charles Jackman, to bring in the *Ayde*, and for that yt fell caulme we came to an anker in the entring thereof, being ebbing water abowt 9 of the clock at night, the master, his mate, and Charles Jackman going then to supper, gave charge to the company to looke well owt for yse, driving towards the ship, willing them to prevent yt in tyme; and before the master had half supped, one of the company came to the master to know whether they shold watche half watche or quarter watche. The master gave order to watche halfe watche, charging them to loke well owt for yse; but the watche neglecting their dutyes, there came driving thwart the halse of the ship a great pece of yse, and the weather being caulme did ly uppon the cabell $\frac{1}{4}$ of an howre before we could be clere of yt fretting the cable in suche sort, that yf yt had put us from our anker we had byn in danger of rocks lying not far from us. God be honored, there chaunced no hurt of yt. Notwithstanding, I thought yt good and my duty to say something unto the watche of their negligence therein, bycause the charge of the vyage did depend upon the savegard of the *Ayde* being the Admirall, whereuppon I rebuked one Holmes, a quarter master, and Hill, bote swayn mate, charging them they shold aunswere their negligent loking to so greet a charge, but they with one other called did will me to meddle with that I had to do, demaunding whether I had commissioned to speake or deale therein, and this did Hill, and willed me to get me to my cabben, and wold not be checked at my hands. I aunswered them, whither I had commyssion or not, I wold tell them their duties, and go to my cabben when I did see cause, and thus with multiplying of words they abused me very much, which I was fayn to put up at their hands. The master can beare no rule amongst them, bycause he is not cowntenanced by the General, and therefore all things hath fallen owt the worse with us, and that hath caused me to speake more earnestly in this cause; for weyther the boat swayn, nor any officer yet hitherto hath byn obedient to the master, and the disobedience of the officers, doth cause the company allso to disobey and neglegt their duties. We had not byn above 2 howres at an anker, but that there came very muche yse driving inwards towards us: at which time, I being still abrode, and the masters mate allso, I said to him, yt were good to way our anker to prevent the danger of the yse; and presently he called vp Charles Jackman, and they caused the company to way the anker with the winde easterly, a smale gale; and after they had purchased home their said anker, there came yse uppon us, but they setting sayle before the anker was catted, the yse stroke the flok of the anker through the bow of the ship, that the water came in fercely, in so muche that we had water in hold 4 fote above the sealing within an howre or les. And our pumps being unready, could not free the ship of yt, but kept yt still at a stay, the leak being stopped as well as they could with beffe and other provisions. And thus we remayned pumping and freeing of the ship with buckets from 12 a clock at night, being the 30 of July, until 9 a clock in the morning, the 31 said, at which tyme we were come into harbour. And then provision was made to beare the ship over of the one side, and the hole mended with lead untill we may come better to yt.

There came into the Cowntesse Sound in company with us and in our sight, the *Hopewell*, the *Ffrances of Foy*, the *Armonell*, the *Emanuell*, the *Salomon* of Weymouth, and the *Bear*. The *Judith* and the *Michael* came into this sownd the 21 of July, and for the space of 3 wekes before they continually were tormented up and down within the streicts amongst the yse, and could not by any meanes get this place nor clere themselves of the yse: the *Judith* being bilged with yse in the bowes, having 2 great holes made in her, every howre loking when they shold perish therewith, but God delivered them, geving them fayre weather to work for their savegard.

The fyrst of August the Generall did order to make tents uppon the iland of the myne for the myners to succour them in their working there, and then began their work.

The second said, the Generall with 2 pynnasses, passed to Beares Sownd, to bring prooffs of the owr there, and to vew what quantity there was to be had, and returned agayn at night, being distant from the Cowntesse Sound 9 leags.

The said, at night, the *Gabriel* came into the Cowntess Sound and Master Hall in her to vew whither the streicts were clere of yse, and left the *Tho. Alin* in a sownd nere Oxford mount untill his return thither agayn.

The fyrst of August the *Fraunces of Foy* toke in 2 pynnasses, ladings of owre, and the 2 day as much.

The said the Generall, with 4 pynnasses and boates with a men, soldiers, and marriners, and Denham with him, went to Jonas Mownt, to seke for owr, and brought sundry samples, whereof as yet no assay is made, but of the riche owr that Jonas fownd the last yere, we could not light of any suche.

The 8 of August the *Thomas Alin* and the *Gabriel* arrived here towards night, by whome we could n(re of the *Thomas of Ipswiche*, the *Anfraunces*, and the *Mone*. I pray God send us good newes of them.

The 9 said, the Generall with the *Gabriell* and the *Michael*, with mariners, myners, and soldiers, departed towards Beares Sound to get owr, for that the myne in the Countess Iland fayled.

The said, the most part of the myners and soldiers were removed to a place called Fentons Fortune, being at the entrance of Countesse Sound to the eastwards. And yt was reported that there were a 1000 tunnes to be had there; but Master Denham, at his returne from thence, this present at night, sayeth he can not see how 40 tunnes will there be had, and that with great travayle to bring yt to the sea side.

The 11 sayd, the master, Robert Davis, Thomas Morice, master of the *Fraunces of Foy*, and I in company with them, travyled with a pynnas to the northwards of the Cowntesse Sound, about 4 myles alongst the coast, and there fownd a myne of black owr, and allso an other of red and of sundry sortes of both, of which sorts we brought ensamples, whereof Denham made proof; and the 13 said Capten Fenton and Denham passed thyther, liking the place very well, and aswell our mariners as the mariners of the said *Fraunces* were there set to work, and by the 15 said we had gotten aboord the *Ayde* of the black sort and some of the red abowt 15 tunnes.

The 15 said, towards the evening, the *Gabriell* and the *Michael* came to the Countesse Sound, both laden with owre from Beares Sound, and the 16 said discharged yt into the *Ayde*, theire lading was adiudged to be abowt 50 tunnes of owr.

The said, all such myners and soldiours as were sent from the Countesse Sound to Fentons Fortune, were removed to the myne that we found to the northwards, which was better liked than yt of Fentons Fortune, where, in the tyme they were there was but 60 or 70 tunnes of owre, they being myners and souldiours that wrought their 6 dayes 60 persons.

The 16 said, the Generall and Denham with him, is gon to a sownd called Dyers Passage, which is uppon the souther land of the Cowntess Sound, to vew a myne there, fownd by Andrew Dyer, and to make assayes thereof.

The said, God called to his mercy Philip, who had charge of certayn apparell brought in by the Generall for the marriners and myners, and allso one of the bark *Denys*, men called Trelos, one allso owt of the *Armonell*, and an other owt of the *Fraunces of Foy*, all buryed uppon Winters Furnace this present day.

The 18 sayd, the *Gabriell* and the *Michael* departed hence to Beares Sound to lade owr and to bring yt hither to the *Ayde*.

The 19 said, the *Solomon* of Weymouth departed towards Bears Sound to take in her lading of owr.

The 20 said, the *Beare* departed towards Dyers Passage to lade there.

The 19 said, the *Hopewell* departed towards Dyers Passage to lade there.

The 21 said, the *Busse* of Bridgewater departed to Dyers Passage to take her lading of owr there.

The 21 of August, the *Fraunces of Foy* was full laden, part of the owr of the Countesse Iland, and the rest of the owr of the myne to the northwards of the Countesse Sound, carrying in all tunnes by estimation 140, whereof 70 from the Countesse Iland, and the rest as aforesaid.

The 19 said, Capten Fenton came to make complaint to the Generall of the boatswayn, and others of the *Aydes* mariners, for disobeying him in certayn service to have byn don for the furderance and dispatche of the ships lading at two severall tymes, his speches tending to due punishment for the same, and after long recitall of their abuses, did loke that the Generall shold have ayded him therein, and to have commanded due punishments for their deserts. The Generall not taking order, therefore Master Fenton and he did grow to hoat speches, by whome eche others credit came by him, and he denying the same, left their former matter, and fell to reason uppon the same with many hoat woords, in somuche that in the end, the Generall affirming he preferred Master Fenton to be the Queens servant, and he denying, alledging that the Generall did not well to rob them that did prefer them both to that service; and then at Master Ffentons departure, he said he had offred him great disgrace in that he wold not punnish the offenders which he complayned of, but rather did animate them against him in neglecting of yt, which he could not take in good part, being his lieutenant generall, and recommending them to do nothing but their duties in their Maiesties service.

The 22 said the *Gabriel* arrived here at the Countess Sound being ladden with owr from Bears Sound, and discharged yt a boord the *Ayde*, bringing tunnes 25 by estimation.

The said, here at the Countess Sound arrived a pynnas of the *An Fraunces*, wherein Captayn Best came, leaving the *An Fraunces* and the *Mone of Foy* at anker in a sownd nere the Queens forelande, and they reported that they had not sene the *Thomas of Ipswich* this 14 dayes, with the said pynnas they came costing up allongst the sowth coast to seke us, and did seke us in Jackmans Sound and Yorks Sound and passed up as far as Gabriels Iland and returned hither this present, bringing them samples of owres, much like that of Winters furnace, and doth purpose that Denham shall make tryall thereof, and fynding yt good, they will lade of yt, having great plenty of yt as they report, they have by report passed great troubles sins they departed from us, by dangers of yse, and rocks, I pray God send us good newes of the *Thomas of Ipswich*.

The 23 said the Generall, Captayn Fenton (his lievtenant), Gilbert York, and George Beste, gentlemen, assembled themselves together, Christopher Hall, and Charles Jackman, masters, with them, for causes touching their instructions, and amongst other matters, did call in question the abuses of the boteswayn and one Robinson used towards the Generalls said lievtenant, and after yt had byn argued of amongst the said Commissioners, the Generall referred the punnishment thereof to them to determyn; then they called the said offenders before them, who acknowledged their abuses, and uppon their submission, as allso affirming they did not know Capteyn Fenton to be the Generalls said lievtenant, they were pardoned and forgeven.

The 23 said of August, the *Michael* arrived here laden with owr from Bears Sound bringing tunnes 25 by estimation and discharged yt aboord the *Ayde*.

The 24 said the *Sollomon* of Weymouth arrived here laden with owr of Bears Sound and with owr taken in her before her departure hence, all tunnes by estimation 130 tunnes, whereof Bears Sound tunnes 60, and of the Countesse Iland Sussex myne 60 tunnes, and Wynters furnace tuns 10.

The Generall departed this present towards Bears Sound in a pynnas and will return hither agayn before he go up into the Streicts.

The said Captayn Beste departed with his pynnas toward the Queens foreland to a sownd where the *An Fraunces* and the *Mone* resteth and stayeth his comming. The said *Ffraunces* and *Mone* by their marriners reports were almost laden with owr before their comming hither, the samples thereof hath byn proved and are reasonably well liked of Denham, and therefore I here order is taken that the *Moone* shall discharge all her owr into the *An Fraunces*, and that the said *Mone* shall take in all such here as the *An Fraunces* hath discharged there a land which was provided for Captayn Fenton and his company, and as wynde and weather shall serve to come hither with the same, and at Bears Sound she shall have her lading of owr provided.

The said the *Thomas Alin* departed hens, having taken in here 100 tunnes of owr had at the north myne called the Countesse of Sussex Myne, and the rest of her lading she is to take in at Beares Sound, and to that end she is gon thither where she is to lade 60 tuns more.

The 26 at night the Generall returned from Bears Sound with the pynnas that he departed from hens. And the 27 in the morning he passed with the same up into the Streict as well to discover mynes as allso to take of the people yf he may conveniently have them.

The 27 said at night the *Thomas Alin* arrived here from Bears Sound being fully laden.

The 28 said in the morning the *An Fraunces* arrived here from a sownd called being nere the Queens foreland and laden with owr of that place.

The said at night, the Generall returned with fowle weather and the winde easterly with rayne and snow and so continued till the 30 towards night.

The 31 said in the morning we wayed and made sayle from Countesse of Warwick Sound with the *Ayde*, the *Thomas Alin*, the *Bear*, the *Salomon*, the *Armonell*, and the two barks, and for that yt fell caulme, we ankered all that night at the mowth of the sownd, being all night caulme and the *Fraunces of Foy*.

The fyrst of September 1578 in the morning the *Gabriell* and the *Michael* did put into Bears Sound to lade there.

The said the Generall with a pynnas departed towards

Beares Sound to provide 10 or 12 tunnes of lading more for the *Ayde* and to send yt owt to us with boats and pynnasses.

The said the *Ayde* and all the other ships aforesaid wayde, the winde northerly, bearing alongst towards Bears Sound with a small gale, and about none ankered thwart of Bears Sound.

The 28 of August before, God called to his mercy Roger Littlestonne the Generalls servant, who by the judgement of the surgian had the horrible disease of the pox.

The last of July at night, God called to his mercy Anthony Sparrow, one of the quarter-masters of the *Ayde*.

The *Fraunces of Foy*, the *Armonell*, the *Thomas Alin*, the *Beare*, the *Salomon* came all laden owt of the Countess Sound, the *Ayde* lacked 10 or 12 tunnes but laden of sundry mynes as before is said.

The *An Fraunces*, the *Hopewell*, and the *Judith* arrived with us thwart of the said Bears Sound the fyrst of September and kept under sayle by us.

The first of September said we receyved tunns of owre into the *Ayde*, and all the myners this present at night were ready to come aboord from thens.

The said at night the winde chopping up to the N.W. a small gale and the sea growing thereby, forced us to way and made sayle, bearing of S.W. untill we came into 23 faddoms, and then ankered agayn, staying for the comming of the Generall, and abowt 2 howres after, our ship did drive, our anker being broken, which caused us to set saile agayn and did beare of W. and W. and by S. and afterwards did lye a hull, staying for the Generall, the winde still growing of great force at N.N.W. caused us to set our fore-saile agayn, bearing of sowth towards the foreland the second day of September and towing our gondelo at starn, she did split therewith and so we were forced to cut her of from the ship and lost her and then we did strike our sayle and spooned before the sea S.E. untill the Queens foreland did beare of us, the Generall is condemned of all men for bringing the flete in danger to anker there, thwart of Beares Sound onely for 2 boates of owre and in daungering him self allso, whome they iudge will hardly recover to come aboord of us, but rather forced to go with the barks or the *Emanuel* of Bridgewater into England; of the whole flete, there is now in our company, or to be seen but 6 sailes.

Master Hall went aland after the ship came first to an anker thwart the said Bears Sound, and did geve him counsaill to make hast a boord before night: God send him well to recover us and all his company.

The *Ayde* hath lading of owr in her as followeth:—Of Bears Sound tunnes by estimation 110; of the Countess of Sussex myne, tunnes 20.

The *Thomas Alin*, owr in her as followeth:—Of the Countess of Sussex myne, tunnes 100; of Beares Sound owre, tunnes 60.

The *Hopewell*, owr in her as followeth:—Of Dyers Passage or Sound, tunnes 140.

The *Fraunces of Foy* hath our laden in her as followeth:—Of the Cowntess of Warwiks myne, tuns 50; of the Countess of Sussex myne, tunnes 80.

The *An Fraunces* hath owr in her as followeth:—Of the Queens foreland, tunnes 130.

The *Mone of Foy* hath owr in her as followeth:—Of the Queens foreland, tunnes 100.

The *Beare Leycestr* hath owr laden in her—Of Dyers Passage, tunnes 100.

The *Judith* hath owr laden in her as followeth:—Of the Countess of Sussex myne, tunnes 80.

The *Gabriell* hath owr laden in her as followeth:—Of Beares Sound, tunnes 20.

The *Michael* hath owr laden in her as followeth:—Of Beares Sound, tunnes 20.

The *Armonell* hath owr laden in her as followeth:—Of

Fentons fortune, tunnes 5; of the Countess of Warwicks myne, tunnes 5; of Winters furnace, tunnes 5; of the Countesse of Sussex myne, tunnes 85.

The *Emanuel* of Bridgewater hath owr laden in her as followeth:—Of the Countess of Sussex myne, tunnes 30; of Dyers Passage, tunnes 20; of Bears Sound, tunnes 60.

The *Salomon* hath owr laden in her as followeth:—Of the Countess of Warwicks myne, tuns 10; of the Countess of Sussex myne, tunnes 60; of Beares Sound, tunnes 60.

Forasmuch as the Countesse of Warwick myne fayled being so hard stone to breke and by iudgement *yelded not above a hundreth tunnes*, we were driven to seke mynes as above named and having but a short tyme to tarry and some proofs made of the best owr fownd in those mynes abovesaid, men were willed to get there lading of them and every man so employed him self to have lading, that many symple men (I iudge) toke good and bad together: so that amongst the fleets lading I think much bad owr will be found.

If the owr now laden doth prove good, at the mynes and places abovesaid is plenty thereof, but gotten with hard labour and travayle: uppon the Countesse of Warwick's Iland Capteyn Fenton hath hidden and covered in the place of the myne all the tymber that came hither for the howse, and divers other things, to whose note I refer me.

Allso he hath caused to be buylded a little howse uppon the same iland and covered yt with boords to prove how yt will abyde or stand untill the next yere and hath left in yt sundry things.

The second said of September, the Queens foreland bearing from us to N.W. and by north, there passed by us these ships bearing to seawards we lying a hull: the *Hopewell*, the *Fraunces of Foy*, the *Beare Leycestr*, the *Armonell*, and the *Salomon*, the *Armonell* at that instant lost her boat and one man; the *Salomon* lost her boat before her comming by us. All which ships the 3 present in the morning was owt of our sighte homewards bound lying to seawards S.S.E. with the winde at N.W. a great gale of wynde.

The second said at night came unto us our pynnas with 8 mariners in her who came from Bears Sound that morning, and bearing over with the S. coast with 18 mariners in her, landed uppon certayn ilands to loke to seawards for us, and after them came the Generall in the *Gabriell* and in their company the *Judith* and the *Michael*, our men at that instant aland and loking for us, did scry 2 ships one under sayle and the other at hull, whereof we in the *Ayde* was one and the *Armonell* the other, she under sayle and we a hull, allso betwene us and them was the *Mone of Foy*, our men which were landed as beforesaid embarked them selves agayn in theyr pynnas an did beare after the *Gabriell*, the *Michael*, and the *Judith*, and did put aboord the *Gabriell* and *Michael* all the 18 mariners and then being somewhat nearer the *Judith* did put a man allso aboord her: and having order before of the Generall, the mariners remayning in the pynnas did beare from the *Judith* towards the *Mone of Ffoy* willing them to remayn with her, but they having a bold pynnas with sayles afterwards espying us a hull, but not knowing us to be the *Ayde* did owt sayle the *Mone of Foy* and at the closing up of the evening we made the said pynnas to be the *Michael* and the *Moone* to be the *Gabriel* and sometymes lying spooning before the sea and sometymes thwart remayning their comming up at length we fownd yt the pynnas as abovesaid: then they bringing us newes that the Generall was comming in the bark abovesaid and in the company of the ships allso aforesaid with the *An Fraunces* allso, the said night we did ly a hull and did hang owt lights for them all night long to show him and burnt a pike of wylde fyre to the end they might the better fynde us we hoping to have had them a boord long before day; but when day was come, we loking owt for them could not see any of them but the *Mone of Foy*: then we iudging they had overshot us or did afterwards spone before the sea 3 or 4 howres, and the *Thomas Alin* then being to seawards and wyndwards of us came bearing toward us and after we had spoken with them, they allso iudged them to be a hed of us and then we made our sayle with our corses and foretopsaile, the winde at N.W. a great gale, and we sayled S.S.E. and towards night the winde came at W.S.W. and we sayled allso S.S.E. the winde somwhat slacked our lesser still keeping company with the *Thomas Alin* and the *Mone of Foy*.

The second said allso our mariners of the pynnas declared that they at their comming over from Bears Sound did see the *Emanuel* of Bridgewater in great danger to be lost to the leewards of the sownd and did strike their sayles uppon the last of the flud to anker as they did iudge amongst the rocks, and then yt was not likely they shold ride to escape all the next eb, the winde at N.N.W. and a very great gale: God be mercifull unto them.

The said allso they declared that the captayn of the *An Fraunces*, George Beste, was with his pynnas in Beare's Sound laden with owr and the number of myners and mariners in her about 30 persons: they rowed with the said pynnas towards the *Michael*, but whither they boorded her, they cannot tell, and at that instant the *Michael* had the *Thomas Alins* pynnas at her starn, which the master said he wold cut of yf she did hinder him his comming owt as yt was thought she wold do: and afterwards our said men did see the *Michael* withowt any pynnas at her starn, and thereby do iudge that the *An Fraunces* pynnass and her men remayned in the said sownd and are in dowt of their getting their ship.

The 4 said still keping our course homewards S.S.E. the winde at N.W. a reasonable bearing gale: in the morning our company did hale up our pynnas which we towed at her starn to clere the water owt: the sea thrust her up with great force against the starn of the ship whereby she perished, and so they did cut of the tow ropes: she came up with such force, that yf she had byn strong as she was but weak, she mought have put the ship allso in danger striking in some plank; the blow was such that a company were commaunded to loke whether we had hurt thereby or not, but God be thanked we had none.

The 5 said at night in a storme we lost the company of the *Mone of Foy*.

The 6 said Thomas Batterby God called to his mercy.

The 10 said, being in the latitude of 53½, about 2 of the clock after midnight, our mayn yard did break a sundre in mydds which to recover in we did beare rome with our fore-saile before the winde, the winde at S.W. and presently did put owt 2 lights and shot of a pece to geve the *Thomas Alin* knowledge of our mishap, but yt shold seme they loked not owt for owr light nor pece, but still carry all their sailes and in the morning we could not see her: the sayd yard was peryshed 5 or 6 dayes before striking of yt tarrying for them at which tyme yt gave a great crak, but we could not finde where yt was, nor what yt was that craked.

The 11 said yt was amended and strengthened with a plank and anker stocks and woulded with ropes, and then we brought a new mayn saile to the yard: and about 7 of the clock at night we did set saile with yt with a reasonable gale of winde and immediately yt being but weakly fisshed gave a great clak and therewithall we stroke yt agayn and so rested with it all that night.

The 12 said yt fell caulme and then we fished the said yard and woulded yt with ropes in sundry other places and so strengthened yt very strong so that we had the use of yt agayne.

The 14 said at 3 of the clock at afternone, the winde at sowth S.E. began very fiercely and so encreased all that night growing to a terrible storme contynuing untill the 15 said to 8 a clock but altered uppon sundry points increasing that yt was not sayle worthy, whereuppon we were forced to spone before the sea withowt sayle and at the end of the second watche, the seas was so terribly grown that one sea came so fast after the other, the one carrying up her head and an other came with such force that yt brake in all the starn of the Generalls cabbin and did beare down with yt the cowbredge head of the said cabben, striking allso one Fraunces Austin from the helme, who called to the company for help fearing we shold have perished, but withall spede yt was amended, God be praysed, and we by his Godly providence wonderfully delivered.

The 17 said God called to his mercy George Yong myner.

The 19 said being in the latitude of 52 degrees we encountred with the *Hopewell* being to leewards of us they declared that the *Beare* and the *Salomon* were to weatherwards of us, and that they were seperated in the great storme from the *Armonell* and the *Fraunces of Foy*: the *Hopewell* lost her boat and a cable and an anker at her comming from the streict.

The 21 said we had sight of 3 sayles being in the latitude of 51, whereof 2 was to leewards of us and one to weatherwards, we did suspect them to be men of war by their working, and therefore we did hale close by the winde to speak with the weathermost ship, and being inowgh in the weather of the leeward ships did ly les in the winde untill the weathermost ship did come within our knowledge, and then we did fynde her to be the *An Fraunces* at the shutting in of the evening and did lose sight of the other 2 sayles, but we iudge them to be of our company, the winde was then at N.W. and by W. by the *An Fraunces* we had understanding the Generall to be in the *Gabriell*, and was seperated from their company the 14 said in a storme, they iudge them to be a head of us: the *Judith* and the *Michael* they left in company together, which they judge to be a starn and allso the *Mone*, they spake with her and left her a starn allso. And the *Busse* of Bridgewater they left at an anker to leewards of Beares Sound amongst the rocks. God send good newes of her, she was left in great perill.

Owt of the *An Fraunces* we received men of ours this instant 22 said.

The 23 said we lost the company of the *Hopewell* and the *An Fraunces* in a storme, which began the 22 at 6 a clock at night and continued till 8 of the clock the 24 in the morning, the winde at west and west N.W.

The 24 said God called to his mercy Water Krelle and Thomas Tort.

The said we sownded and had 70 faddems oosy sand, whereby we iudged us to the northwards of Silly, and afterwards sayled sowth east all that night, the winde at north stormy weather.

The 25 said God called to his mercy Thomas Coningham.

The 27 in the morning we had sight of the Start, 5 leags of, God be praysed therefore and make us thankfull for delivering us from innumerable dangers this present vyage.

The said, God called to his mercy Corneyles Riche a Dutchman.

The 28 of the said God called to his mercy John Wilmet.

FINIS.

STATE PAPERS SUBSEQUENT TO THE THIRD VOYAGE.

STATE PAPERS SUBSEQUENT TO THE THIRD VOYAGE.

[*Colonial*, 102. *Dom. Eliz.*, cxxvi, No. 22.]

OCTOBER 29TH, 1578. MINUTES TO MR. LOCKE ABOWTE MR. FURBISHER VIAGE.

After our very harty commendations. Whereas the shyps imploied in the viage of *Meta Incognita* are nowe retorned all home in saffetie wth Mr. Ffurbusher, and forasmuche as we are informed yt in this voyage dyvers new places and mynes have byn dyscovred. We have thought yt necessarye to require you to have a care in these matters, and to call before you the generall, and the captaynes, masters and pilotes of the shyps, and to demand of them account in wryting severallie of their doinges and procedinges in this voyage, wth discourse of the thinges happened in the same, And also to demand and take of them such platts and cartes of descriptions of the countries and places as they have made, and to forbyd them and others to publish or gyve out to others any platts or descriptions of the same countries.

And also we requyre you to have dew consyderation of the state of the shyps and goodes now retorned home, and to sett suche order therin as best may be for the saffetye of the goodes, and the commoditie and credite of the companie of venturers, and avoydans of unnecessarie expenses. And furdermore, wee doo ernestly pray and requyre you throughlye to consyder of the state of the workes at Dartford, that withe all expedition sum good prooffe and triall may be had of the trew valew of the ewr brought home, aswell in this voyage as in the other before; and that we may be certified therof from you, for that her Matie hathe very great expectation of the same.

The Commyssioners.

Indorsed.

[*Colonial*, 100. *Dom. Eliz.*, cxxvi, No. 20.]

THE QUEENS AUTHORITY TO MICHAEL LOK TO COLLECT OF THE ADVENTURERS THEIR SUBSCRIPTIONS.

After our harty commendacions. Fforasmoche as the shipps now come home wth oure lovinge frende Martyn Furbusher have brought doble the quantitie of ewar that was expected, wherby the charges of the ffraight therof, and of the maryners and mynars employed in the voyage are doble the rate sett downe at the begynnyng therof, as it is certiffyed to us by the Commyssioners therunto appoynted for the payment wherof and discharge of the said men, it is requysyt to collect of the venturars presently the sum of vj^{m} pounds of money. And forasmoche as it is greatly needfull to use all dylygens for the present spedye collection of the said sum of money, aswell for the avoyding of excessyve great charges wch grow theruppon daylye untill the said men be paid and ships discharged, as also for the performans of dewtye and mayntaynans of credite of the companye. This is therfore to wyll and require you (being thresorer appointed) presentlye, wth all the dyllygens that you can, to collect and receave of the venturars in this voyage the severall sums of money dew by them for the rate of their venture, according to a cedule of their names and sums herewithall under the handes of the said commyssyoners. And in case that you shall fynd any of the venturers to be remysse in payment, and doo not presently pay his part and dewty as aforesaid (wch we trust shall not happen), then doe you thinke meate that you gyve knowledg therof unto the Lord Mayor of London, and to Sr W. Cordell, Master of Records, whome we have appointed to be assystant unto you in that case, according to the tenor of our letters directed unto them in that behalfe.

Michael Lok.

[*Colonial*, 101. *Dom. Eliz.*, cxxvi, No. 21.]

After our very harty commendacions. Wheras our loving frynd Michael Lok is appointed presently and spedely to collect and receave of the venturars in the voyage of Mr. Ffurbusher, according to a cedule of their names delyvred to hym, a good sum of money for the payment of the maryners and discharge of the ships now come. And for that it may happen sum of them wyll not make ready payment of their partes, or wyll refuse to pay the same, wch thinge would be a hynderans to the rest by great charges dayly groweng theron untill the maryners be paid and the ships dyscharged.

Therefore we have thought good to requyre you twayne to be assystant to the said Michael Lok in this case, and uppon his information or complainte unto you to be made to calle before you suche parsons as shalbe found slak in payment, or shall refuse to pay their partes as aforesaid, and to perswade them eyther to pay the same presentlie, or els to comaunde them, as so dyrected by us, to appeare before us presentlye to shew cause why they doe not make payment accordingly. And so—

My L. Mayor.

Cordell.

[*Colonial*, 107. *Dom. Eliz.*, cxxvii, No. 8.]

DECEMBER 1578. M^{M} FROM MY LORDS TO CERTAYNE GENTLEMEN FOR THE PAYMENT OF CERTAYNE SOMMES DUE BY THEM FOR THEIR ADVENTURE IN MR. FURBISHERS VIAGE.

After our harty commendacions. The Quenes Matie being geven to understand that the myners, maryners and others imployed in the late vyage under our loving frynd Martin Furbusher, gentilman, are not yett paid all their wagys for their sarvys in the sayd voyage, but doo lye styll at the great charges of all the venturars, for lak of payment of the money dew by dyvers of the particuler venturars, althoughe her Matie and many of the venturers have paid their partes dew for the same. And for that uppon thaccount taken it......appere that for yor part therof you are to paye the sum of (*blank*). She hathe therefore geven us expresse commaundement to require you amongest others and straytly to charge you in her name to geve order for the payment of the sayd somme in London unto the handes of Thomas Allen, tresorer therunto appointed, wthin ten days after the receyt herof wthout......ffor that otherwyse yt is ordered that suche as shall......or fayle to make payment at the daye limyted shall be quyt exempted from all maner of benefytt and priviledg that may grow unto them by their former ventures made in the said voyages. And thus we hartely bid you Farewell.[1]

In the countrie.

The Erle Pembroke - - -	£172	10	0
The Countesse Pembroke - - -	- 28	15	0
The Lord Hunsden - - - -	- 85	0	0
Sr Henry Wallop - - - -	- 57	10	0
Sr John Brockett - - - -	- 77	10	0
Mr. William Pellham - - -	- 135	0	0
Anthonye Jenkinson - - -	- 57	10	0
The Ladye Anne Talbot - - -	- 10	0	0
	£623	15	0

In the Court.

- - - - -	£		

In London.

Nexte weke.	Sr Thomas Gresham - -	- 180	0	0
	Sr Leonell Ducket - -	- 91	5	0
	Mathe Fyld - - -	- 57	10	0
	Edmond Hogan - -	- 115	0	0
	William Harington - -	- 28	15	0

[1] [And—Farewell], expuncted MS.

Christmas.	Mr. Thomas Randall	45	0	0
	„ William Paintor	57	10	0
	„ Jeffrey Turvile	57	10	0
	„ Richard Bowland	57	10	0
	Mrs. Anne Kynnersley	86	5	0
	Mathew Kynersley	28	15	0
	Robert Kynersley	57	10	0
	William Bonde	115	0	0
	William Burde	20	0	0
	Thomas Owen	28	15	0
	William Ormshaw	28	15	0
	William Dowgle	28	15	0
		£1123	15	0
	Sr Wm Wintar	40	0	0
	Christofer Andrews	5	0	0
	Robert Martin	5	0	0

[*Colonial*, 110. *Dom. Eliz.*, cxxvii, No. 12.]

DECEMBER 8, 1578. THE EXCLAMATION OF THE MARRINERS, ETC., FOR THEIR PAYMENT FOR SERVICE UNDER MR. FURBISHER.

My dewty remembryd unto your honar. This is to syngnyfy unto yor honar that we, commynge home wth out mony where hit was declaryd beffore we came that we wolde brynge hit wth us, they keppe a gretor store nowe then they dyd before, and wyl beleve nothyng that we do saye. If hit maye please your honar that suche order may be taken that those wc was taxyd by your honars maye be recevyd wth the reste that owght to paye who be in the corte by somme one yor honar maye please to apoynt. And that hit maye please your honar to sende the messenger wt the letter to those in the syete (city) that they maye paye presently, and I shall geve my attendance there to receve hit and to paye hit ageyne accordynge to order, ffor lyvynge at the corte is great chargys, and all moste be put to accownt. All so there is a great dell of ffreyt to paye: no shippe p'd but one, wch is caulyd the *Beare*, Lester, wch is Mr. Lockes shippe, and she is holy payde, as your honar may se in his accownt of the mony wch he dyd receve; hit is 350*li.* the laste parsell. There is other that wolde be p'd as well, as he Crystmas beynge so nere every man cryythe out for mony. I wyshe all myght be payde before the tyme and hit be possybell, desyerynge your honar to helppe at a pynche, or elce I wolde I had my mony and another had my offece. This I take my leve, commyttynge your honar to the Lorde, who blesse you and kepe you for ever. Wrytten at my howsse in London this viij December in anno 1578.

Yours to commaunde,

Thomas Allen.

To the Right honorable Sr Francis Walsingham, knight and principall Secretary to ye quenes highnes geve these.

[*Colonial* 111. *Dom. Eliz.*, cxxvii, No. 16.]

DECR. 11, 1578. FROM MICHAEL LOK TOWCHYNGE THE ADDITAMENTS.

Right honorable. This berar the messenger wyll report unto you what he hathe done wth the venturars for their money. We have not yet receved anye but of Wylliam Ormshawe. We hope the rest will come shortlye. This messinger sayethe he must have his ffees. I know not what to answere him thereon but as yor honor will appoint.

The great workes at Dartford stand still untill additament come from the northe or the west; that of the northe wyll come shortlie I hope, that of the west is not yet sent for, bycause the commyssioners had not byn togetheres sins I was at the court, but to morrow Sr Thomas Gresham and others of them wilbe in towne as I am informed, but when they meete I think they cannot do moche for Mr. Edgecome's dytament, wthout sum speciall letters to hym from the courte. And in the meane tyme I think it very needfull that letters were wrytten to hym to send a ton therof by land wth the very first spedy convayans, for that we are very certaynelye assured by Jonas and Denan that that is most good and most fitt to work wth our ewre, and the like surans have we by one Goodyere an English workman, who hathe wrought in my hows these iiij or v dayes on divers small sayes of our ewr by appointement of Sr Leonell Duckętt, whose report yor honor shall know wthin ij or iij dayes. And thus I commit yor honor to Almighty God. From London this Thursday xj Dęcember 1578.

Yor honors most bounden

Michael Lok.

To the right honorable Sr Francis Wąlsingham, knight, her Maties principall Secretarie.

at the Court.

[*Colonial*, 112. *Dom. Eliz.*, cxxvii, No. 20.]

DECR. 15TH 1578. MR. LOCKES ACCOUNT.

Right honorable, I have receved presently yor letter wherin yr honor dothe write me of informations gyven against me to detayne in my handes the companyes money and their goodes, for answere therof I can saye no more, but that I have none of their money in my handes, and for prooffe therof I referre me to myne accounttes,[1] wch I am redye to showe in particulers, whensoever the commyssyoners and audytors wyll take a tyme to paruse the same. And syns that Mr. Allyn was appointed to be tresorer I have not receved one peny of money of any of the venturars, but onely xxviijli of my Lady Martin, wherof I paid out xxli unto Denam for his journey into the northe, and the rest wth a more summe is paid outt for divers petty parsells wch grow dayly uppon the workes of Dartford and amonges men for their sarvyce. And I am fully determyned not to receave one penny of money nor other matter of any of the venturers but to gyve over myne offyce unto Mr. Allyn, althoughe dayly I doo styll take payne to passe all accountts wth all men, and wyll doo styll to bring this busynes to a good end, the best I can. And I have not receved of any of the venturers any one parcell of wares syns Mr. Furbusher retorned home into England nor before he went on the voyage, but onelye of iij or iiij of them, summe munition or tackeling for the ships, wch stode for money for their venture outwardes, wch is answered in thaccounttes, butt nothing at all have I receved of any of them for this their dewtye for the ffraight of the ships nor wagys of the men. And I have no goodes in my handes belonging to any of the venturers in particuler, but I have my howsse full paystered of the goodes of the companye dyscharged out of their ships come home, wch is tackeling of ships, monytion, vyttells, and many od things, wch is all by inventarye receved under the handes of the masters and offycers of the ships, wch goodes I am ready at all tymes to delyver into the handes and charge of Mr. Allyn when soever it pleases hym to receave ytt.

Herein have I wrytten to yor honor the trewthe of my doinges wch I wyll justyfye. I beseche yor honor to stand my ffrynd as you shall see cause of defect by my doinges. And bycaus that sclanderous tonges wyll not be stopped by wordes, I make no answere to them, but abyde the tyme when God shall make my doinges knowen wherby he shall stop them for me. And I comytt yor honor to Almighty God. From London this Monday xv December 1578.

Yor honors most bounden

Michael Lok.

To the right honorable Sr Francis Walsingham, knight,

her Maties principall Secretarie.

at the Court.

[1] Two volumes of these accounts are in the Miscell. of the Exchequer, vol. 60, 61.

[*Colonial* 123. *Dom. Eliz.*, cxxvi, No. 57.]

THE NORTH-WEST VOIAGE. A BRIEF REPORTE OF THE ACCOMPTE OF MICHAELL LOCKE CONCERNING THE CHARDGES OF IIJ VOIAGES INTO THE NORTH-WEST PARTES UNDER THE CONDUICTE OF MARTIN FURBUSSHER, TOGUITHER W^{T} THE CHARGES OF BUILDINGES AT DERTFORD.

Divers sommes of money receved by Michaell Locke of the adventurers, viz., for the furst voiage, viij^{c} lxxv^{li}; the second voiage, v^{mt} cccv^{li}, and the iij^{de} voiage, vj^{mt} vij^{c} iiij^{xx} iij^{li} xv^{s} . . .	xij^{mt} ix^{c} lxiij^{li} xv^{s}
Divers sommes of money receved of the said adventurers for buildenges at Dartford	m^{l} iiij^{xxli}
Divers sommes of money recevid for paiment of fraightes . . .	v^{mt} vij^{c} lxxviij^{li} xv^{s}
Sum totall of the receiptes. .	xix^{mt}viij^{c}xxij^{li} x^{s}
Wherof	
Allowed in the said accomptes for buildeng repaireng and furniture of shippes, victuals, implements, wages, paiment of fraightes and buildenges at Dertford, and divers other thinges as in the accomptes maie appere .	xv^{mt}ciiij^{xx}vij^{li} x^{s} iiij^{d}
And then remaineth to be accompted for .	iiij^{mt}vj^{c}xxxiiij^{li} xix^{s} viiij^{d}
Wherof	
Due by Thomas Allen, Threasurer of the voiage for money by him recevid of the said adventurers	viij^{c}iiij^{xx}ij^{li} x^{s}
Divers other persons for their adventures yet unpaide	ij^{mt}v^{c}xxxv^{li} xiij^{s} iiij^{d}
Michaell Locke for money supposed to remaine in his handes . . .	m^{t}ccxvj^{li} xvj^{s} iiij^{d}

Whereof he demandeth allowaunce of m^{t}cc^{li} for his attendance and charges sustained in the causes of the said iij voiages.

Tho : Neale, Audit.

On the back occurs—Articles to be inquired of by Mr. Thomas Neale and Mr. — Baynham, Auditor^{rs} appoynted to take the accompt of the northwest viage.

What som the whole adventure in the sayd viage dothe amount unto.

What sommes the adventurers in that viage have payd of the same.

To whom the same was payde.

What is behinde by the sayd adventurars.

What Michaell Lock is forther to be charged wthall for wares sowlde pertaynyng to the compagnye.

What hathe been paydd to Mr. Th. Alin the 2 Marcheant as Threasorer, and howe the same hathe been ussede and what remaynethe in his handes.

What is dewe by the compagny for fraight of shippes and otherwyse and to whom the same is dewe.

[*Colonial*, 126. *Domestic Eliz.*, cxxvi, No. 35.]

AN AUNSWEARE TO MR. LOCKES REQUEST FOR 1200^li^ W^ch^ HE DEMAUNDETH OF THE COMPANIE OF THE NORTHWEST VOIAGE FOR HIS SERVICE FOR IIJ YERES.

1576. The first yere.

1. Ffirst, where the saide Locke demaundeth for three yeres charges and for warehouses and for kepinge of house. It is thought y^t^ the first yere is not to be brought into this reckoninge, for that he was then in the service of the companie of Muscovia, and that yeare also the companie of the Northwest voiage had no need to use warehouses or anie meetinges ; ffor in the said first yere there went forthe but two pinasses, and the adventure was but 875^li^ wch was all lost and spent. Therefore, if the said Lock be allowed iij^li^ for the hundreth, it is verye muche for 875^li^ wch is 20^li^ £20 0 0

1577. The second voiage.

2. This yere went out the *Ayde*, wth two pinasses, and the adventure was 1075^li^. And if the said Locke be allowed for 200^li^ after iij^li^ for the hundreth, wch is 96 : that is, wth the most, he tarying at home and not travailinge. Moreover, in this yere he was not muche troubled wth house roome, servantes, or greatt dyett. But lett there be allowed him above his provision towards his charges and servants, xx^li^ . £116 0 0

1578. The third voiage.

3. In this third voiage there went out ten ships, wch the said Lock had to deale wth all the *Ayde*, the *Judith*, and two pinasses, and the adventure this yere was 7000^li^, wherof the said Lockes adventure was 2030^li^. Beinge allowed for 5000^li^, after iij^li^, for the hundreth 250^li^, and for his servantes, three in number, at xiij^li^ vj^s^ viij^d^ the peeee, wch is 40^li^ (thoughe in the said Lockes accompt nothing is putt downe for their charges), and for meetinge of the commissioners diverse times after, 10^li^ the monthe, for iiij monthes, wch is x^li^ ; this may be thought sufficient, ffor the commissioners did not eate often wth him, and but a fewe of them at a time £230 0 0

Touchinge the interest of money taken upp by the said Locke by exchange.

4. It is not thought meet that he shoulde be allowed anie thinge, because it is verely thought he tooke not upp anie monie for the use of

the said companie, but rather for him selfe, because he was so greate an adventurer in the voiages aforesaid: at the leaste, if he did take upp anie it could not be verie muche. This demaunde being so greate, it is like he knoweth for whom he tooke upp so muche money, who are to repay the same to him, and no reason that other adventurers w[ch] have paide their money should pay him interest.

5. Ffor boate hire, to and from the Courte at sundrie times in two yeres, the said Lock may have allowance of x[li], w[ch] is reasonable.

6. Ffor the time of the buildinge at Dartforde, for his riding charges, and keepinge three horses about sixe monthes after, xv[s] the weeke, w[ch] amounteth to xviij[li]. As for his mens and his owne diats are allowed before: yet, because riding charges be great, he may be allowed for him selfe and his men xxij[li] more . . . . 40[li]

7. Ffor the said Lockes charges to the Courtes, and following hir Ma[tie] in progresse, a certain estimat cannot be made thereof: it being uncertain howe often and howe farre he did ryde; but it is to be supposed not farre, because hir Ma[tie] was ever in hir progresse when the ships were absent in their voiages, at what time Locke had least to doe wth the Courte. And his ridinge in this case and time could be but part of two progresses in the two last yeres, for w[ch] he may be allowed xx[li] £430 0 0

[*Colonial*, 114. *State Papers. Eliz.*, No. 4, Vol. 129.]

13 JANUARY, 1578-9.

The Second Minute for this purpose.

After o[r] hartie comendacions, &c., albeit that not longe sithence uppon informacion given unto the Quenes Ma[tie], that the myners, maryners, & others imploied in the late voiage under o[r] loving freind, Mr. Martyn Furbusher, weare unpaide of their waiges for their service in the said voiage (and thereby do lye still at the great increase of charges of all the adventures). We wrote o[r] letters by her Ma[ts] precise commandem[t] to [yo[r] L.] emonge others, to make paiem[t] of [clxxij[li]], w[ch] remayned unpaid of the somme that [yo[r] L.] assented to contribute to the said adventure to the hands of Mr. Thom[s] Allin, thresourer, appointed for that p'pose wth in x daies after the receipt of o[r] said letters: yet the same not wth standinge her Ma[tie] is eftesones given to understand that the said somme gevon by your L., as also lyke sommes geven by other the adventorers, remayne yet unpayed, whereby as the inconvenience aforesaid, onelie by the wante of paiement of suche money, is increased. So is her Ma[tie] the more moved to mislike thereof (Her Highnes & diverse of us & others having paid o[r] parts according to o[r] promisse). Theise are therefore once againe to praye yo[r] L. to geve present order for the payment of the sayd some, for besides her Ma[ts] good contentacion, that hath alwaies bene well effected to the voiage: though somme men, uppon misliking, can be contented to withdrawe themselfs, & to be exempted from the adventure & all privileges of the same, as was mencioned in or former letters that all suche should be as refused to make paiement by the daie lymitted: yet is it not thought in any wise resonable, howe soever the thing shall fall out; but that they should make satisfaction of so muche as they have promised, for without these promises the voiage had never bene taken in hand. And thus trusting that for the respects aforesaid, we shall nede in this case to write no more; we bidd yor L. hartely farewell.

(*Endorsed.*)
Md To the adventurers.

[*Colonial*, 119. *State Papers.* Vol. 129. *Domestic Eliz.*, No. 9.]

My dewtie remembred. Hit maye pleise youre honor to undarstand that Mr. Furbusher doyth moche myseuse me in words, saynge, I have complayned to the consell of hym. And that I have saide that all ys nothing worth at Dartforde, and howe he hayth receyved so moche monye and donne wth it, he pleaseth wth all, where in I ame une oneste and have no onestie in me. I do remebar I dide declare unto yor own honor, wiche was wryten in my byll of debts at the lower ende, that he dide reseve 86li of Mr. Frances, 50li of my L. of Warwicke, 7li of Mr. Turwill, wiche I most nedes declare, be cause I have gyven no quyttans for yt. And youre honor axed me what he hade donne wth all. I ansured, I colde not well tell; but, as I dide here p'ade frawght unto some of those shippes wiche he dide frawght in the west contrye, & some of the myners, he sayth those shippes dide hym the best serves when owre other shippes dide rune awaye. Sr, he wyll werye us all, and he have the brydell to moche. Sr, under youre correction, I do not thincke yt amyse, and yt be youre honors pleasure, that there weare comysion frome youre honors that Mr. Furbusher accounte sholde be nowe presentlye awdited wth these same auditors, for Mr. Lockes wylbe downe this daye, and so shall youre honor see boythe the accounts to gether, and what ys owyng by them. And that ordar may be gyven by youre honors, that all superfluus charges may be cutt of whyth spede or else yt wyll rowne one styll to great charge. I wolde I weare dyscharged, rather then I wyll be thus raled at for my paynes. This I take my leave of youre honor, commytting youe to the All Mightie. Wrytten at my howse in London, the xiij of Janewary, 1578.

Your honars to commaunde,
Thomas Allen.

(Endorsed.)
To the right honorable Sr Frances Walsingham, knight,
and prensepall secretory to hir Matie, geve these.

[*Colonial*, 120. *State Papers*, Vol. 129. *Domestic Eliz.*, No. 11.]

MR W^{M} BOROUGH, TO S^{R} F. WALSINGHAM. 14 JAN^{Y} 1578-9.

After my dewty unto yor honoure, dewly considered, this daye being Wensday, the xiiijth Janewary, I receved yor letters bering date at Richemond the xijth of this present, wherby I understand it hathe ben informed yor honour that I shuld owe for myne adventure in this last voyage wth Mr. Frobisher, the som of lvij^{li} x^{s}, the wch yo^{u} requyre me to paye out of hand, or els to repayr presently to the court, wher I shuld understand her Maties furder pleasure.

True it is, right honorable, that at the setting forth of this last voyage to Meta Incognita, Myghell Loke, then treasurer, and cheefe dealer for the same voyage, bought of me (to serve in the same voyage) a shipp called the *Judeth*, of burden about lxxv tons, for the som of 320^{li}, the wch he condicioned to pay me in Maye last, wher of I allowed him then lxvij^{li} x^{s} for my adventur, in the sayd last voyage. Afterwards (in June last) I recevid of him lxxxx^{li}. So I accompt to have recevid 157^{li} x^{s}, (and he then rested dettor to me 162^{li} 10^{s}), the wch rest I cold by no meanes get of him, unlest I shuld have recovered it by lawe, wch I was loth to doo.

Since the commyng home of the fleet of that voyage, and the charges of the same knowne; and, theruppon, every man allotted his porcion thereof, according to his adventure, my parte (as I understand) comyth to lvij^{li} x^{s}, the wch I accompt to be payd out of the sayd 162^{li} 10^{s}. And yet remaynyth dew to me, 106^{li}.

Thus, as appearith, I have payd my porcion fully wth the first, and ought not to have ben brought in now as a dettor. But, seeing Mr. Lok, his dealing towards me herein, hat ben wth so small credit or honesty, that neyther he wold paye me the money that hath ben so long time dew to me, nor yet accompt my porcion of charges to be payd as before I have shewed; but hath geven report unto yor honour that I shuld yet remayne dettor for the sayd som of lvij^{li} x^{s}. I will no longer credit him. And therfore doo besech yor honoure, that by yor good meanes I maye be appointed payment of my rest, 106^{li}, wch hath ben dewe to me so longe since.

I wold have attended uppon yor honoure according to yor order; but, true it is, that I have ben ever since Christmas very much trobled wth an extreme payne in my hed, so as I have ben forced to keepe my howse, and yet am not clear of the same, but hope of amendment shortly. Thus I commyt my cawse to yor honoure, and yor selfe to the tuicion of the Almyghty, who blesse the Quenes Matie wth longe most helthfull, happy life and rayne. Amen.

Lymehowse, the xiiijth of Janeuary, 1578.

Yo^{r} honours most humble to use and commande,

W. Borowgh.

(Endorsements.)

To the right honorable, Sir Francis Walsingam, knyght, principal secretary to the Queenes most excellent Matie, at the Court.

14 Januarie, 157

From Mr. Willm a Burrough.

He hathe payd the 5... 10s demanded, in a shipp Mr. Locke had of him for Furbishers last viage: & there remayneth due to him more for that shippe 106li.

[*Colonial*, 121. *State Papers*. Vol. 129, No. 12.]

MICHAEL LOKE TO S F. WALSINGHAM.

Right honorable,—

I have receved yor letter, wherin I am charged to paye ixc xli to Mr. Allyn, for my part of the fraight of the ships retorned since wth Mr. Furbussher, in this third voyage. For answere whereof, it may please yor H: to be advertysed, my part of that fraight cometh to iijc xvjli vs wch I have paid, as by myne accountt dothe appere, wch accountt the Awditors are now in hand wth all, and by them yor H: shalbe advertised very shortlys, bothe of thatt, and of all the rest of my doinges in the companies busynes. And for more part of the said ixc xli it is sett downe that the right honorable th'erle of Oxford, ys to paye iiijc lli accordinge to the order and rate of all the rest of the venturars, wherfore it may please yor H: to call on his H: for the same sum. And yf that his H: be not satisfied of this matter, I am to be ruled by yor H: and others, uppon vew of the bargayne wch I made wth his honor, wch he hathe under my hand and sealle, for I will not doo any wronge wyttingly to any man lyving, especiallye to his H: to whome I doo owe bothe dewtye and reverence.

And thus for this tyme I take my leave humbly of yor H: and commytt the same to almightie God. From London the xiiij Januarie 1578.

Yor H: most bounden

Michael Loke.

(Endorsements)

To the right honorable Sr Francis Walsingham Knight her Maties principall Secretarie.

at the Court.

14 Januarie 1578.

From Mr. Michaell Locke

Towchyng the money wch he was written unto to pay to Mr. Allen for his adventure in Mr. Furbishers viage.

[*Colonial*, 132. *Domestic Eliz.*, cxxix, No. 44, I.]

THE 20 JANUARYE 1578. THE ANSWER OF ME MICHAELL LOK UNTO THE WORSHIPFULL THE COMYSSIONERS AND AUDITORES OF MYNE ACCOUNTES UPON THE SECONDE AUDITE THEROF.

In the month of Auguste laste 1578 my iij bookes of accountes of the iij voyages made by Martine Ffurbusher for the northwest partes were audited by sufficient parsones therunto appoynted, who uppon the particular examinacon of the same, dide certifye under their handes writtinge that uppon those accounttes dide remayne dewe unto me the summe of ix^c iiij^xx ix^li iiij^s vj^d of money, besydes all my stocke in venture wth the companye, wch was about iiij^m li in all thes voyages. First audite of myne accountes.

Afterwardes M. Furbusher beinge retorned home, he of his owne evell disposed mynde dislyked of myne accounttes, and made greate complaynt of the audite therof and procured you the new comyssyoners and new auditores to revewe and examine better all myne accounttes and doinges wch you have done. And now by that wch you have sene well provide agayne, you doo fynde my said accountes to be juste and trewe as they wer befor in all partes excepte onlye in iij poyntes following, to the whiche I doo now answere. Seconde awdyte of myne accountes.

Ffirste you wold dysalowe me a somme of 96^li wch I make paid to the shippe *Beare lester* for vittels of men that were passingers therin, wherunto I answer that I dide paye that some of mony and agreater summe unto that shippe befor her departure from London, and I dide knowe by a covenaunt of agrement made in the charter partie, that the said shippe was to carrye xx men passingers, for the wch was dew eyther that somme of monye or so muche vittells as should suffice for them. And I did know that shippe dide carrye from London to Hawiche more than xxx men w^thout any manner of vittalls put into the shippe for them, and they fedd only of the shippes owne provissione, and I dide know that thes vittelles were denyed her in the Tames, and I dide not knowe that any vittells were put into that shippe for them at Harwiche, but I was informed that none would be ther delivered therfor although this payment be sett doune and allowed unto me in account, yett is yt but uppon a good accounte, for mysreconynge is no payment for I have recevid no mony for yt of the company uppon that account made, but now yt standeth stille as mony paid uppon the fraight dew to that shippe. The j objection for 96li. paid.

The second poynte is matter of yor dislykinge of the order of my said accounttes, in that you saye I ought to have charged myselfe uppon the fotte of that myne accounte for suche debts as were owinge for the goodes bought and were not then paid: to this I answere that in dede yf I had so done ther wold not so great a remayner of ix^c iiij^xx ix^li have appered to be then dew unto me by the somme of vj^c li wch I hade not then paid, The ij objectione for the order of myne accounts.

but I fynde no cause neather in reason nor in justyce that I should have made suche manner of accounte forasmoche as no day of payment was assigned but present mony and for the same by appoyntement of the commissioners I was become bounden unto those men that were unpaide, and they daly and howerly calleynge and exclaiminge on me for the mony and thretening me to prissone for yt, from wch how hardly and chargablye hether to I have kepte myself, bothe I and my frindes doo knowe and felle; wherin I coulde fynde no helpe of any of all the companye wch trouble might have bene avoyded yf the companye accordinge to equitie and right had payed me that some of mony that therwthall I might have paid the creditors.

The iij objectione for my stoke.

The thirde poynte is an objection made uppon my stoke that I have in venture wth the Company in these iij voyages; to the whiche I nede not answer, for the account sheweth yt planly, that I and my parteners have in stoke and venture in all the summe of £4,680 of money wch is all paide excepte £450 wch muste be paide shortlye, wherof God graunt good successe and then all this ware wilbe turned into peace.

The great sumes of mony wch I have disbursed for the Company.

Morover you doo nowe see by myne accounttes that over and besydes all the said summe of mony for any stoke and venture and over and above all that I dide receave of all the companye I dide disburse of myne owne mony for the Company as followth: In the firste voyage more then vjc*li*, and in the second voyage xiijc*li* wch afterwardes I dyde lette stande stylle for my stoke and venture therin.

And in the thirde voyage iijc*li*, and in the buildinges and workes at Darteforde viijc*li* besydes other great sumes dysbursed in other meane tymes wch myne accounttes cannot declare particularlye, bycause the same are made up wthout dystinction of dayes or tymes but onlye one daye at thende of every yeres accounte.

All wch said summes of mony I have forborne longe tyme to my great troble and charges of interesse, and I was not repayd the same untille of late, about amonthe laste paste, as myne accounttes doo declare wherof I know that you wille have dew consideracone as reason requirethe.

This answere may suffice unto you that be wise and reasonable, and as for Mr. Furbushers faustye in this matter, yt deservethe none answer at all.

THE 26 JANUARYE 1578. MICHAELL LOK SALUTETH THE WORSHIPFULL COMYSSIONERS AND AUDITORS OF HIS ACCOMPTS OF THE IIJ VOYAGES OF C. FURBUSHER.

Ffor as muche as wordes are but winde and are easely forgotten of those whome they towche not, and yet beinge misplassed by ignorauns of the speker, or mishard by negligence of the hearer, or misconstrued by the malitious, they do breed oftentymes contention and displeasure wthout cause, therfor I have thought good to set downe my mind in this writtinge wherby you maye the better consider therof and so resolve uppon that wch reason and equitie will requier.

Nowe that you have gonne throughe all myne accountes and have particularlie examined the same, you do fynd that I have made them justelye and trulye, and have not charged them wth and matter wrongfullie nor falselye. And by the course of them you maye parceave my great labour cost and troble had wth this busynes in these iij yeres voyages.

Myne accounttes found right and trew as at the first awdite.

And by conclusion of thes accomptes you do fynde that I have a great somme of mony of myne owne in stoke and venture in thes voyages, and that I have disbursed and paid from time to tyme great somes of monye uppon the credit of my selfe and my frindes for the companyes busynes, to furnishe those voyages. And that in my handes no monye dothe remayne dew unto the companye.

My great venture in these voyages.

And wheras in thende of that myne accompt I do set downe a summe of xij^{c}*li* of monye in divers particuler parcells followinge, wch I have paid and disbursed in the companyes busynes in these iij yeares, and have not had any allowance therof, nor have sett downe anye demande untille nowe at thend and conclusion of these myne accountes to saye,

The charges in iij yeres busynes.

1. Ffirst for my ridinge charges in iij yeres to the courte and abroade for collection of monye of the venturares and other busynes of the company . . . Summe £120

2. for my ridinge charges to Darteford in viij monthes solicitinge the buildinges thear . . . Summe £60

3. for my boathire to the corte at Grenwhiche and to the shipes and other places about the foresaid busynes in these iij yeares . . . . . . Summe £20

4. for the table diate of the commissioners, auditores, captaynes and others, dalye meatinge at my howse about the busynes of these 3 yeres . . . Summe £150

5. for interesse of money taken uppe frome tyme to tyme to furnishe this busyness and dispatche of the shippes on ther voyages in dewe tyme for lacke of the venturars mony Summe £250

Summe £600

6. for charges and travayle of my selfe, my servauntes, and howshold, to followe this busynes, and takinge charge of thaccomptes and howserome of the goodes in these iij yeres Summe £600

Summe £1200

uppon wch said parcelles some of you the commissioners do make dowbt of the spendinge of the mony of some of these parcells and of the dewtie of other some of them, wherefore hereunder I do declare unto you my reason and prooffe of dewtie for the same.

The firste begyninge of thes voyages.

At the firste begininge of these voyages for the discovery of Cathay, etc., Martine Ffurbusher did procure the same to be taken in hand of a good mynd towardes his contrye and comodite towardes him selfe, by the good likinge of the right honorable the Lorde Burghley, Lord Highe Treasorer of England, and others of her Maties honorable privie counsell, whose letters he brought in that behalfe, dated in December 1574, directed to the Company of Muskovie for their lycens by their priveledge to doo the same, wch first they refused to graunte, of whose busynes I had then the chiefe charge and whole doinges, whereby I entered into knowledge of the matter, and althoughe (accordinge to my dutie towardes the Company of Muscovie, knowinge the inconveniences that therby might growe unto their trade of marchandiz) I did also dislyke of this motion for a tyme; yet afterwardes, uppon consideracion of my dutye towardes my contrye, and knowinge by myne owne knowledge (as my manifold writtinges therof wille witnes) the greate benyfitte that therby might growe to the same, and perceavinge the corrage and knowinge the aptnes of Martine Furbusher (by former acquayntance wth him, and uppon newe conference had wth him) to execute that attempte, I did so enterelie joyne wth him therein, that through my frindshippe wth the company I obtayned of them a previledge and lycens to followe that attempt, datid the thirde of Ffebruarye 1574, wch I have, and so gave out my selfe openlye for a chiefe frynde and followar of the matter, wherby many men wer brought into a good lickinge of the matter, wch before could fynde no trace therof. And hereuppon I used M. Furbusher as my fellow and frinde, and opened unto him all myne owne privàte studies and labores passed in twentye yeares continuans befor, for knowledge of the state of the worlde, and shewed him all my bookes, cartes, mappes, instruments, so many as cost me v^{c} poundes of mony, and writtinges, and my nottes collected therof. And dalye instructid him therin to my skyll, and lent him the same to his owne lodginge at his will for his better defence in talke thereof wth other men. And to be short, dalye increased my good will towardes him, makinge my howse his howse and my purse his purse at his neede, and my credite to his credite to my powre, when he was utterlye destitute boath of mony and credite and of frindes, as his letters unto me and his protection of her M^{tie} dothe witnes, when he was first lodged at the house of one Browne in Flete Street, and afterwardes to have my better helpe and relief removed nearer to me to the howse of a widowe named Mrs. Hancokes in Marke Lane, who cane bare witnes of the same, wth others more that I cowld name. Herewthall M. Furbusher was a glad man, and hoaped of great good fortune towardes him, and towld me great matters of venturars that he would procure to furnishe this matter, wheruppon to begine the matter I made a writtinge dated the 9 of Februarie 1574, for the venturars to sett downe their some of monye wth their owne handes, and for the better incorraginge of others

M. Lok joynethe wth M. Furbusher.

I first sett doune my selfe for one hundreth poundes, wheruppon divers others followid in the cittie to the some of v^{c}*li*, and afterwards M. Furbusher carried that in writtinge to the court (for befor that tyme no handes wold be hade there), and there he had the handes of divers of her M^{ties} honourable privy councell to the [sum] of iij^{c}*li*, wherin the right honorable the Lorde Burghley sette downe a condicion that a convenient parson should take charge of this sarvice, and afterwardes divers other parsones did sett downe divers sommes of monye of small value. And more venturars could not be gotten for that time, wheruppon thenterprice was geven over for that yere.

The attempt over-throwne first yere, 1575.

And now Mr. Furbusher was become a sade man, for that by this meanes his credite grew dalye in questione, and more & more dislikinge grew of his dealinges ; yet he contenewed styll abowt London and the Court, hoapinge and solicitinge what he could agaynst the next yeare. And my good will and good word contynewed still towardes him as before, wch did him no hurte, but in the meane tyme coste muche monye for thinges provided and charges bestowed for the same voyage.

The next yere beinge anno 1576, the enterprice of the voiage was revyved agayn, and the question beinge asked of the venturars, they said they wold contynew their good will and venture sett downe the yere before. Hereuppon, M. Furbusher was alyve agayne, and solicited the furtheraunce of the matter by the helpe of Mr. Burde, then costomer of London, and Mr. Alderman Bonde, now deceassed, at whose howse we had divers conferences of the maters.

Th attempte revived next yere, 1576.

Now, in procedinge to the preparacion of this voyage questione grewe amongest the venturars, accordinge to the noate of condicion sett downe by my Lord Treasorer, who should take charge of the mony to be colected of the venturares, and who for the provicion and furniture of the shippes, and who in the conducte of the voyage wth the shippes at sea; yt was aunswered, that for the monye I would not meadell; and theruppon Mr. Hogan and Mr. Borow was named; but Mr. Borrowe uterly refused, and would not medell. Wheruppon Mr. Hogan tooke paynes for a tyme and receaved suche mony as he cowld gette, and percevinge the travell therin or ells no voyage at all, had bine made that yere neyther, and for the provicion and furniture of the shippes M. Furbusher did sett doune divers noats of divers thinges wch grewe to a great somme, wch the said 3 or 4 of the venturares dide correct, and did ordayne dalye what they wolde have donne, and I daly keapte regestar in writtinge of all there agreementes, and accordingly I did see all thinges accomplished and executed, and tooke charge of the accompts of all thinges; but now the greatest matter remayned still in doubte, and not satisfied amongest the venturares, wch was who should take charge of conducte and commandement of the shippes beinge alreadie at the see; for that M. Furbusher had verie littell credite at home and muche lesse to be credited wth the shippes abroade: this matter was the cause

Order of the busynes.

A treasorer.

A chefe governor of the ships at sea.

of the overthrow of the voyage in the yere before, and this matter also now was like to overthrow it this yere, and did cause most of the venturares to keape backe their moneye in thend; but, to forther the voyage, I dide helpe this matter the best that I cowld, and I steped in wth my credite for his credite to satisfie all the venturars that he should deale honestlye and lyke a trewe man wth the ships in the voyage; but this would not sarve their torne. Wheruppon, afterwardes, by their consent, I devised a writtinge wherin was joyned wth him in comission Christofer Hall and Owen Griffen, Mrs of the shippes, and Nicholas Chaunsler, marchaunt & purser of the voyage, who were knowne for trustye men, wth out whose consent he should not comaund nor carrie the shippes, but accordinge to the comission geven them by indenture under their handes and seales, wch I have to showe. This did satisfie most of the venturars, but all this dilligens would not bringe in mony to furnishe owt one quarter of the shippinge intended for the voyage. Wheruppon the shippinge was dyminyshed, and insteade of iij shippes, we could scante furnishe two small barkes and one lyttell boate wherwth all he went wth cost xvjc*li* of mony wherof ixc*li* came out of my purse alone, as thaccomptes doo wittnes, besydes other thinges not declared in myne accomptes. And now this was the begininge of my travell and service done to the Company in theis iij voyages.

Lytle mony of the venturares but M. Lok ixc*li*.

M. Furbusher retorned home.

Now, when Martyn Furbusher was retorned hom againe, in October 1576, wth his strange man of Cataye, and his great rumor of the passage to Cathai, he was called to the courte and greatly embraced and lyked of the best. And uppon his great informacione of many great matters of this new world, yt pleased her Maties Honorable Privie Counsell to directe their letters and comissiones unto Sr William Winter, Mr. Thomas Randall, my selfe, and others, to calle unto us M. Furbusher and Christofer Hawle, and to take accompt of them of all their doinges in this voyage, and to take knowledge what were requisite to be donne in the followinge of this discoverie made for the passage to Cathai, for another voiage the next yere, and to certifie their honors therof.

Hereuppon manye mettinges were at my howse and sometymes at Sr William Wintares howse, and certificat was made by the comissioners to their honors of good lykinge of the passages to Cathai.

The first stone of ewer brought home.

In this meane tyme happened to be discovered the riches of a mynerall stone brought home by chaunce by Mr. Furbusher and delivered to me, wherof I caused prooffe to be made by skilfull men, and was sertifyed the same to be of a myne of golde, wherof I gave knowledge to her Matie accordinge to my dutye, wheruppon muche marvale was made and muche enquire and triall made by others of more credite, by whome att the last it was confirmed to be trewe, and so was certifyed; wheruppon M. Furbusher was called to knowe what quantitie therof was to be had in that new worlde; he aunswered, that ther was inough to lade all the shippes of her Ma ^{ie} and theruppon now was prepared the second voyage anno 1577, wth muche greater preparacione then was purposed before for that her Matie would be a great venturar therin.

Second voyag, 1577.

And here uppon daly grew new busynes and new venturars and new collectiones for greater matters then befor and dalye new comyssiones and new directiones from her Maties Honorable Councell, wthout whose knowledge no great matter might be donne. And stille the chief charge comitted unto me as treasorer as many of their honors letters to me directed in that behalfe doth wittnes; wheruppon many assembles of the Comissioners and others wer made at my howse dalye, as the regesters of manye of their meetinges and agrementes of the busynes can declare. And thus now may you see how and by whose comaundem^{t} my travaylles, trobles and charges in this busynes was contynewed and encreased stille unto this daye.

The thirde voiage, 1578.

Thes foresaid matters beinge well considered, and callinge to yor rememberaunce the great busynes wch ther uppon followed in the thirde voyage made in the next yeare followinge, wch was anno 1578, wch remayneth stille befor your eyes. I trust you will thinke my foaresaid demaundes of xij$^{c li}$ for my charges layd out of my purse in this busynes to be verye reasonable.

M. Lokes answere for the charges paid.

And never the lesse, bicause you shall know that I do not demaund ytt wthout ground of dutie, I will now answer unto your doubtes and make proofe wth a good conscience, as followithe :—

No. 1. Ridinge charges.

Ffor my ridinge chardges I have sett downe cxx^{li} for these iij yeres, wch is xl^{li} by yere, in wch tyme I have travyled to the courte in the countrie, and in progresse not so fewe as c dayes yerlye, for collection of mony and comissions for busynes, sometyme wth one man, sometym wth twayne, wth my horses and companye that hath drawne to me about this busynes, hath cost me some tymes xx^{s}, sometymes xv^{s}, and when least xiij^{s} iiij^{d} a daye.

No. 2. Dartford.

For my ridinge chardges to Darteford, I have set downe lx^{li} in the tyme of viij monthes, yt is well knowne I was there everye weke sometymes iiij dayes, sometymes iij dayes, and most comenly ij dayes wth iij horses, sometymes iiij or vj by the workemen carried wth me for the buyldinges where I was lodged in a comyn yn as I am stille when I go thether and all my folke also, when they be there by cause. C. Furbusher hathe nowe taken uppe the lodginge that is at the worke houses, leste I and my men wch have charge of the thinges there, should lye neare unto o^{r} chardges to loke to the salfe garde of the thinges there, wch hath cost me wth the workemen, resortinge to me dalye about bussynes xl^{s} adaye xxx^{s}, and never under xx$_{s}$ everye daye that I tarried there.

No. 3. bote hyer.

Ffor my boat hire to the Court at Grenew^{ch} and to the shippes and other places about this busynes, for these iij yeres, I have sett downe xx^{li}. The nomber of jorneys I cannot remember, they were so many;

but well I do remember that everye jorney to Grenew^{ch} dide cost me and my men, and his meate theare gevinge attendance everye daye v^{s}, and when least iij^{s} liij^{d}, and manye dayes makinge two jorneys thether in one daye.

No. 4. table dyat of the comyssioners, etc.

Ffor the table diatt of the comyssioners, auditores, captaynes, shippe masters, and other daly metinge and resortinge to my howse, about the busynes in these iij yeares, I have sett downe the somme of cl^{li}, wch is after l^{li}, by the yere, one wth another. What resort hathe byn there at daly about this busynes, the men themselves can wittnes makinge my howse as their howse, and my table as their table, and what yt hath cost me above myne ordenarie, I do knowe and ffeale, and those that have experience of howshold chardges in thes dayes, also can consider; but sure I knowe that, yf I were allowed iij^{cli} for thes iij yere yt would not recompense my charges therof.

No. 5. Interesses paid for mony for the venturars.

Ffor the interesses of mony taken upp from tyme to tyme to furnishe this busynes and dispatche of the shippes on their voyages in dewe tyme for lacke of the venturares mony, I have set downe but ccl^{li}, wch parchaunce somme of you maye thinke to be verye muche, not havinge byn acquanted befor wth my deallinges in the busynes, nor havinge consydered by myne accomptes what great somes of mony I have disbursed and paid for the Companye from tyme to tyme for the furniture of the voyages above all y^{t} wch I receaved for them ; yet I praye you thinke of my doinge thus faborably, as the rest of my doinges dothe geve you cause, that I will not willingly charge the accomptes wth any more somme then that I thinke may stand wth equitye and good concience. Trewe it is, that I have a great somme of mony for myne owne stoke and venture in the seconde and third voiages, wch would have stoped a great hole in the paymentes of those ij voyages, wherby you maye suspecte that I neaded not to have taken uppe so muche mony for the sarvice of the rest of the venturares ; but verrie trewe is this also, that it were to hard dealinge wth me to make my mony stope the gape for other mens dutyes, and make me stylle bare theyr burden ; and reason would that yf anye of all the venturares be favorabley borne wthall for the payment of this monye, I should have the most favor of all others therein, consideringe my charge and my care and travell for all ; but for playne proffe of my uppright dealinge in this matter, yt shall appeare unto you by wytnes of the brokers by whose handes I have taken upe the mony from tyme to tyme for the sarvice and furniture of this busynes, that the interesse wch I have paid for mony taken uppe, therfor wthin these iii yeres hath cost me of my purse the somme of v^{cli}, besydes c^{li} more wch I payed for the surans of v^{cli} wch remayned in the seconde voyage, at I cannot telle whose venture, but at the least lighted uppon myne owne venture and coste, wch is togeathers vj^{cli} of reddie monye paid out of my owne purse, wch is not charged in any of myne accomptes, but only now in thend of myne accomptes I have sett downe ij$^{c\,li}$ for the companyes parte of that v$^{c\,li}$ of interest paid the other rest being iij^{c}l^{li}. I do bare and paye of myne owne purse, wch I thinke may suffice for the interest of myne owne stoke yf it were not paid to thaccompte so sone as you would have it, but so sone as reason required that it should be.

All these foresaid sommes of mony sett downe do amount to the somme of vj^{c}*li.* of redye mony wch I have paid out of my purse by extraordinarye charges in the sarvice donne for the busynes of the company, wherof I trust you do now see good cause to be owt of dowbte of my dutye therof, and so to allowe me the same.

No. 6. M. Loks owne charges for his travayll iij yeares.

Also I have sett downe at thende of myne accompt the somme of vj^{c}*li.* of mony, wch I demaund of the ordinarie charges and travayle of my selfe, my sarvantes and howshold to followe this busynes and takinge charge of thaccompts and howse rome of the goodes in these iij yeares, wch is after the rate of cc^{li} by yeare, wherin I trust I shall not neade to saye muche, consideringe that all of you be men of good reason, and can consider what belongeth to the mayntenaunce of suche a famely as I have.

And somme of you by y^{or} owne experience have founde, and all of you have seene, what a chargable travayle and great troble bothe I and all my hows hold have had in the executione and followinge of this busynes in thes iij yeares, and also all of you do know that wheras I was well placed in the busynes of the Companye of Moskovia, wch I did execute quietlye, and for the doinges therof I had of them a pention of cc markes by yeare, bysides my howse rent ffree and other thinges well worthe to me cc^{li} by yeare; I have forsaken and geven over that office and assured trade of quiet lyvinge purposly to follow this bussines of the companye wth better effecte, according to my good wille and mynde desirous to sarve them to the best of my skille and power, wch I trust shall not be evill bestowed on them nor evell recompensed towardes me, havinge now torned all my goodes into the stones at Darteforde, and left to my selfe a howse full of children, wch maye bagge their bread yf the stones at Dartford be but stones.

The awtoritie and comaundement geven to M. Lok to follow this busynes.

Thus, I trust I have satisfied you consernynge my dutie of the vj^{c}*li* sett downe for monye paid owte of my purse for the extraordinarye charges in matters of the Companyes busynes, and also conserninge the other vj^{c}*li* for the ordinarye mayntenaunce of my selfe and famely during this tyme that I have sarved the companye and followed their busynes accordinge to the office and charge comitted and comaunded unto me from tyme to tyme by the letters and comissiones of her Maties most honorable privye councell and accordinge to the agreements and directiones sett downe by the comissioners in writtinge remayninge by me wch I have faithfully and dutifully executed from tyme to tyme to the best of my skille and power, wch sommes of mony you ought justly to allowe unto me, and maye as lawfullye so doo by yo[r] owne discretione w[th]out expectinge any further auctoritie as you and other comissioners by their owne discretiones have all redye paid and allowed unto C. Furbusher more then viij[c]*li* for his sarvice not so well bestowed as myne.

C. Furbusher paid and allowed him viiic*li*.

The objection for the Q. Maties monye.

But yo[u] maye object against me that when I had receaved the Quenes Ma[ties] monye dewe towardes the fraight of the shipps and wages of men come home, I might have paid it owt unto them. To this I aunswer that in dede I might have so done and so I did the moste parte of it, but in all that tyme that I was in the courte suter for hir Ma[ties] monye and for others of the venturars w[ch] was more then 3 wekes tyme continewally, C. Furbusher was at Darteford solicitinge the workemen their to make some good proffe of their worke w[ch] thinge beinge so greatlye longed for at the courte as w[th]out that ther was no money to be had amongest the venturars, and havinge so evell successe at Darteforde as cowld not be worse then was reported, the matter grew to so great discredit, as I could not parceave any hope where to receave any more mony amonge the venturares to parforme this busynes, and here w[th]all my debt beinge great for mony taken for the Company for the furniture of this busynes and my venture knowne to be so great therin as all that I had was worth, my credit decayed w[th] the discredit of these workes, so as I could not prolonge my debte any longer tyme, but was forced to paye the somme wth suche mony as I hade left me, w[ch] was dewe to me by the Company. And heruppon all matters growinge to miserie throughe scarcitie of mony because the rest of the venturares would not paye their dewtyes. And C. Furbusher lackinge now the mony he was wont to have still at my handes for the askinge, and I havinge no mony now for him to paye his men, he entred into great stormes and rages wth me lyke a made best, and raysed on me suche shamefull reportes and false sclaunders as the whole court and cittie was full therof, w[ch] did me great hurt, and did muche more hurte to the state of the companyes busynes, w[ch] is not yet recovered, but wilbe shortlye, at w[ch] tyme his false talles wilbe retorned uppon his owne heade, but in the mean tyme his slaunderous reportes have byn made agaynst me wth suche vement wordes of affirmation to be trewe, that through their sound of matter for the venturares profite and vantage, they are yt credited to be trew amongest them in the court, and so will remayne stylle untyll you do scertifye them of the truth of my doinges uppon yo[r] audite made of myne accomptes.

C. Furbushers rage againste M. Lok.

M. Loks great stoke and venture in thes voyages.

Moreover you doo see that uppon my good hoape and desier of the good successe of this busynes I have put in stoke and venture in the same all the goodes that I have in the world w[th]out exceptione, and for the accomplisshinge of the same to the companyes desire, I have gaged all that I have and have pressed my selfe and my frindes w[th] all the credite that I could make, and have spent all my tyme and oppressed my selfe wth continuall labour and troble therein, wherby maye appeare that thear hathe not byn any lacke of good wille nor dewtie on my parte, for the good successe of this busynes, wherfore yf any evell successe should happen in this busynes of the ewre at Darteford, wch I trust shall not yet is not that any way to be imputed to me, whose innocentie therin my goodes bestowed and ventured therin shall witnes and my writtinges delivered to her Maties honorable privie counsell of my dewtifull sarvis donne in that behalfe shall declare, but yf any suche mischief should happen, wch God forbed, the same wer rather to be imputed and layed on Martine Furbusher, who therin hathe comytted great abuses agaynst the companye, as in a paper of artickells therof shall appeare in dewe tyme, and uppon Jonas and Denham who be the chiefe workemen thereof. And on them the same were to be ponished sharplye as men who have byne the fynders and bringers of that ewer wch is brought, and causers of the cost bestowed for the fetchinge and workinge of the same, but I trust no suche cause shalbe geven.

M. Lok not giltie in the busynes at Darteforde.

And now I praye you waye uppright1y my former doinges and consider wth equitie my present state, and give not eare to the false reportes and sclanderous clamores latly raysed and sprede against me by Martine Furbusher w^th^out any foundacion of trewth, but defend my cause as my inocentie deserveth. And certifye her Maties honorable privie councell planly the effecte of my doinges in this sarvice and busynes, as you do fynd it by myne accomptes, that their honors maye be satisfied of me. And that I maye satisfie the worlde by the tryall of my doinges wch I will justifye. And yf you thus do, yo^u^ shall do justice, and I shall give thanks. And yf you thinke otherwise of the premises, I referre me to that wch shall stand wth reason and equitie.

Request of M. Lok to certifye the truthe of his doinges.

The 18 of Ffebruarie 1578.

And nowe to conclude this my aunswer unto yo^u^, I must saye that yo^u^ have delt verrie hardly with me in that you have suffered myne accomptes to lye dead and not touched ever syns the xviij of Januarie last untill yesterdaye, by wch meanes muche suspicion and clamor is growen agaynst me withowt just cause. Albeit yo^u^ maye saye in trewthe that in this tyme the chief commissioners have byn so occupied abowt the busynes of Duke Cesimirus, and the awditors abowt their busynes in the tearme tyme that yo^u^ cowld not assemble togethers, yet when some of yo^u^ did assemble, ones to make aunswer to the busynes of Captayn Ffenton in the west countrye and agayne to make inventorie and praysment of the shipps to be sould, some thing more then is donne myght have byn donne in myne accomptes, yf the awditors would or could have come to yo^u^, in whose absence yo^u^ would do nothing in myne accomptes.

The comyssioners hard dealinge wth M. Lok.

Also I maye saye that yo^u^ have donne me great wrong in writting

The comyssioners would not heare the auswere of M. Lok.

yesterdaye unto her Maties honorable privie councell yor aunswer uppon the conclusion of myne accomptes wthout having had anny maner of consideracion of my demaundes sett downe in my book of accomptes delivered to you for my great chardges paid and sustayned for the doing of the companyes busynes in these iij yeres voiages, and chieflye in that I having made this present book of myne aunswer uppon yor awdit of myne accomptes and making some of you acquaynted therwthall privatlye bifoare, and yesterday laying it open byfoare you to consyder, you would not vowchesafe to here it readde, which justyce requyrethe to be donne, alledging that the tyme was spent and you cowld not tell when to mete agayn any more. And tharfore (in post haste) you would make aunswer uppon myne accomptes and referre my demandes to their honors.

Mr. Auditor Neall controllethe not thaccountes.

And wheras Mr. Awditor Neale of privat affection hathe set down in the letter of yor aunswer that myne accomptes are wthowt controlement, suerly he dothe me great wrong therin for myne accomptes are controlled by billes and quyttaunces wch I have showed and delivered to the awditors in presens of you the comissioners and of Captayn Furbisher, in whose presentts they have byn examyned wth my bookes of accomptes[1] for the proffe of all the emptions of the goodes and the payementes of monney sett down therin wch said bills and bookes have byn now iij monthes in custodie of Mr. Neale, Mr. Furbisher dayly comyng to his howse as hys famyliar frynd, who in all this tyme might have controlled the same, and would have donne it, yf they cowld have found anny matter. And yet never the lesse yf all thesse awdytinge and reawditinge be not sufficient to justifie my trewe dealinges in myne accomptes I will take no vantage therbye, but let them be awdited and controlled agayn by some others, and I will abyde the uttermost tryall of anny man that can controll anny part of my doinges in myne accomptes. And for all the payementes therin contayned I will bring before them the partyes themselves of whom the goodes were bought and to whome the monny was paid (yf they be lyving) or ells a sufficient testimoniall from them by othe or wyttnesses by notarie.

M. Lok disbursed ijmli for the Companye.

And whereas by the examynacion of myne accompts yt dothe playnly appeare that I did paye and disburse of myne owne for the companyes busynes, the somme of ijm poundes of monney at dyvers tymes for yers and monthes of tyme wch is repaid me but wthin these ij or iij mounthes last past, you the comissioners and awditors have not made anny manner mention in this yor last letter unto her Maties honorable councell of that good torne donne by me for the companye wch equytie requiereth you should have donne.

And thus it may appeare that you have a thorne in yor owne foote wch dothe somwhat prick you wch now you would pull owt and put into

[1] Misc. Exchequer, Qu. Rev., vols. 60, 61.

myne who am not able to cure it so well as yorselves, but I praye you put yorselfe into my place and then do to me as yorselves would be donne to. And so shall God prosper us all. Do as you would be done to.

[*Colonial*, 131. *Dom. Eliz.*, cxxix, No. 44.]

FEB. 18, 1578. FROM MICHELL LOCKE CERTEFIETHE THE AUDITORS AND COMMISSIONERS PROCEADINGE WTH HIM ABOUT HIS LAST ACCOMPTE.

Right honorable,—Hereinclosed I send to yor honor the letter of the report of the Commyssyoners and Audytors uppon their last awdyte of myne accounttes, wherein they have delt very hardly wth me bothe in wordes and in deedes, but I trust I shall fynd their honors of her Maties councell bothe reasonable and good unto me, according to my trew delynges in thaccounttes and paynfull sarvyce in their busynes. Wherof to the end their honors may be better informed I send to yor honor herewthall a large declaration in wrytynge wch I directed unto the commyssyoners, purposinge therby to gyve them knowledge therof to their satisfaction, but for lak of tyme convenient they have not read ytt, for ttheir meetynges hathe byn so selldome and the busynes in thaccountts hathe byn so tedious unto them to paruse that they waxed wearye before their tyme and so have knytt up the end in hast as yor honor may see and would not read myne answere but referred it to their honors. Yor honor was my fyrst and chief frynd at myne entrans into this troblesom and chargeable busynes, and I would be right sorye that any of my doinges should move you to repent, and I trust there is no suche cause. Wherfore, I beseche yor honor to stand styll good unto me, and to gyve me styll yor favor and good countenans, and to thinke of me as of a trew man, for so wyll I trye my sellf in all my doinges, and wyll abyde the uttermost tryall of any man that can controll any part of my doinges in myne accounttes. This booke of myne answere uppon this awdyte of myne accounttes may seeme to yor honor tedious to reade, wherfore ytt may please yor honor that one of yor men may read ytt and report unto you the effect therof and afterwardes that yt may be used for the information of my Lords of her Maties Councell, as my cause shall requyre. And I wyll not be furder tedious unto yor honor at this tyme, but onely agayne I crave yor favour towardes my trewthe. And I commytt yor honor to Almightye God. From London the 18th of Februarie 1578.

Yor honors most bounden,

Michael Lok.

To the right honorable Sr Francys Walsingham, knight,
her Maties chyef Secretarie.

Colonial, 103. *Dom. Eliz.*, cxxvi, No. 33.

THE HUMBLE SUTE OF THOMAS BONHAM.

Ffirste the said Bonham firnished a shippe called the *Thomas of Ippiswiche*, beinge of the berthen of viii^{xx} or thear aboutes, and victualed the same, and ffirnished hir with all thinges necessarye to the somme of above iij^{c}*li.*

Item, the said shippe was so beaten by weather in hire viadge as *cli.* will not suffice to repayer hir, besides continuall charges of victualles for ye mariners sinste hir comminge home.

Item, the pilate being apoynted by the companie of adventurers, and by Mr. Ffurbusher, so as I ame not to be charged withe hir retorne without ffraight, the same being happined onely by the ffaulte of theim.

My humble sute therfore is, y^{t} by yor honorable good meanes some spedie order may be taken y^{t} thextreame charges I have bene at for not paymint of enie ffraight maybe presently releved withe the satisfaction of siche somes of monye as yor honor, withe ye reste of hir Majesties moste honorable counsell, shall thinke mete and requisite for the sayd shipes ffraighte.

Indorsed. Thomas Bonhams sute. Towchyng allowance to be yealded him for Furbishers viage.

Colonial, 135. *Domestic Eliz.*, cxxx, No. 21.

MARCH 28, 1579. FROM SIR THOMAS GRESHAM AND OTHER TOUCHYNGE Y^{E} ORDER THEY HAVE TAKEN FOR THE PAYMENT OF THE MARINERS.

Ytt may lyke yor good LL. to understand that we haue receaved yor LL. letter of the 26 Marche, wherby yor LL. plesure is, that we shall appoint iiij men for the sale of the other ships and other thinges remaynynge, for the payment of the men not yett paid. Accordinglie, we haue appointed men therto, who shall see the same executed as spedelie as may be, but that beinge done, is but a verie small matter to pay that wch is owinge to the men that are unpaid, and the ffraightts of the shyps owinge. Wherfore yt may pleise yor LL. to gyve order, that suche of the venturars as have not yet paid their partes towardes the said fraights and wagys may pay the same forthwth, for that otherwyse wee see not other present remedye for the same. The names of the venturars wch have not yet paid their full part is in a wrytinge hereinclosed, to whome ytt may please yor LL. to gyv suche straight order, as that they may pay their partes owinge, as the other venturars have done, or ells to be secluded from all benefyte that may grow to them by their former ventures made, and other ventures hereafter to be made, and to lose all that wch alreadye they have disbursed.

Also yt may lyke yor LL. to understand that this daye we have had conferens wth one John Barton, gentleman, who semethe to have experiens of myneral workes, who hathe offered to make a proffe of the ewr at Dartford, at his owne charges in the great workes at Dartford, and theruppon will procede in the work of all the ewr there, to have for his industrie, vppon the valew of the ewr after this rate; to saye, for everie ton yeldinge the valew of xx ponds money clere above all charges, he to have ten shillings for his paynes and industrie, and yelding the valew of xxx ponds the ton clere of all charges, he to have xx shillings; and yeldinge the valew of xl ponds the [ton] to have xxx shillings for his industrie; and he to work ytt at his charges, and wyll work xvc tons by yere, to whom (yf yor LL. lyke of ytt) we have consented he shall make prooffe ymediatelie after Ester, wherof yt may plese yor LL. to advertyse us yor plesures, for that we doo staye the man in towne uppon yor LL. answere to be had.

And for the procedinge of the great workes at Dartford by Jonas, we thynk ytt very requysytt to procede in the same, and therto is needfull of dyttamentts to be provyded for them, and other necessarie charges at Dartford for the working of the same, wch in all would ask the suum of vc pondes untill the workes wyll mayntaine ytt sellf for the reste, for the wch money we have nott any means heere, for that the former cessementts wyll not suffyce for the fraights and wagys of men yet unpaid, neyther doo we know how to provyde the same but by a new cessementt uppon all the venturars, and the charge and accountt therof to be commytted to a severall man.

Thus humbly we take our leave of yor LL., and commytt the same to the tuition of Almightie God. From London the xxviij Marche, 1579.

At yor LL. commandements,

Thomas Gresh'm.	Lionell Duckett.
John Dee.	Martin Frobisher.
Thomas Allen.	Edwarde Fenton.
Xopfer Hoddesdonn.	Gylbert Yorke.
Michael Lok.	Mathew Fyeld.

To the Right Honorable our very goods Lordes the LL. of Her Maties Honorable Privie Councell.

[*Colonial*, 138. *Domestic Eliz.*, cxxx, No. 42.]

THE VENTURARS W[TH] HIM NOTE PAYDE AT THE 25 APRILL 1579.

	li.	*s.*	*d.*
The Lord High Tresorer	065	00	00
The Lorde High Admerall	065	00	00
The Erle of Sussex	065	00	00
The Earle of Warwicke	065	00	00
The Earle of Lecester	011	03	04
The Lorde Hondeston	085	00	00
S[r] Francs Knowles, Tresore	032	10	00
The Earle of Oxforde	450	00	00
The Earle of Penbroke	172	10	00
The Countesse of Warwick	057	10	00
The Countesse of Penbroke	028	15	00
The Lady Ann Talbott	010	00	00
S[r] W[m.] Winter	040	00	00
S[r] Johane Broquete	077	10	00
Mr. Phallapp Sydney	067	10	00
Mr. Edward Dyer	067	10	00
Mr. Willm Pelhame	135	00	00
Mr. Thomas Randolphe	085	00	00
Johne Somers	067	10	00
Symonde Boyere	028	15	00
Antony Jenkenson	057	10	00
Jeffry Turvill	007	00	00
Richerd Bolande	027	10	00
Mathewe Kenersley	028	15	00
Robarte Kinersley	057	10	00
William Burde	020	00	00
Thomas Owene	012	15	00
Christopher Androwes	005	00	00
Robart Martine	005	00	00
Marten Furbysher	270	00	00
Sum of all	2167	03	04

Thes whos names be under wrytten be suche as adventured in the second viage, and not in the thirde, except the Countesse of Sussex, who was no venturer in the second vyage, and S[r] Lionell Duckett who hathe adventured the moytie in the third viage accordinge to the some adventurid by in the second viage the w[ch] moste be referred unto the consideratione of your Honors.

The Countesss of Sussex	135 00 00
The Lady Ann Talbote	062 10 00
Sr Lyonell Duckett	067 10 00
Sr William Winter	500 00 00
Willm Burde	250 00 00
Christopher Andrwes	062 10 00
Robart Martyne	062 10 00

There is also owinge by Mighell Locke for the foote of his accompte 1217*li.* 19*s.* 04*d.*, the consideracion whereof moste be in like case referrid to the determynatione of your Honors.

There is also a reare acompt of Mr. Locks for dyvers marchandizes and victuales, etc., retornid in the shipps, and by him sould unawdited.

Even so in leke case the whole and full acompt of Mr. Furbysher as yete to awdite to bothe wch acompts I cann saye nothinge untell the same be fynyshed and by the comyssioners throughly seane.

Indorsed.—A note of the accompt towchynge the northwest viage.

[*Colonial*, 140. *Domestic Eliz.*, cxxx. No. 47.]

After or hartie commendations, whereas for want of the paiment of suche somes as are due by sundrie the adventurers to the northwest in the late voiage made by Mr. Frobisher, not onely manie that served in the saide voiage be yet unpaide and undischarged, but also the ewre brought home remainethe untried and so unprofitable Her Matie hath caused an order to be sette downe by my LL for the aunsweringe of the saide sommes whereunto her pleasure is that so manie as be behinde hand in their paiments, and intend by continewinge in the societie of this companie to reape the benefitte that may happely growe thereof, shall subscribe their names in the testimonie that they will see the sommes due by them paide to such person, and within such time as is expressed in the saide order. And to that ende we are willed to sende to you, as we do by this bearer the same order to be by you subscribed in case you meane to continue an adventurer, otherwise purposinge to venture no more, to require you to subscribe to one other bill wch this bearer also hathe to exhibite to you, thereby testifyinge yor refusall to be for the presente anie longer an adventurer of this companie. By a note wch this bearer hath under the hands of the Commissioners appointed to regarde the accounts of this Companie, you may see what is behinde to be by you paide, wch if you shall like to see paide accordinge to the order, then are you by a note of yor hand to signifie the day of yor subscription to the order, that accordingly the sommes wch you are to pay may be looked for and receaved here by the Threasurer of the Companie. And so we bid you hartely farewell from the courte the of Aprill, 1579.

Yor lovinge frende.

[*Colonial*, 109. *State Papers. Dom. Eliz.*, cxxx. No. 16.]

THE VENTURARS Wch HAVE NOT PAID TO M. LOK, BUT MUST PAIE TO Mr ALLEN.

	Wages second voiage. 1577.	Buyldings Dartford.	Third voiag outwards.	Fraight retorne. 1578.
The Lord Highe Treasurer	*li*	*li*	*li*	*li* 118
The Lord Highe Admirall	*li*	*li*	*li*	*li* 118
The Erle of Sussex	*li*	*li*	*li*	*li* 118
The Erle of Warwick	*li*	*li*	*li*	*li* 118
The Erle of Leycester	*li*	*li*	*li*	*li* 11 3 4
The Lord of Hunsdon	*li*	*li* 10	*li* 17 10	*li* 57 10 0
Sr Fraunc̄s Knowlls	*li*	*li*	*li*	*li* 57 10 0
The Erle of Oxford	*li*	*li*	*li*	*li* 450
The Erle of Penbrook	*li*	*li*	*li*	*li* 172 10 0
The Countesse of Penbroke	*li*	*li*	*li*	*li* 28 15 0
The Countesse of Sussex	*li*	*li* 10	*li* 67 10	*li* 57 10 0
The Countesse of Warwick	*li*	*li*	*li*	*li* 57 10 0
The Ladie Anne Talbot	*li* 5	*li* 5	*li*	*li*
Sr Henrye Wallope	*li*	*li*	*li*	*li* 57 10
Sr Thoms Gresham	*li*	*li*	*li*	*li* 230
Sr Leonell Ducket	*li*	*li* 5	*li* 33 15	*li* 28 15
Sr Willm Wynter	*li*	*li* 40	*li*	*li*
Sr John Brocket	*li* 10	*li* 10	*li*	*li* 57 10
Mr Phillip Sidney	*li* 10	*li*	*li*	*li* 57 10
Edward Dier	*li*	*li* 5	*li* 33 15	*li* 28 15
Willm Pelham	*li*	*li* 10	*li* 67 10	*li* 57 10
Thomas Randolph	*li*	*li*	*li* 27 10	*li* 57 10
John Somers	*li* 10	*li*	*li*	*li* 57 10 0
Symon Bowyer	*li*	*li*	*li*	*li* 28 15 0
Anthony Jenkynson	*li*	*li*	*li*	*li* 57 10 0
Jeffrey Turvile	*li*	*li*	*li*	*li* 37
Willm Paynter	*li*	*li*	*li*	*li* 57 10 0
Richard Bowland	*li*	*li*	*li*	*li* 57 10 0
Mathew Kyndersley	*li*	*li*	*li*	*li* 28 15 0
Robert Kyndersley	*li*	*li*	*li*	*li* 57 10 0
Mrs AnneFraunc̄s Kyndersley	*li*	*li*	*li*	*li* 86 5 0
Willm Burd Mercer	*li*	*li* 20	*li*	*li*
Willm Ormeshawe	*li*	*li*	*li*	*li* 28 15 0
Thoms Allen	*li*	*li*	*li*	*li* 57 10
Richard Young	*li*	*li*	*li*	*li* 57 10
Willm Bond	*li*	*li*	*li*	*li* 115
Thoms Owen	*li*	*li*	*li*	*li* 28 15 0

	Wages second voiage. 1577.	Buyldings Dartford.	Third voiag outwards.	Fraight retorne. 1578.
Willm Dowgell -	- li	li	li	li 28 15 0
Anthony Marlor -	- li	li	li	li 28 15 0
Christopher Androwes	- li	li 5	li	li
Robert Martyn -	- li	li 5	li	li
Martyn Furbisher -	- li	li 20	li 138	li 115
	li 35	li 145	li 382 10	li 2855 13 4 not red
				2923 1 8 received
				li 5778 18 0

38
145
382 10
2855 13 4

3418 3 4 Not recd by Mr Lok
72 1578
The Venturars money not paid to Mr Lok but to Mr Allyn.

[*Colonial*, 124. *Dom. Eliz.*, cxxx, No. 18.]

THE HUMBLE PETITION OF MICHAEL LOK FOR CHARGES DYSBURSED.

To the right Honorable the Lordes and others of Her Maties moste Honorable Privie Councell.

In most humble dewtye besechethe yor Honors, yor most humble suppliant Michael Lok, that wheras by the manyfold comandementts of yor honors, and by the dayly directions of the comyssyoners appointed for the voiages lately made by Martin Furbusher, yor said suppliant for the space of these iij yeres hathe taken the charge and dewtyfully to his power hathe followed and seene executed all the busynes therunto appertaininge, according to the orders to him appointed in that behalf from tyme to tyme, wch he hathe to shew to his continewall great paynes and trouble, and his very great charges and expenses. And of all his doinges in the premisses he hathe made dew and trew accountt, wch is awdyted and certiffied unto yor Honors, in wch accountt yor said humble suppliant hathe sett downe the sum of xijc*li* by hymn expended and layd out of his owne purse for dyvers particular charges, for the followinge of the said busynes in the said tyme of iij yeres, as therin dothe appeare; wch said awditors would not allowe unto yor said suppliant uppon his said account sayenge that they had none auctoritie therto, but doo referre the same to the consideration of yor Honors. Now yor Honors said humble suppliant, most humbly besechethe yor Honors to have consyderation of the premysses, in respect of his dewtyfull sarvyce done therin, and his trew dealynges in his accountt made, as also for that most trewlye he hathe expended and layd out of his owne purse for the said busynes the said sum of xij^{c}*li* sett downe in thaccountt, and also iiij^{c}*li* more not sett downe in thaccountt, as he wyll make dew proffe besydes the great sum beinge ij^{m}ij^{c}l*li* of money wch he hathe paid in the said voyages, for his owne stok and venture therin, whiche is all the goodes that he hathe in the world wthout exception ; wherby now hym sellf and wyfe and xv children are left in state to beg their bread hensforthe except God turne the stones at Dartford into his bread agayne, and that yo$_{r}$ Honors be good unto hym in this his humble sute, accordinge to his dewtifull trew meaninge in this his sarvyce done. And yor said humble suppliant and his children, according to their bounden dewtye, shall pray to God contynewallye for the encreas of all yor Honors estates with all prosperitie.

[*Colonial*, 125. *Dom. Eliz.*, cxxx, No. 19.]

MICHAEL LOK HATHE PAID FOR DYVERS CHARGES FOR THE AFFAYRES OF THE COMPANYE IN THE IIJ YERES OF THE IIJ VOYAGES OF MARTIN FURBUSHER, GENT., FOR THE NORTHWEST PARTS, AS FOLOWTH :—

For my rydynge charges to the Court at Hampton, Wyndsore, Rychemond, and other places in progresse to attend on Her Maties most Honorable Privie Councell, for comyssyons, directions, and money collections of the venturars in iij yeres, at xl^{li} by yere .	cxx^{li}
For my ryding charges to Dartford, and for the bylldings and workes theare, and to other places, daylye, for necessaries in these viij monthes . . .	lx^{li}
For my botehyer to the Court at Grenewiche to attende on the Councell for the comyssyons, and money collections of the venturars, and to the ships, and other places *li*	xx^{li}
For the table dyatt of the comyssyoners, awdytors, captaynes, and others of these voyages, at my howsse often and daylye in these iij yeres, at l^{li} by yere *li* .	cl^{li}
For interesses of money taken up from tyme to tyme to furnishe thes iij voyages and dyspache of the ships in dew tyme, for lak of the venturars money *li*	ccl^{li}
S'mm vj *li*	

For the ordynary chages and travayll of my sellfe and my sarvants and howshold to follow this busynes and take charge of thaccountts, and howserome of the goodes in these iij yeres voyages at cc^{li} by yere . vj^{c}*li*

S'mm of all xij^{c}*li*

Thus moche money xij^{c}*li*, and more hathe ytt cost me out of my purse, wherof nothinge is yett allowed me in accountts. Allow me what reason and equitie requyrethe.

And Michael Lok hathe in stok and venture for hym sellf and hys chyldren wch he hathe paid . . ij^{m}clij^{li} x^{s}

And in the name of John Dee . iiij^{xx}xvij^{li} x^{s}

S'mm . ij^{m}ccl*li*

Besydes the stok and venturre of the Right Honorable the Erle of Oxford, wch is . . . ij^{m}iiij^{c}xxx*li*

By me, Michael Lok.

[*Colonial*, 105. *Dom. Eliz.* cxxvj, No. 56.]

THE VENTURARS W^{CH} HAVE NOT PAID THEIR PARTES FOR FFRAIGHT AND WAGES THE LAST NOVEMBER, 1578.

Off the Court.

The Lord High Treasurer .	cxv^{li}	50		
The Lord High Admirall .	cxv^{li}	50		Ffor third voiag
The Erle of Sussex . .	cxv^{li}	50	200	venture outwardes.
The Erle of Warwick .	cxv^{li}	50		
The Lord Hunsdon . .	lvij^{li} x^{s}	25		xxvij^{li} x^{s}
Sr Ffraunces Knowles .	lvij^{li} x^{s}	25	50	
The Erle of Oxford . .	iiij^{c}l^{li}	200		
The Erle of Pembroke .	clxxij^{li} x^{s}	60	275	
The Countez of Sussex .	lvij^{li} x^{s}	25		lxx^{li} x^{s}
The Countez of Warwick .	lvij^{li} x^{s}	25	40	
The Countez of Penbroke .	xxviij^{li} xv^{s}	15		
Sr Henrie Wallope . .	lvij^{li} x^{s}	25		
Sr John Brocket . .	lvij^{li} x^{s}	25		xx^{li}
Mr. Philip Sidney . .	lvij^{li} x^{s}	25	125	x^{li}
Mr. William Pelham .	lvij^{li} x^{s}	25		lxxvij^{li} x^{s}
Mr. Thomas Randolphe .	lvij^{li} x^{s}	25		xxxvij^{li} x^{s}

Edward Diar .	. xxviij^{li} xv^{s}	15	40	xxxviij^{li} xv^{s}
John Somers .	. lvij^{li} x^{s}	25		x^{li}
Symon Boyer .	. xxviij^{li} xv^{s}	15	65	
Martyn Ffurbisher .	. cxv^{li}	50		clv^{li}
Anthonye Jenkynson	. lvij^{li} x^{s}	25	115	
Jeffrey Turvile .	. lvij^{li} x^{s}	25		
William Paynter .	. lvij^{li} x^{s}	25		
Richard Bowland .	. lvij^{li} x^{s}	25		
John Dee . .	. xxviij^{li} xv^{s}	15		

Sum ij^{m}cxvij^{li} x^{s}

Off the Cittie.

Sr Thomas Gressham	. clxxx^{li}	65	90	xxxiij^{li} xv^{s}
Sr Leonell Ducket .	. lvij^{li} x^{s}	25		
My Ladye Martyn .	. xxviij^{li} xv^{s}	15	105	
Mathewe Kyndersleye	. xxviij^{li} xv^{s}	15		
Robert Kyndersleye.	. lvij^{li} x^{s}	15		
Mrs. Anne Fraunces Kyndersley	lxxxvj^{li} v^{s}	35		
Mathew Ffield .	. lvij^{li} x^{s}	25		
Edmund Hogan .	. cxv^{li}	50	115	
William Bond .	. cxv^{li}	50		
Thomas Owen .	. xxviij^{li} xv^{s}	15		
William Borrowe .	. lvij^{li} x^{s}	25		
William Ormeshawe.	. xxviij^{li} xv^{s}	15	100	
William Dowgle .	. xxviij^{li} xv^{s}	15		
Anthonye Marlo^{r} .	. xxviij^{li} xv^{s}	15		
William Harrington	. xxviij^{li} xv^{s}	15		
Michael Lok .	. iiij^{c}lx^{li}		220	

Sum j^{m}iiij^{c}lxxxvij^{li} x^{s}
and ij^{m}cxvj^{li} x^{s}

Not receved fraight iij^{m}v^{c}v^{li}
outwardes venture vj^{c}x^{li} x^{s}

Sum iij^{m}cxv^{li} x^{s} not receved
Ffor the second voiag.

———— x^{li} My Ladie Anne Talbot
———— xl^{li} Sr William Wynter
————xx^{li} William Burde
———— v^{li} Christofer Androwes
———— v^{li} Robert Martyn
Sum vj^{c}x^{li} x^{s}

ACCOMPT OF MICHAEL LOK, TREASURER, MADE THE LAST OF NOVEMBER, 1578, OF MONNEY RECEIVED AND PAID BY HYM SYNS HIS LAST ACCOMPT, AUDITED IN AUGUST 1578.

Receiptts.

Off the Quens Matie for fraight, and wages . .	mcl^{li}
Of Therle of Leycester, for fraight . . .	clxxj^{li} x^{s}
Of Therle of Oxford, part for fraight . .	v^{c} lxxxv^{li}
Of Mr. Secretarie Walsingham, for fraight . .	ij^{c} xxx^{li}
Of hym, for venture owtwardes . . .	lxij^{li} xiij^{s}
Of Mr. Secretarie Wilson, fraight . . .	cxij^{li} x^{s}
Of S. Thomas Gresham, venture owtwards . .	c^{li}
Of hym, for parte of fraight . . .	l^{li}
Of Thomas Allen, fraight	lvij^{li} x^{s}
Of Christopher Hadson, fraight . . .	lvij^{li} x^{s}
Of Richard Young, fraight	lvij^{li} x^{s}

Sum receved, xxv iiij$^{xx li}$ iij^{s}

Payments.

To Michael Lok, rest of his accompt . .	ix^{c} lxxxix^{li}
For the workes at Dartford, above all receved .	iiij^{c} lxxx^{li}
To the Erle Leycester, ewr from Bristow . .	clxxij^{li} x^{s}
To the shippe *Hoapwell*, part of fraight . .	c^{li}
To the shippe *Anne Fraunces*, part fraight . .	c^{li}
To the shippe *Tho. Allen*, part fraight . .	clxij^{li} x^{s}
To the mynars and men, part of wages . .	ij^{c}l^{li}
To the shippe *Beare Leycester*, fraight . .	iij^{c}l^{li}

Sum paid, xxv^{c} iiij^{xx} xix^{li}

By me, Michael Lok.

Colonial, 137. *Dom. Eliz.*, cxxvii, No. 10.

AN ORDER SETT DOWNE BY THE QUEENES MA^{TIES} EXPRESSE COMMANDEMENT, TOUCHING THE SUPPLYING OF SUCH SUMMES OF MONEY AS ARE DUE BY THE ADVENTURERS IN THE NORTHWEST VOYAGE, OTHERWISE CALLED META INCOGNITA.

The Q. Matie being given to understand that diverse of those that were adventurers in the late viages performed by Martin Ffurbusher, gent., into the northwest partes (not wth standing sondrie admonitions given by letters directed unto them from the lords and others of the privy councell), that they shold bringe in such summes of money as were due by them at tymes and daies limited by the said letters, have not brought in the said summes accordingly; wherby diverse mariners continewed a long season undischarged, and the fraught of the moste parte of the shippes employed in that voyage unpayed to th' utter undoing of diverse of the owners of the said shippes, and greatly to her Ma[ties] dishonor, being an adventurer in the said voyage, and having payed all such summes of money as were due by her. Ffor redresse wherof her Ma[tie] doth therfore order that all such adventurers as have not yet payed in such summes of money as are by them due, shall wth in ten dayes after notice given to them of this her Ma[ties] pleasure bring in and deliver into the hands of Thomas Allen, appoynted to be the treasorer for this purpose, the moytie of such summes as are by them yet due, and th' other moytie wth in a moneth after, w[ch] if they shall not observe that than they shall not only be forthw[th] excluded out of the company, but also loose the benefitt of such summes of money as they have alreaddy putt in, being a matter agreeable wth lawe and justice for not observing the rules of societie. And for that it may be knowen out of hand who meane to continewe the said adventure by making payment of such sommes as are by them due accordingly, as is above mentioned, and who shall refuse: yt is by her highnes thought meete, and so ordered that such as meane to continew the same shold subscribe there names to this order, as thereby binding themselves to the payment of the summes by them due, as above is expressed. And that such as shall refuse to subscribe the same therby to bind themselves to the payement, but meaning to adventure no more money in the said voyage shalbe herafter utterly excluded in suche sorte as is above specified. And to thintent that no man shall pretend ignoraunce what he ought to paye at this present, the bearer herof hath a scedule conteyning the names of all such as have adventured wth the summes by them payed, and what summes are remayning due to be payd, subscribed by such commissioners as have had authorysie to have regard thereto.

Also, The L. Treasorer.
The L. Admirall.
The L. Chamberlain.
The E. of Warwycke.
The E. of Leycester.
The L. of Hunsdon.
Mr. Thresorer
Secret. Walsingham.
Mr. Secret. Wylson.

[*Colonial*, 141. *Dom. Eliz.*, cxxvi, No. 36.]

THE NAMES OF THEM THAT BE LEFTE OWTE OF THIS LAST BILL AND THOSE THAT BE SETT LESSE IN THIS LAST BYLL THEN IN THE OTHER THESE NAMES FFOLLOINGE.

	£	s.	d.
Thomas Randolphe	10	0	0
Jeffraye Turvell	20	0	0
Jhon Dice	28	15	0
Sr Lyonell Duckett	28	15	0
My Ladye Martyn	28	15	0
Mathew Ffelde	57	0	0
Edmond Huggan	115	0	0
Wylliam Burroo	57	0	0
Wylliam Harryngton	28	15	0
Mychell Locke	460	0	0
	835li	0s	0d

[*Lansdowne*, xxx, No. 4, fol. 12. *Colonial*, 153.]

THE OFFER OF MICHAEL LOK FOR THE NORTHWEST EWR AT DARTFORD.

He requirethe to have lycense for iij yeares to serche for myneralls by the patent made to William Humfrey or by the mynes Royall. And libertye to work them at his pleasure at his owne charges, gyving therfore v of the c, of the clere gayne that shall grow therbye.

To have the use of the workhouses at Dartford for iij yeres, and libertye to work the northwest ewr that is there at his pleasure, at his owne charges.

To have a man assigned to be of his councell, and take account under hymn in all that shalbe done. And all the clere gaynes that shall grow by this workes, he is content shalbe gyven to the payment of the debtts owing by the companye, for the northwest voiages of Captaine Furbusher.

That he may have recompenses of the Royal Majestie for his land bought and recovered from hym, the which recompens he is content shalbe emploied in these workes, to be repaid hym in account of the workes, withall his owne travayll and industrie.

That he have a protection of her Majestie for iij yeres, and a quietus est of his accountes, and a clere discharge for all his bondes, and all the debtes of the companye of the voiages of Captain Furbusher.

That may have commyssion to collect the goodes of the companye that lye scattered, and to recover the debtes owing to the companye, and set downe what is owinge to them and to take account of the state of the companye.

And after that he hath paid and sattysffied the creditors of the companye for the debttes owing to them, he shall have freelye the leasse of the workhouses at Dartford, with all the companyes interest therein.

A letter to be wrytten from my Lords of the Counsell unto the Commyssyoners, to examyne all the workmen for the addytaments used with the northwest ewr, in the small sayes made in Meta Incognita, which shewed clene gold.

Commyssyoners.	Workmen.
Sir William Wyntar.	John Baptista Agnello.
Sir William Pelham.	Jonas Shutz.
Mr. Thomas Randolphe.	Robert Denam.
Mr. Dyar.	William Humfrey.
Mr. Dee.	Humfrey Cole.
Mr. Yonge.	
Mr. Hogan.	D. Burcot is deade.
Mr. Lok.	
Mr. Palmar.	

[*Lansdowne MS.*, xxx, No. 4, fol. 10. *Colonial*, 152.]

THE OFFER OF MICHAEL LOK.

All the northwest ewr brought home by Captayne Furbusher is estemed xij^{cc} tons.

For the which ewr Michael Lok shall paye v^{l} the ton, which amountethe vj^{ml} of money.

The first payment to begyn at the end of one yere, and then to pay every monthe c^{l} of money untill the vj^{ml} be payd.

To have better suerty than by self.

And for suretye of this payment he shall fyrst receve 1 ton of ewr uppon his owne bond, and afterward shall gyve suretye for the ewr as he recevethe yt.

The vallew of the lesse to be knowen.

And he shall have the whole leasse of the mylles and workhowsses at Dartford, and benefyt therof in suche state as the same is taken of the Quenes farmar, and as the same now ys.

He shall have freelye all the implements and furnyture for the workes now beinge at Dartford, and all the myneralls and mettals that are there being wroughte.

The excepting all other former grants.

He shall have lycens for terme of the said leasse to serche and myne for ewrs and mynerallss in all groundes which are not already opened and myned, and therout dygged the quantyte of fyve ton of ewr within all the Quenes Majesties domynyons, except the priviledges of the stannerie of Cornwall for their tynne ewrs, and the same ewrs and myneralls to take and carye away and use at his pleasure, compoundyng with the ownars for the brekynge of their ground. made by her Majesty or by any her progenitors.

And, nevertheles, this Lycens shall not restrayne any other man to serche and dygge also for any ewrs and myneralls in any other place not beinge within fyve myles of the place that shalbe dygged and followed by vertew of this Lycens.

He shalbe clerely dyscharged and kept harmeles, quyet of all maner debtes and demandes of all men, for all the busynes of the Companye done before this day.

The Quenes Majestie shall have libertie to take agayne into her handes this grant and contract at the end of vij yeres, payenge and recompensyng the charges done and domage to be sustayned therebye by the arbitrement of vj indyfferent parsons.

Lansdowne MS., xxx, No. 4, fol. 11.

Or ells yf the forsaid offer be not lyked, then Michael Lok shall cause to be wrought all the said northwest ewr, for the account and use of the companye of venturars.

And shall make yt worthe v[li] the ton at the least, and better yf yt wilbe clere of almaner charges from hens forthe to be done.

And the company of venturars shall gyve hym the awcthoritie to governe, command, and direct all the workes.

And shall pay hym x[s] a day for his owne charges and travayll, out of the sayd valew of the ewr. And shall provyde a stok of money iiij[c]£, to buye and provyde addytaments and to begyn the workes. And shall appointe a man to be of counsell of his doinges and to kepe the money and to take thaccounts daylye of all that passethe. The Company hath lost over much money

And he shalbe clerely dyscharged and kept quyet of all maner debtts and demandes of all men, for all the busynes of the companye done before this day.

And after that all this northwest ewr is wrought as aforesaid, Michael Lok shall have the state and right of the said leasse of the sayd mylles and workhowsses at Dartford for the rest of the yeres therein then to come.

And allso the Lycence to serche and myne and work all ewrs and myneralls as aforesaid, duringe the rest of the yeres of that leasse for his owne account and use, payeng to the Quenes Majestie fyve shillinges money for every tonne of ewr that shalbe dygged and melted by vertew therof.

Colonial, 104. *Dom. Eliz.*, cxxvi, No. 47.

18 NOVEMBRIS, 1578. AN OFFER THEN MADE AT MOSKOVY HOUSE BY JONAS SUTE BEFORE MR FEILD, MR LOCK, AND ANDREW PALMER.

A tonne of ewer	viij^{li}
vj^{c} of copper rerquisite from Keswicke	xx^{s}
Of lead, i^{cwt}	x^{s}
Of lead ewer, vi^{cwt}	xxiiij^{s}
Wood for roste, *di* (2^{cwt}) m^{t}? (500)	v^{s}
Coles for meltinge	xx^{s}
Ffees and wages a tonne	xx^{s}
Ffor extraordinary charges	xvj^{s}
Sum, xiij^{li} xv^{s}	

Hereof Jonas will deliver gold and silver nett to the valewe of xxiij^{li} xv^{s}

Indorsed. Mr. Palmers note touchynge Jonas offer abowt Furbishers ewre. 18th Nov., 1578.

Colonial, 103. *Dom. Eliz.*, cxxvi, No. 34.

ALL THE STOK OF THE VENTURERS IN ALL THE IIJ VOYAGES.

Sum of all the stok of all the venturars.	All the venturars.	wherof	Michael Lok and his children.
For the first voyage, anno 1576	£875		£100
For the second voyage, anno 1577	£4275		£1075
For wagys at retorne therof	£1030		£225
For byldinges at Dartford	£1105		£260
For the third voyage, anno 1578	£6952 10s.		£1755
For fraight & wagys at retorne therof,	£3347 10s.		£845
Sum	£17585		£4270
For the second rate of fraight	£2575		£650
Sum all	£20160		£4920

And note that of the forsaid summe of £4270 of his venture, the Erle of Oxford became partner wth him for £2000 in suche order and maner as hym selfe was and is venturar.

And over and besydes the said summe of £4920 of his venture Michael Lok dyd pay of his owne purse for the furnyture of the first voyage £700, whiche was restored to him in account of the second voyage.

And he dyd also paye of his owne purse, for the furnyture of second voyage, £400, whiche is now latelye repaid hym in accountt.

And he did also pay of his owne purse, for furnyture of the third voyage and byldyng at Dartford, £700, whiche is nowe latelye repaid hym in accountt.

And, more over, he hathe taken great paynes and travayll, and byn att very great charges and expenses in doinge the companies busynes in all these iij voyages, and hathe not yet charged anye of his accounttes withe one penye for the same, knowinge that the venturars wyll consyder of it withe reason.

And now, Michael Lok havinge done all the premysses in sarvyce of her Matie and the venturars, he is openlie sclandered by Captaine Furbusher thus to be :—

A false accountant to the companye.

A cossener of my L. of Oxford.

No venturer at all in the voiages.

A bankerot knave.

Wherfore most humblye he besechethe yor Lordships to direct yor letters unto the commyssioners of the busynes and the awditors of his accounttes to certyffye yor Lordships what he hathe done in the premysses.

[*Colonial* 122. *Dom. Eliz.*, cxxx, No. 17.]

THE ABUSES OF CAPTAYN FURBUSHER AGAYNST THE COMPANYE. AN^O^ 1578.

In the first voyage he brought home by chaunce a stoane of riche ewre, and being examyned by Sr William Wynter, Mr Randall, Mr Hogan, and the rest of the Comissioners, what quantitie was to be had, he said that in that countrie was inoughe therof to lade all the Quenes shippes, and promised to lade the shippes of the seconde voiage ther w^th^all, wheruppon the seconde voiage was prepared, and comyssion geven him to bringe of the same. And Jonas, Denham, and Grigorie, were sent with him for the same; but he performed nothinge at all, & brought not so muche as one stoane therof; for ther was none to lade, as Jonas and the rest do witnes, but laded the ships wth other mynes founde by chaunse. No. 1. The ewr promysed was not brought.

In the seconde voyag he retorned the shipps laden wth stoanes of strainge ewr found by chaunce there, sainge they were of gold myne worth iiii^xx^ poundes a tonne, wch is not yet so founde; and also he brought some stoanes of redde ewre and yellow ewer of Jonas mount, verye riche of gold, as D. Burcot witnessed, and the stoanes are yett to be seen. And promised to the comissioneres that ther was mountaynes therof, and he would lade all the shippes therw^th^al in the thirde voiage, No. 2. The ewr promised was not brought.

wheruppon the thirde voiage was prepared wth so great chardg; but he brought home not one stoane therof afterwards that is yet found.

No. 3. Superfluous shippes and chargs.

In the thirde voiage he promised to lade all the shippes wth the ewr of Jonas mount, and other so riche ewre as the best of the second voiage was, and carried owt a nomber of ships for that purpose, and a c. men to inhabit there under culler of the Frenche mens preparacon to that countrie, and besydes the nomber appoynted to him by the Comissioners, he carried mor 4 shippes and a c. men more for his owne purpose, wthout the knowledge of the Comissioners, wch now rest uppon the charge of the Companye, and he brought home those ships laden wth none of the ewre that he promised, but wth other strainge ewr, wher he could fynd yt, wch he said was better then the best that was brought the yeare befor, wch is not yt so found.

He carried 4 ships, and c men, wthout comission.

No. 4. He would not place C. Fenton there.

Also he promised to the Comissioners and had speciall charge by comissione first to plant C. Fenton and the c. men to inhabit in that new land, wheruppon the great preparacon was made; but afterwards, before his departure from London, he dislyked that enterprice, and diswaded the same cullerablie, and when he came there he would not helpe them therin one jote, not so muche as for 50 men wherwithall C. Fenton would have tarried there, he feringe that C. Fentons deede therin woulde dashe his glorye, and because he toke the victualls of that provicion to victuall his owne 4 shippes taken wth him extraordinare, wch went from hens unvictualled uppon his promisse made them to victuall them, as Captayn Fenton and others witnes.

No. 5. He made no discourie of passage.

He promised and had comission to send the two barks this yere to make some discoverie of the passage for Cathai, wch he might have donne; but when he came at Meta Incognita, he would do nothinge at all therin as Hawll & Jakman wytenes, but made all his endevour to lade his owne shippes, and the rest home agayne wth ewre.

No. 6. His owne men evell officers in the shippes.

He hathe byn still verrie costlye and prodigall in the furniture of the shippes and men for the voyage, and his owne men beinge shipped for officieres of the shippes have made verie great spoile, wast, and pilfrye of the goods in the shippes, for the wch he must give account.

No. 7. He mayntayned D. Burcott's doings.

He did practyse to advaunce D. Burcot into the place of Jonas, & mayntan Burcots false proffes made of the ewre, to thend he might be sett on agayn in this third voyage, as the Comissioneres and Denham canne witnes.

No. 8. He vittelled the shipe *Ayde*.

He was sent to Bristowe to dispatche the ships, the *Ayde* on the thirde voyage, wherin he was made victualler of the shippe, for the whiche sarvice he had money before hand, but he dide so evell vittell the same, as wheras the Companye allowed him to vittell her wth fleshe 4 daye in the weke, he sarved the men therof onlye 3 dayes, and 2 dayes in the weke, and the rest of the weke wth evill fishe, and that wth scarsetie wherbye manye of them died, as the men do reporte.

He was sent into the west countrie to provide the 120 myners for the voiage, for whose furniture he received money of the Companye by fore hande, for their wags *li*240, and for their weapons *li*120, but therof he paid these men uppon their wags, to some xx[s], to some xiii[s] iiii[d], and to some nothinge the man, as the accounts declare. And what weapones they had, or he for them, as yet is unknoune. But in the west countrie is spreade agreat clamor that those mynares beinge prest by comissione many of them were afterwards chaunged by favour for showmakeres, taylores, and other artificers, no workemen, and were furnished to see at the charge of the townes and villages in maner of a subsedye as it is reported openlye. *No. 9.* He dealt doble in the myners provision.

He toke the shipe the *Sallomon* of Weymoth, in the west countrie, w[th]out knowledge of the comissioneres, by force of Her Ma[ties] generall comission to him geven, and therby caused the owner, Hew Randall, to furnishe her, and to be with him in this sarvice of the thirde voyage, promisinge hime victualls and other great matters, w[ch] he performed not, as Hewe Randall dothe saye. No. 10. He toke the shippe *Salomon* by comission.

He led all the shipps this yere to a wronge place of Meta Incognita, throughe his obstinate ignorance, wherby they were all in great danger to perrishe, as Hawll, Davis, and the rest of the shipps masters will witnes. No. 11. He led the flete of ships to wrong place.

He, beinge at Meta Incognita, did refuse conference and counsell of all others, and said his instructiones, geven by her Ma[ties] Honourable Privie Councell, were but the device of Fenton and Lok, and never reade by the Councell, though their hands were at the same, as Captayn Fenton and the other captaynes, and Hawell do witnes. No. 12. He denied the Councells comission.

And when the shippes were mored salf in harbor in the countey of Warwicks Sounde, where they should lade, and from whence they should have departed orderlye, he beinge at Beares Sound, comaunded all the shippes (w[th]out anney advice or discretion) to come thether to take him and his men in, w[ch] place beinge no harbor, but wilde see, a storme of weather happened w[ch] put all the shippes to see to save them selves, w[th] losse of all their boates and pynnesses, and other spoile, leavinge him there behind them in the barke *Gabriell*. And so they came home in suche disorder as is openlye knowne. No. 13. He caused the great disorder of the ships retorne home.

He is so arrogant in his governement, as Hawle, Jakman, Davis, nor the other of the masters wille no more take charge of ships under him, and so imperious in his doinges as some of the Comissioners are werie of his company, and manney of the venturares mynded to medle no more wth him. No. 14. His arrogancie.

He drew his dagger and furioslye ranne uppon Jonas, beinge in his worke at Tower hill, and threatned to kill him yf he did not finishe his worke owt of hand, that he might be sett owt againe on the thirde voiage, wheruppon Jonas did conseave so eavell nature in him, that he No. 15. He drew his dagar on Jonas.

made a sollempe vowe he would never go to see any more wth him, wch hath byn no small domage to the Company in the ewre brought home the thirde voyage.

No. 16. He drew his dagear on C. Fenton.

He drew his dager on Captayne Fenton at Darteford, uppon a quarrelous humor, and wolde haue mischefed him uppon the sodayne, yf Mr. Pelham and others had not bine present.

No. 17. Litle trewthe in his talke.

He is so full of lyinge talke as no man maye credit anye thinge that he doth speake, and so impudent of his tonge as his best frindes are most sclanndered of him when he cannot have his wille.

No. 18. He sclandered M. Lok, to the great domage of the Companye.

He hath raysed lately such sclannderous reportes against Mr. Lok, and geven suche vehement false informaciones of iiim *li*, and other greate somes of money to remayne in his hands dew to the Company, to paye the shipps fraights and mens wages, as hath well lyked some of the venturares, which hoaped therby to be forborne of the payment of their owne parts of money dewe for that purpose, wherby littell money cane yett be had of them of the *li*3,400 dewe by them to dyscharg that dutye whiche hath caused the Company to spend m*li* of monye in vayne, for chardgs of the shippes and men synes they came home, and by that meanes for lake of payment of their dewtye, a sclanderous rumoor is spreade over all the realme, to the great discredite of the Company.

No. 19. He paid wages to men against comaundment.

He did paye wages to the men of the shipe *Thomas* of Ipswiche for v monthes, wheras the Comissioners did agree and comannd to paye them but for iiij monthes.

No. 20. He brought men into wags wthout order.

He hathe brought into wages of the Companye so many men, and suche men as he lyste, and many of them at suche wages as he lyste, wthout regard of their sarvyce or deserts wherof he is to geve accountt for that many of them are dead, and gone awaye.

No. 21. The men in the *Ayde* make great spoyle.

He hath plased styll in the shippe *Ayde*, now in the Tames, a nomber of men at the Companyes charges, wherof many are suche disordered men, bothe of their tonges and of their hands, as are the cause of moche sclander to the Company, and great spoyle done in their shippes, and yt have but small dutye of wages owinge to them, when their accounts shalbe examined particulerlie.

No. 22. He hath not distributed the *cli* to the men.

He receaved *cli* of mony by Her Maties order, at retorne of the seconde voiage, as of Her Maties gyfte and reward to be distrybuted amonge the marineres and other men wch sarved in that voyage, but no distribution is made therof as yett, as the men doo complayne.

No. 23.

To conclude, yf his doinges in thes iij voyages be well looked into, parchanse he wilbe found the most unproffitable sarvante of all that have sarved the Companye therin.

THE SCLANDEROUS CLAMORS OF CAPTAINE FURBUSHER AGAINST MICHAEL LOK. 1578.

He hathe made false accountts to the Companye, and hathe cossened them of iii^{m}*li* of money.

He hathe cossened my Lord of Oxford of m*li*.

He hathe not one grote of venture in these voiages.

He is a bankerot knave.

THE ANSWERS OF MICHAEL LOK.

All these forsaid clamors are proved to be false sclanders, aswell by the new awdyte made of M. Loks accountts as also by the open knowledge had of all his doinges certiffied to Her Maties Honorable Privie Councell.

And now, yf any evell successe should happen in the work of the ewr now layd at Dartford, wch I trust shall not happen, yet wear not that to be imputed anye wayes unto M. Lok, whose innocentie therein is proved by his great goodes beinge ij^{m}vi^{c}*li* of money bestowed and ventured therin, and by the testimonie of the Comyssioners certiffienge the first proffe of the work made in the second voiage, but rather yf any such myschyef should happen, wch God forbyd, the same wear to be layed on Captaine Furbusher, whose great abusses therin are before declared, and on Jonas and Denam, being the workmen therof, as men who have byn the fyndars and bringars of that ewr wch is brought, and causars of the cost bestowed for the fetchinge and workinge of the same, and on them the same weare to be punished sharplie, but I trust no suche cause shalbe gyven.

1578.

THE ABUSSES OF CAPTAYNE FURBUSHER AGAINST THE COMPANIE.

A DESCRIPTIVE CATALOGUE

OF

RELICS

OF

FROBISHER'S EXPEDITIONS

TO

THE ARCTIC REGIONS IN THE YEARS 1576-7-8,

DISCOVERED BY

MR. CHARLES F. HALL, OF CINCINNATI, U.S.A.,

WHILE ON AN

EXPEDITION IN SEARCH OF TRACES OF SIR JOHN FRANKLIN AND HIS SHIPS' COMPANIES IN THE YEARS 1860-1-2, AND BY HIM TRANSMITTED TO THE BRITISH PEOPLE

THROUGH THE ROYAL GEOGRAPHICAL SOCIETY OF LONDON.

DESCRIPTIVE CATALOGUE

OF

FROBISHER RELICS.

NOTE. *The references by letters following descriptions in this Catalogue are to the accompanying map of Kod-lu-narn.*

A.

Note. The relics under A obtained Sept. 22nd and 25th, 1861, July 15th and 16th, 1862. My companions, Esquimaux. They are from the ruins of a small house near the centre of the Island Kod-lu-narn.* (Vide B, small map of Kod-lu-narn.) Said house was of lime and stone. The foundation and walls, and also floor being laid in lime cement, are still in a good state of preservation.

On returning home in the fall of 1862, I saw for the first time in my life Hakluyt's work. Read on page 634, edition of 1589, as follows—"But before we took shipping we builded a little house in the Countess of Warwick's Island," etc. (Thomas Ellis, author.)

Also read, p. 91, vol. 3, ed. 1598-1600, "this day (30th August, 1578) the masons finished a house which Captain Fenton caused to be made of *lyme and stone* upon the Countess of Warwicks Island," etc. (George Best, author.)

A 1. Round box containing several stones cemented together with lime.

A 2. Stone and lime cement.

A 3. Ditto.

A 4. Ditto.

A 5. Ditto.

A 6. Two stones, lime cement (one of these with moss upon its edge).

A 7. Two stones, and lime cement.

A 8. Ditto.

A 9. Stone, and lime cement, moss upon its upper surface.

A 10. Small round box containing lime cement—flint stones and fragments of tile.

A 11. Same as A 10.

A 12. Lime cement and burnt flint stones.

A 13. Stone, and lime cement.

A 14. Four small stones, and lime cement.

* Kodlunarn is the Countes of Warrick Yland of the map.

A 15. Stone, and lime cement.

A 16. Ditto.

A 17. Lime cement, partly reduced to powder-flint.

A 18. Lime cement, mostly reduced to powder.

A 19. Lime cement, burnt flint stones, oxide of iron, fragment red stone.

A 20. Lime cement and small stone.

A 21. Three pieces lime cement.

A 22. Lime cement with small stone.

A 23. Three pieces lime cement.

A 24. Powdered lime cement with burnt flint stones.

A 25. Stone, and lime cement.

A 26. Thick moss with lime cement at its base, stem of dwarf willow in the moss.

A 27. Sod and dwarf willow (in large box by itself), from over the foundation wall of house of lime and stone.

B.

B 1. Semisphere of iron, found under east embankment (G) of ship's trench (A A). *The exact spot where I found this at* F.

B 2. Sand that was fast cemented to bottom of B 1 by oxide of iron.

C.

C 1. Stone covered with black moss of ages found on one of the embankments (G G) of the ship's trench (A A).

C 2. Same as C 1.

C 3. Ditto.

C 4. Ditto.

C 5. Three stones from ship's embankments (G G).

C 6. Two stones from ship's embankments (G G).

C 7. Same as C 5.

C 8. Twelve stones from reservoir embankments (C).

C 9. Small stone with white moss, from reservoir embankments (C).

C 10. Small stone with black moss of ages, from ship's embankments (G G).

C 11. Four stones from ship's embankments (G G).

C 12. Small stone with black moss of ages, from ship's embankments (G G).

C 13. Two stones with black moss of ages, from ship's embankments (G G).

C 14. Two stones; one, quartz, has upon it a spot of black moss of ages.

C 15. Stone with black moss of ages, from reservoir embankments (C).

C 16. Stone with black moss of ages, from reservoir embankments (C).

C 17. Three stones from ship's embankments (G G).

C 18. Two stones from ship's embankments (G G).

C 19. Stone from ship's embankments (G G).

D.

D 1. Long box—wood, dug out of base of ship's trench (A A).

D 2. Same as D 1.

D 3. Same as D 1.

D 4. Sand and wood dug out of base of ship's trench (A A).

NOTE. Frobisher left the timber of his intended fort on the "Countess of Warwick's Island." Vide *Hakluyt*, p. 91, vol. 3, edition of 1600.*

E.

NOTE. The history of *Frobisher's Second Voyage as written by George Best*, in referring to the natives (Esquimaux) building their "poor caves and houses which serve them for their winter dwellings," says, "From the ground upwards they build with whales' bones, for lack of timber, which, bending over one another, are handsomely compacted in the top together, and are covered over with seal skins, which, instead of tiles, fence them from the rain," etc. *Pinkerton*, vol. 12, p. 522. Is not this reference indirect proof that Frobisher had "*tiles*" as covering for the house or "fort" which he took out with him in 1578 ?

See "Notes framed by M. Richard Hakluyt, of the Middle Temple, Esquire, given to certain gentlemen that went out with M. Frobisher in his north-west discoverie," wherein the word "tile" or "tyle" occurs several times. *Hakluyt*, p. 636, edition 1589.

E 1. Fragments of tile; some dug from under ship's embankments (G G) and trench (A A), the rest picked up on same side of the island. Two or three pieces of oxide of iron.

E 2. Fragments of tile, charcoal, sea-coal, flint stones, oxide of iron, picked up on the island of Kod-lu-narn.

E 3. Fragments of tile, few pieces of charcoal, and oxide of iron. Some of the tile dug up at ship's embankments (G G), remainder picked up on the island.

E 4. Fragments of tile, few pieces of sea-coal, oxide of iron, slag, coke, flint stones, small bone covered with moss, small stones. All picked up on the island.

E 5. Fragments of tile, flint stones, coke, sea-coal, charcoal. Some of the tile dug up at ship's embankments (G G).

E 6. Fragments of tile, charcoal, small roots, dug up from the ruins of blacksmith's shop. (?) (d)

E 7. Fragments of tile, oxide of iron, sea-coal, picked up on the island (Kod-lu-narn).

E 8. Fragments of tile dug up from ruins of blacksmith's shop. (?) (d)

* See page 272.

E 9. Three fragments glazed tile dug from under ship's embankments (G G).

E 10. Two fragments same as E 9.

E 11. Nine fragments same as E 9.

E 12. Four fragments same as E 9.

E 13. Two fragments tile and gravel stone united by moss.

E 14. Two fragments tile with moss of ages upon them.

E 15. Fragment tile and stone united by moss of ages.

E 16. Same as E 15.

E 17. Three fragments tile (two with glazing).

E 18. Fragment of tile with moss.

E 19. Fragment of tile.

E 20. Fragment tile and stone united by moss.

E 21. Fragment of tile found embedded in the coal deposit, etc., on Ek-ke-lu-zhun.*

E 22. Fragments tile, sea-coal, flint stone, oxide of iron. All these covered with the moss of ages. From that portion of Kod-lu-narn between ship's trench (A A) and reservoir (C).

E 23. Ten fragments tile (nine glazed).

NOTE. Nos. 13 to 20 inclusive, from vicinity of ship's trench (A A).

F.

F 1. Oxide of iron. Some pieces found in the centre as the nucleus, the scales lying around. Found on the ground, most of it near the head of the ship's trench (A A). Some at "Best's Bulwark" (E).

F 2. Large piece of coke, small piece of charcoal in one of the protuberances. Found on Kod-lu-narn.

F 3. Coke dug from under ship's embankments (G G).

F 4. Oxide of iron and gravel, found on the ground south-east side of island at H.

G.

G 1. Lime stone found in Kod-lu-narn.

H.

H 1. Sea-coal, coke, fragments of tile, oxide of iron, lime stone, small piece of bone with moss upon it. All as picked up on Kod-lu-narn.

I.

I 1. Burnt stones, charcoal, fragments of tile, dug up from beneath ruins of blacksmith's shop. (?) (d)

I 2. Sod, with charcoal, from ruins of blacksmith's shop. (?) (d)

I 3. Fragments of tile, charcoal and earth from ruins of blacksmith's shop. (?) (d)

* Ek-ke-lu-zhun is a bay on the east side of Countess of Warwick's Sound

J.

NOTE. In box marked J, velvet lined.

J 1. Fragment of tile and four gravel stones united by moss of ages.

J 2. Fragment of pottery found near "Best's Bulwark" (E).

J 3. Small piece of cord, apparently of hair, found deeply embedded in the coal deposit of Ek-ke-lu-zhun.

J 4. Four fragments glass, apparently of a jar or bottle, found on the ground near ship's trench—the exact spot marked I.

J 5. Piece oxide of iron with moss of ages upon it, found near ship's trench (A A).

J 6. Piece of wood dug up from base of ship's trench (A A).

J 7. Sea-coal, with moss of ages upon it, found near "Best's Bulwark" (E).

J 8. Piece of pottery found near "Best's Bulwark" (E).

J 9. Fragment white pottery, (?) black glazing outside and inside, found near "Best's Bulwark" (E).

J 10. Choice specimen of tile, covered with moss of ages, from Kod-lu-narn.

J 11. Sea-coal, covered with moss of ages, from coal deposit at Ek-ke-lu-zhun.

J 12. Stone, covered with moss of ages, from top of ship's embankments (G G).

J 13. Flint stone, covered with moss of ages, found near the head of ship's trench (A A).

J 14. Fragment of tile, glazed, apparently portion of human figure represented upon it: leg and foot *in relievo*. Largest piece of tile found. Dug up from beneath one of the ship's embankments (G G).

J 15. Stone with lime cement from ruins stone house (B).

J 16. Probably one of the ears or knob-handles of an earthen jar. From near "Best's Bulwark" (E).*

J 17. Flint stone, with moss of ages upon it.

J 18. Chip found deeply embedded in coal deposit on Ek-ke-lu-zhun.

J 19. Burnt flint stone with lime cement, from ruins stone house (B).

J 20. Charcoal of coarse grained wood, apparently of thrifty growth, found under stones and sods by the ruins of blacksmith's shop. (?) (d) The grain of this charcoal indicates it to be from the same kind of wood as that found at the base of ship's trench (A A). Vide Box D 1.

K.

K 1. (In keg). Sea-coal, flint stones, wood chips, the latter found deeply embedded in the coal. All in this keg precisely as gathered from coal deposit Ek-ke-lu-zhun.

K 2. Sod of moss with sea-coal.

* See page 148.

K 3. Sea-coal overgrown with moss of ages.

K 4. Four pieces coal covered with moss of ages.

K 5. Three pieces coal and one of coke with moss of ages.

K 6. Coal with moss of ages upon it.

K 7. Three pieces coal with moss of ages upon them.

K 8. Small pieces of coal enveloped in moss.

K 9. Two pieces of coal with moss of ages upon them.

K 10. Small pieces enveloped in moss.

K 11. Two pieces coal with moss of ages.

K 12. Same as K 11.

K 13. Three very small pieces coal united by moss.

K 14. Fifteen bits of wood excavated from coal deposit.

NOTE. All the above from Ek-ke-lu-zhun, except coke in K 5, which is from Kod-lu-narn.

L.

L 1. Two pieces coal from Ek-ke-lu-zhun.

L 2. Coal from near "Best's Bulwark," (E) Kod-lu-narn.

L 3. Three pieces coal from Ek-ke-lu-zhun.

L 4. Same as L 3.

L 5. Five pieces coal from Ek-ke-lu-zhun.

For Ek-ke-lu-zhun, *vide* Chart "Countess of Warwick's Sound."

M.

M 1. Sod with coal intermixed.

M 2. Two sods with coal intermixed.

M 3. Moss interlocking and covering coal.

M 4. Earth, first layer beneath coal.

M 5. Earth, second layer beneath coal.

M 6. Earth, third layer beneath coal.

NOTE. All under M from the island Ni-oun-te-ling.*

N.

N 1. Flint stones found embedded in coal deposit on Ni-oun-te-ling.

N 2. Flint stones found embedded in coal deposit on Ek-ke-lu-zhun.

N 3. Same as N 2.

O.

O 1. Red stone found on top of the coal deposit on Ni-oun-te-ling.

P.

P 1. Two pieces iron pyrites from above Countess of Warwick's Sound.

P 2. Two pieces, apparently mineral (iron), "like to sea-coal," found at Ek-ke-lu-zhun.

* Ni-oun-te-ling is a small island on the east side of the Countess of Warwick's Sound.

Q.

Q 1. Walrus rib with heavy moss upon it. Another bone with moss, ound on the *Esquimaux Deserted Land* (Frobisher's *North Foreland*).

R.

R 1. Wood model of an anvil made by an old man Esquimaux named An-na-wa, Oct. 15, 1861. I have three other models of like fashioning made by three Esquimaux, one apart from another. These, I am confident, from Esquimaux traditions, are models of a relic of Frobisher's expedition. Some six years ago the Esquimaux cast the anvil of which these are models into the sea from point x of Oo-pung-ne-wing (see Chart of Countess of Warwick's Sound), an island three miles distant from Kod-lu-narn. I have endeavoured to recover it at lowest tides, new and full moon, but the ice has probably carried it away. Only the strongest Esquimaux could lift it.

S.

S 1. (In small square box). Thick sod, grass, moss and coal and flint stones intermixed. Loose coal, flint stones and sand put into the box to fill up. Contents of this box gathered from coal deposited on Ni-oun-te-ling.

Note. About one ton coal at Ni-oun-te-ling.

NOTES.

In relation to sea-coal, Hakluyt, vol. 3rd, ed. 1598-1600, p. 91, has the following:—"And the great cause of this leakage and wasting was for that the great timber and sea-coal, which lay so waighty upon the barrels, brake, bruised and rotted the hoops in sunder."

A truthful description of "Countess of Warwick's Sound."

Nine recognisable physical facts in a few words.

"The 29th July (1577), about five leagues from Bear's Sound, we discovered a bay, which, being fenced on each side with small islands lying off the main, which break the force of the tides and make the place free from any indrafts of ice, did prove a very fit harbour for our ships, where we came to anchor under a small island, which now, together with the sound, is called by the name of that right honorable and virtuous lady, Anne Countess of Warwick. And this is the furtherest place that this year we have entered up within the straights, and is reckoned from the Cape of Queen Elizabeth's Foreland (Cape Resolution of Resolution Island, C. F. H.), which is the entrance of the straights not above thirty leagues." *Pinkerton*, vol. xii, p. 522.*

The Reservoir (?) or Frobisher's "Gold Diggings," vide c, small map of Kod-lu-narn. Read *Hakluyt*, ed. 1589, p. 626.

"In this isle (Countess of Warwick's), our General (Frobisher)

* See page 137.

thought good for this voyage to fraight both the ships and barkes with such stone or gold mineral as he judged to countervail the charges of his first and this his second navigation to these countries.................. It (stone or gold mineral) riseth so abundantly, that, from the beginning of Aug. to the 22nd thereof (every man following the dilligence of our General, we raysed above ground 200 tunne, which we judged a reasonable fraight for the ship and two barks, in the sayd Anne Warwick's Isle."

For what I recognise excellent descriptions of Bear's Sound (channel), see *Pinkerton*, vol. xii, pp. 521 and 555, and *Hakluyt*, ed. 1589, p. 635.

Ek-ke-lu-zhun (where a coal deposit is) is about ten miles east of Oo-pung-ne-wing.

Oo-pung-ne-wing and Ni-oun-te-ling are about three miles from Kod-lu-narn (Countess of Warwick's Island).

On the small map of Kod-lu-narn will be noticed E (Best's Bulwark). At this point I found considerable oxide of iron, several pieces of pottery and sea-coal.

In *Pinkerton*, vol. xii, p. 527, read the following:—"On Thursday, the 9th Aug. [1577] we began to make a small fort for our defence in the Countess's Island, and entrenched the corner of a cliff, which, on three parts, like a wall of good height, was encompassed and well fenced with the sea, and this was called Best's Bulwark, after the lieutenant's name, who first devised the same. The above description of cliff (E) is truthful."

C. F. HALL.

New York, Feb. 7, 1863.

INDEX.

INDEX.

T. RICHARDS, PRINTER, 37, GT. QUEEN STREET.

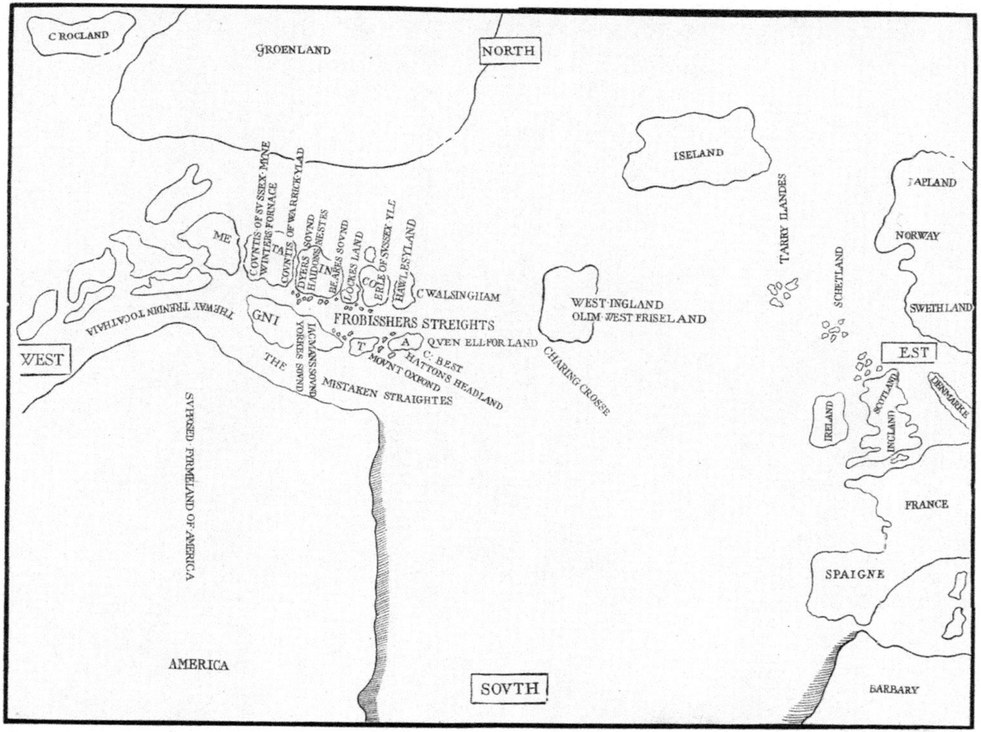
C ROCLAND
GROENLAND
NORTH
ISELAND
TARRY ILANDES
IAPLAND
NORWAY
SWETHLAND
SCHETLAND
EST
SCOTLAND
INGLAND
IRELAND
DENMARKE
FRANCE
SPAIGNE
BARBARY
WEST·INGLAND
OLIM·WEST FRISELAND
CHARING CROSSE
ME
COVNTIS·OF SVSSEX·MINE
WINTERS FORNACE
TA
COVNTIS OF WARRICK·YLAD
DYERS SOVND
HAIDONS NESTES
IN
BEARES SOVND
LOCKES LAND
CO
ERLE OF SVSSEX YLE
HAWLESYLAND
C WALSINGHAM
FROBISSHERS STREIGHTS
GNI
T
A
QVEN ELLFOR LAND
C: BEST
HATTONS HEADLAND
MOVNT OXFOND
IACMANS SOVND
YORKES SOVND
THE
MISTAKEN STRAIGHTES
THEWAY TRENDIN TOCATHAIA
WEST
SVPPOSED·FIRMELAND OF·AMERICA
AMERICA
SOVTH

Printed in Dunstable, United Kingdom

91222494R00241